Mediterranean Tourism

As a region, the Mediterranean attracts more tourists than any other in the world. However, until now no substantial scholarly work has been dedicated to the study of the region, its tourism activity and the implications of this industry for the region's socioeconomic development. This volume compiles original contributions from twenty-one leading authorities, both scholars and practitioners, and constitutes the first comprehensive effort to examine Mediterranean tourism as a coherent unit of analysis.

Using a critical, interdisciplinary approach, the organization, structure and ramifications of fourteen countries' tourism sectors are examined. Building on this important empirical research, the book identifies the commonalities of these national tourist products, and suggests directions for future strategies to strengthen the position of Mediterranean tourism in a period of international restructuring and globalization.

This book is essential reading for anybody with a scholarly or professional interest in the socioeconomic development of the Mediterranean region.

Yorghos Apostolopoulos is a Research Associate Professor of Sociology at Arizona State University. His research focuses on the epidemological and public health effects of tourist migration and subsequent impacts on sustainable development.

Philippos Loukissas is a Professor and Chairman at the Department of Planning and Regional Development at the University of Thessaly. His research includes issues in planning theory and public policy planning in the areas of tourism and transportation.

Lila Leontidou is Professor at the Department of Geography at the University of the Aegean. Her research focuses on urban restructuring, modernity, social disadvantage, theory and epistemology in geography. She is currently coordinating the European Union project *Border Cities and Towns: Causes of Social Exclusion in Peripheral Europe.*

Routledge advances in tourism
Series editors: Brian Goodall and Gregory Ashworth

Mediterranean Tourism

Facets of socioeconomic development
and cultural change

**Edited by
Yorghos Apostolopoulos,
Philippos Loukissas and
Lila Leontidou**

London and New York

First published 2001
by Routledge
2 Park Square, Milton Park, Abingdon, Oxfordshire OX14 4RN

Simultaneously published in the USA and Canada
by Routledge
711 Third Avenue, New York, NY 10017

First issued in paperback 2014

Routledge is an imprint of the Taylor and Francis Group, an informa company

Transferred to Digital Printing 2006

Editorial material and selection © 2001 Yorghos Apostolopoulos,
Philippos Loukissas and Lila Leontidou

Individual chapters © 2001 the contributors

Typeset in Baskerville by Wearset, Boldon, Tyne and Wear

British Library Cataloguing in Publication Data
A catalogue record for this book is available from the British Library

Library of Congress Cataloging Publication Data
Mediterranean tourism : facets of socioeconomic development and
cultural change/
edited by Yorghos Apostolopoulos, Philippos Loukissas, and Lila
Leontidou.
    p. cm.
    Includes bibliographical references (p.).
    1. Tourism – Mediterranean Region. I. Apostolopoulos, Yorghos.
    II. Loukissas, Philippos J. III. Leontidou, Lila.

G155.M46 M44 2000
338.4'791091822–dc21
                                                        00-032822

ISBN 13: 978-0-415-18023-8 (hbk)
ISBN 13: 978-0-415-75744-7 (pbk)

Contents

List of figures

List of tables

Contributors

Dr Yorghos Apostolopoulos, Department of Sociology, Arizona State University, Tempe, Arizona, USA.

Dr Jaques Barbier, UrbaPlan, Laussane, Switzerland and Rabat, Morocco.

Dr Marc Boyer, Faculte de Geographie, Histoire, Histoire de l'Art et Tourisme, University of Lumiere, Lyon, France.

Dr Matthew Gray, Centre for Middle Eastern and Central Asian Studies, Australian National University, Canberra, Australia.

Dr Khalid Zakaria Imam, Department of Urban Design, Faculty of Urban and Regional Planning, University of Cairo, Cairo, Egypt.

Dr Dimitri Ioannides, Department of Geography, Geology and Planning, Southwest Missouri State University, Springfield, Missouri, USA.

Dr Josep A. Ivars Baidal, Cavanilles Foundation of Advanced Studies in Tourism, University of Alicante, Alicante, Spain.

Mr Kostas Krantonellis, Athens, Greece.

Dr Lila Leontidou, Department of Geography, University of the Aegean, Mytilini, Lesuos, Greece.

Dr Philippos Loukissas, Department of Planning and Regional Development, University of Thessaly, Volos, Greece.

Dr Hyam Mallat, Lebanese University, Beirut, Lebanon.

Dr Yoel Mansfeld, Department of Geography, University of Haifa, Haifa, Israel.

Dr Emmanuel Marmaras, Department of Geography, University of the Aegean, Mytilini, Lesuos, Greece.

Dr Vicente M. Monfort Mir, Department of Tourism Studies, University of Castellon, Valencia, Spain.

Dr Giovanni Montemagno, Institute of Economics, Faculty of Political Science, University of Catania, Catania, Italy.

Dr Robert A. Poirier, Department of Political Science, Northern Arizona University, Flagstaff, Arizona, USA.

Dr Pantelis Skayannis, Department of Planning and Regional Development, University of Thessaly, Volos, Greece.

Dr Sevil Sonmez, Department of Recreation Management and Tourism, Arizona State University, Tempe, Arizona, USA.

Dr Turgut Var, Department of Recreation, Park, and Tourism Sciences, Texas A&M University, College Station, Texas, USA.

Dr Boris Vukonić, Department of Tourism and Trade, Faculty of Economics, University of Zagreb, Zagreb, Croatia.

Dr Yahia Zoubir, Thunderbird, American Graduate School of International Management, Glendale, Arizona, USA.

Preface

The genesis of this edited volume can be traced back to a relatively short break in Greece in my exploration of a possibly permanent stay in my homeland a few years ago. As a complete stranger in both the academic circles and the tourism sector of the country, I was fortunate enough to have met professors Philippos Loukissas and Lila Leontidou. As established academics in the areas of regional development and planning and cultural geography respectively, they both demonstrated a sincere interest in my plans for an edited book on Mediterranean tourism. Like myself, both Philippos and Lila were born and educated in Greece, had experienced the graduate educational systems of the USA and UK respectively, and had lived and worked in these countries for a number of years. These invaluable shared experiences – despite our diverse disciplinary backgrounds – along with our strong conviction in tourism's significance in the economic and cultural development of the Mediterranean region, brought us together.

This work has been long overdue and is intended to fill an enormous gap in the international tourism literature. It concentrates on the structure and organization of Mediterranean tourism and its implications for national and regional economies in the oldest and most touristically-advanced region of the world. Despite undoubted shortcomings or omissions, it constitutes the first comprehensive attempt to examine, both individually and comparatively, the tourism sectors of the fourteen most significant countries of the Mediterranean (in terms of tourism) and of the basin as a whole, during an era of intense competition and restructuring in the production and consumption of global tourism. This comparative book is presented to the audience of academics and practitioners in the midst of reshuffling, shifts in the market share and intense phenomena of economic globalization.

The book overcame several hurdles, underwent numerous phases, and took more than two years to complete. Obviously, this volume would not have materialized without the collective support of all the involved individuals and organizations. It was a great challenge to locate, commission, and coordinate high-calibre international scholars and practitioners in various spheres of Mediterranean tourism. The calibre of their work made the review and revision process a much more enjoyable task.

I would like to extend my gratitude to the educational attachés of the embassies of Algeria, France, Italy, Lebanon, Spain, and Turkey in Athens as well as to the National Tourism Organizations of Greece in Athens and Lebanon in Beirut for their invaluable assistance during the initial phases of the book. Furthermore, I would also like to extend my gratitude to colleagues Dimitri Ioannides, Panajotis Komilis, Vasilis Pappas, Nikos Triantafyllopoulos and Sevil Sonmez for their valuable advice and assistance during various phases of the project. Thanks must also go to Lorrine Basinger, a doctoral candidate of sociology at the University of Miami, for her valuable input and editing throughout the initial phases of the book. I also wish to thank Elizabeth Brown, Sarah Carty, Craig Fowlie, Casey Main, Valerie Rose, and Simon Whitmore at Routledge publishers for their encouragement, guidance, patience, and support during this process. Last, but certainly not least, I would like to thank my two co-editors, professors Loukissas and Leontidou. I am enormously grateful for their generous guidance and support throughout the project.

Yorghos Apostolopoulos
Tempe, Arizona, USA
Summer 2000

Foreword

During the early 1980s, tourism was considered everywhere as 'the gold mine of the twentieth century', bound to become by the year 2000 the world's leading industry. In general terms the forecast has shown its accuracy now that we have arrived at the new millennium. If we look closer at some major destinations such as the Mediterranean coastal areas and its islands, we must admit that the last few years have brought considerable changes to the once-smiling landscape of traditional mass tourism. Recent WTO data show, however, that arrivals in the Mediterranean have been reduced for more than 20 per cent as compared with a decade ago.

What has changed? A rough answer can be provided by any traveller travelling along the Mediterranean shores: the huge investments and mass tourism resorts developed from the 1970s onwards simply do not correspond any more in terms of typology and quality to the requirements of a new international demand. National and local authorities, and all kinds of operators are, of course, reacting towards the ongoing trends. Just in observing the recent promotional campaigns we discover that almost all of them seem to be engaged in the offer of similar products more or less related to ecological and cultural values. From the operational point of view, nothing distinguishes the new labels from the traditional tourism operations – except perhaps the word 'sustainable' being added here and there. What can be felt from the aggressiveness of the above campaigns, set in motion by the Mediterranean operators is that they are addressing the international market as a dispersed troop, within a highly-competitive context. Some have adopted the so-called 'new communication technologies' but if somebody is patient enough to go through the many pages of their web sites, he/she discovers that nothing new is offered, beyond that which was already displayed in the usual tourism leaflets.

Can we explain these simple facts presently put before us? A couple of observations might help to explain them. First, we are still confronted with an indiscriminate mass tourism policy serving primarily the short-term needs of the economy rather than the long-term structural and intergenerational requirements of an evolving leisure sector. Second, because of its national and international pervasiveness Mediterranean tourism, particularly on islands, tends to crystallize many vested interests strongly committed to 'business as usual' local policy-makers facing elections, local land owners seeking short-run speculative profits, and the myriad of constructions, service and transport-makers whose fortunes are closely tied to the tourist season.

On the other side of the horizon, international airline, cruise, hotel and travel interests, are committed to the intensive use of available large-scale facilities. Both groups together produce a policy propensity for elastic short-run growth, irrespective of future perspectives. This mass orientation in the Mediterranean tourism policy further constrains a change in direction and quality in the tourism sector. Finally tourism is not only a complex multi-dimensional process embracing visitors, residents, environment, and so forth together with the system itself, but it is also dynamic and difficult to capture. In addition, its gestation and pervasive impacts occur unevenly on the environment through time and particularly on soft sociocultural systems, highly sensitive to the imbedded complicated interactions involving, for example, the differences in perception between hosts and visitors. Tourism research is obviously affected by all the above ambiguities, often lacking both empirical strength and theoretical rigour. As a result, no universal model of tourism evolution as yet dominates our understanding, nor reliably provides policy-makers with appropriate tools for maximizing short-run economic benefits and minimizing socio-environmental costs. As a consequence, no long-term strategy for achieving intergenerational sustainability is available. All the above, I am convinced, would appear to the reader as a pessimistic but challenging approach to the present realities of the Mediterranean tourism sector. Another reason, if I may say so, to justify the debates raised by this volume, is that no final answers are given here, only hopes and many indications for doing better.

It is enough to satisfy us today in a world where the Mediterranean basin is still at the edge of conflicts and peace. Tourism needs peace and security to grow, as do Mediterranean people. Let me stress this hope and the urgent demand for further scientific thought, a challenging invitation to my old friend and partner Professor Philippos Loukissas.

Pier Giovanni d'Ayala
Secretary General
International Scientific Council for Islands Development, UNESCO

Introduction

The dynamics of Mediterranean tourism

1 Tourism, development, and change in the Mediterranean

Yorghos Apostolopoulos, Philippos Loukissas, and Lila Leontidou

What is the Mediterranean? It's a thousand things together. It isn't only one place but countless places. It isn't only one sea but a series of seas. It isn't only one civilization but layers accumulated over time. A journey through the Mediterranean is the discovery of the Roman world in Lebanon, of prehistory in Sardinia, of Greek cities in Sicily, of an Arabic presence in Spain, of Turkish Islam in Yugoslavia. It is a dive into the depths of centuries, ascendance to the buildings of Malta's megalithic era or to the Egyptian pyramids. It is a meeting with things very old, still alive ... things which coexist with the postmodern.

Excerpt from Fernand Braudel's, *La Mediterranee: L'Espace et l'Histoire*, 1977
(Loosely translated by the first author)

The Mediterranean basin: history, economic geography, and travel

The Mediterranean region constitutes a diverse geopolitical 'system' made up of parts that often have pulled in opposite nationalistic, ethnic, and religious directions, but which ultimately have remained together. The Mediterranean basin[1] is currently comprised of five African, five Asian, and 11 European countries sharing its coastline and totalling a population of 426.1 million in 1999. Eight main cultural and linguistic subsystems, three major religions, land areas ranging from the tiny island-state of Malta (360 km^2) to Algeria's massive 2.4 million km^2, and populations that vary from Malta's 380,000 to Egypt's 67.2 million blend into the Mediterranean mosaic (Figure 1.1).

The Mediterranean ranges broadly in economic growth rates as well as levels of social development. The region presently comprises four major groups of countries, classified on the basis of varying levels of socioeconomic development and political stability (Table 1.1). The first group includes countries with socioeconomically advanced and politically stable democracies. These are the four European Union countries of France, Greece, Italy, and Spain (with various levels of development not only

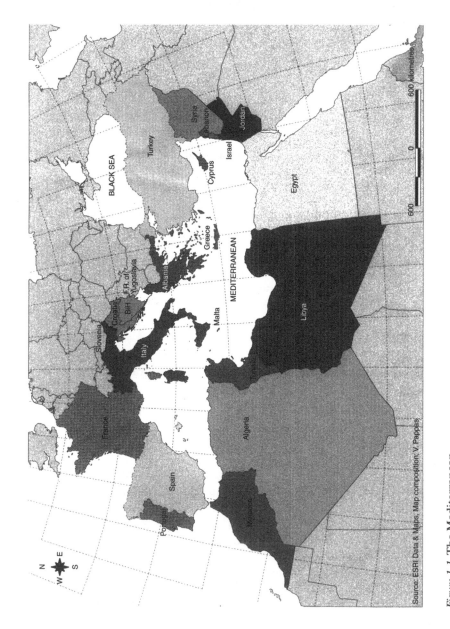

Figure 1.1 The Mediterranean.

among themselves but between regions of the same country) and the former colonies and/or newly established states of Cyprus, Malta, and Israel, as well as Monaco. The second group includes Algeria, Egypt, Lebanon, Libya, Morocco, Syria, and Tunisia, which have diverse types of political regimes and usually rich natural resources, but face serious structural socioeconomic and/or political problems and religious and ethnic tensions. Third, the unique case of Turkey shares characteristics of both previous groups – western orientation, budget deficits, rising external debt, intense socioeconomic polarization and an unstable political system. The fourth group includes the Adriatic coast of the Balkans with the Republic of Albania and four of the five newly established countries in the area of the former Yugoslavia – Bosnia-Herzegovina, Croatia, Slovenia, and the Federal Republic of Yugoslavia, consisting of Serbia and Montenegro. All of these nations have experienced disrupted socioeconomic and political order as a result of civil war and external military intervention. Despite the foregoing differences in levels and styles of development and socioeconomic and political systems, the 21 Mediterranean states share strong cultural ties and environmental resources. As globalization trends expand, local enclaves strive to maintain and preserve unique attributes and individuality, to co-operate against common threats, to share information and experiences, and to maintain regional planning efforts already initiated.

The Mediterranean region has been the cradle of western civilization and some of the world's oldest cultures have developed on its shores, including the Arab, Byzantine, Carthaginian, Egyptian, Greek, Phoenician, and Roman. The Mediterranean Sea has also served as one of the world's chief travel and trade routes – as early as 3000 BC, Egypt attracted travellers interested in seeing the pyramids and sailing the Nile; the ancient Greeks voyaged to Phoenicia and the Black Sea in the east and to Sicily and Marseilles in the west, and journeys to spas, festivals, athletic contests, and sacred places were common. During Roman times, travel was further encouraged by an extensive road network protected by the military, while during the medieval period, the wealthy could travel to France, Italy, Egypt, and the Holy Lands. During the Renaissance, historical cities such as Constantinople, Genoa, Venice, and Barcelona served as centres of commerce and culture.

During the twentieth century, and due to changes in world trade relations and transport technology, the centre of activity gradually shifted from the Mediterranean Sea to the Atlantic Ocean, leaving, by the 1950s, the Mediterranean with the relics of the past. That was the time when the Mediterranean landscape was transformed from the world's centre of commerce and trade to a tourist park (Williams 1997). After the post-World War Two emergence of mass travel, the richness of scenery of its coastal environment, its mild climate, and its impressive cultural heritage, turned the Mediterranean into a significant tourist draw. Today,

Table 1.1 Profile and socioeconomic indicators of 14 Mediterranean countries[a] 1999

Country	Area (km²)	Population (million)	Government	Independence	GDP/c (US$)	External debt (US$bill)	Inflation (%)	Unemployment (%)	World HDI rank	World GDI rank	World Life expectancy	Infant Mortality (per 1000 births)
Spain	507,800	39.1	Parl. Dem.	1492	16,500	90.1	2.0	20.0	11	19	77.7	6.5
France	547,000	58.9	Republic	486	22,600	11.8	0.7	11.5	2	6	78.6	5.6
Italy	301,200	56.8	Republic	1861	20,800	45.1	1.8	12.5	21	23	78.5	6.3
Croatia	56,500	4.6	Pres./Parl. Dem.	1991	5100	8.0	5.4	18.6	77	57	74.0	7.8
Greece	132,000	10.8	Parl. Republic	1829	14,600	40.8	2.3	10.0	20	21	78.5	7.1
Turkey	780,580	65.5	Parl. Dem.	1923	6600	93.4	70.0	10.0	74	58	73.3	35.9
Cyprus	9250	0.8	Republic	1960	13,000[b]	1.5[b]	2.3[b]/87.0[c]	3.3[b]/6.4	24	33	77.1	7.7
Syria	185,100	17.2	RUMR	1946	2500	22.0	21.0	15.0	78	84	68.1	36.5
Lebanon	10,400	3.5	Republic	1943	4500	3.0	5.0	18.0	65	66	70.9	30.5
Israel	20,700	5.7	Republic	1948	18,100	18.7	5.4	8.7	23	22	78.6	7.8
Egypt	1,001,450	67.2	Republic	1922	2850	28.0	3.6	10.0	109	100	62.4	67.4
Tunisia	163,600	9.5	Republic	1956	5200	12.1	3.3	15.6	81	74	73.4	31.4
Algeria	2,381,740	31.1	Republic	1962	4600	31.4	9.0	30.0	82	92	69.9	43.8
Morocco	446,500	29.6	Const. Monarchy	1956	3200	20.9	3.0	19.0	119	105	68.8	50.9

Source: CIA, 1999; *The World in Figures 1999*, London: *The Economist*
[a] = The table includes only the 14 countries the book has focused on.
[b] = Greek Cypriot
[c] = Turkish Cypriot
HDI = human development index (life expectancy, educational attainment, real GDP per capita)
GDI = gender-related development index
Inf. Mort. = infant mortality per 1000 births
RUMR = Republic Under Military Regime

the Mediterranean coast and its numerous islands constitutes a prime destination for international tourists seeking antiquities, culture, history, sun, sea, sand, and fun.

Tourism in the Mediterranean: dimensions and characteristics

Despite various setbacks,[2] travel and tourism continues to constitute the world's pre-eminent industry, producing over $3.6 trillion in gross output and generating over 225 million jobs (WTTC 1997). In 1998, 625 million tourists travelled internationally, with 60 per cent preferring European destinations (WTO 1999). In 1996, European Union member states accounted for 12 out of the 40 leading tourist destinations in the world and for 40 per cent of arrivals, 38 per cent of receipts, and 42 per cent of expenditures (European Union 1999); in 1998, tourism contributed 22.1 million jobs and accounted for 14.1 per cent of GDP in the European Union countries (WTTC and TBR 1999). Aggregate numbers of tourist arrivals in the 21 Mediterranean countries rose from 86 million in 1975, to 125 million in 1985, to over 200 million in 1990, and are estimated to have now surpassed 250 million (Grenon and Batisse 1989; Jenner and Smith 1993; Williams 1997; WTO 1999).[3] At the threshold of the twenty-first century, the Mediterranean region remains the world's main tourist destination, accounting for approximately 40 per cent of all international tourist arrivals and 30 per cent of all tourism revenues (Montanari 1995), while tourism represents 13 per cent of Mediterranean countries' exports and 23 per cent of trade in services, and employs over five million people (WTO 1996). However, the 40-year dominance of the Mediterranean region as a tourist destination has started to fade as it has begun to lose market share (WTO 1999) in favour of other geographic regions (the Pacific or Southeast Asia for example). Table 1.2 illustrates the main indicators of the tourist industries of the 14 countries included in the present volume.

There has always been an uneven distribution of travellers in the region – recently, however, there has been a more pronounced movement from the more mature areas in the north and west to new and fast growing areas in the east and south (with North Africa being the fastest growing region)[4] (WTO 1999). The concentration of tourist activity in relatively few key locations around cultural and physical attractions (i.e., archaeological sites, fragile coastal areas, islands) is accompanied by the concentration of other economic activities as well. Within the 21 countries, the coastal zones of the Mediterranean account for one-fifth of the world's total arrivals (amounting to 125 million foreign tourists), not counting day-trippers, of whom there are almost as many (WTO 1996). Coastal tourism development has become a magnet for additional population concentrations and tourism-related economic

Table 1.2 Main tourism indicators of 14 Mediterranean countries

Country	Tourist arrivals (thousands)	Tourism receipts (US$ million)	Tourist arrivals by air (thousands)	Rooms in hotels and other establishments
Spain	47,749	29,809	34,446	585,606
France	70,000	29,931	N/A	586,944
Italy	34,829	30,427	N/A	N/A
Croatia	4112	2733	631	83,199
Greece	11,416	8182	8946	304,232
Turkey	8960	7809	6699	149,186
Cyprus	2223	1671	1989	36,383
Syria	1267	1107	343	14,717
Lebanon	631	1285	418	10,966
Israel	1942	2656	1553	40,268
Egypt	3213	2564	2281	82,925
Tunisia	4718	1413	N/A	N/A
Algeria	673	20	N/A	N/A
Morocco	3243	1712	1581	63,446

Source: Greek National Tourism Organization (1999); World Tourism Organization (1999)
N/A = not available

activities. In Greece alone, 90 per cent of tourism, 80 per cent of the industrial sector, 35 per cent of fishing and agriculture, as well as infrastructure (i.e., roads, airports, train lines) are located along coastal areas, and 14 of the 50 Greek administrative prefectures (comprising 19 per cent of the land area) contain 73 per cent of hotel beds. In Spain, 76 per cent of the tourist infrastructure is concentrated in five (coastal) regions of the country out of a total of 18. Languedoc-Roussilon at Herault, France had a density of 130 inhabitants per km² versus 14 in the inland province of Lozere, while the Italian town of Lazio near Rome had 701 inhabitants per km², while inland at Rieti there were only 44 (European Community 1994). As a consequence, the international tourist influx in the Mediterranean coastal areas doubled between 1970 and the mid-1980s from 58 to 117 million (Grenon and Batisse 1993). This concentration of activities has put additional pressures on the fragile environmental resources and ecosystems of coastal areas, including coastal, visual and sound pollution, illegal construction practices, as well as urban land use conflicts (Montanari 1995). When, in the 1970s, water pollution became a serious problem, most of the nations of the region formulated the Mediterranean Action Plan which was later transformed to the Blue Plan, in order to reduce pollution and protect the Mediterranean environment from becoming 'a great dustbin' (Grenon and Batisse 1993).

Since the sweeping globalization trends in international trade have not bypassed the tourist industry, it has demonstrated increased dependence on tour operators and has experienced fierce competition from newly emerging destinations. Accommodation patterns and tourist activities are gradually changing as well – hotels represent a smaller percentage of the total supply, whereas second homes, condominiums, time-shares and residential tourism represent a growing portion of the market (see Chapter 17). Rental rooms, usually unauthorized, constitute a major problem for regions in Greece, southern Spain and Italy, as well as for many North African countries. In Greece it is estimated that one out of two rooms operates without a license, and one out of three tourist nights is not registered, resulting in the loss of significant amounts of bed tax revenues and compromised service quality (CPER and UT 1998).

At the other end of the spectrum, management and administrative issues are gaining in importance in both the public and private sectors (i.e., decentralization of regional tourism policies, efforts for spatially equitable distribution of tourism benefits). Countries on the eastern and southern shores can learn from the French, Spanish, and Italian models of regional planning and management, which involve alternative forms of tourism development such as agro-, eco-, cultural, marine, health, and sport tourism. European Union programmes, such as 'Leader', promote rural tourism and provide financial support to small family farms by promoting small-scale development, while protecting the environment. At the same time, there is a growing dissatisfaction with central governments' role in tourism policy and a recognition of the immediate need for the formation of partnerships between the public and private sectors. Funding issues for tourist towns and localities require special attention, especially due to the failure of traditional state funding sources to realize in a timely fashion the importance of carrying capacity during the tourist season. Therefore, the utilization of innovative funding mechanisms such as value capture techniques, fees for museums, archaeological sites, and building permits, or operating licenses, in order to fund planning and research for infrastructure, professional tourism training and research as well for improvements to service quality and sustainability should be seriously considered.

The marked advancements of the less developed Mediterranean countries in socioeconomic and political spheres have diminished their 'exotic' nature as well. This occurred in southern Europe in the 1960s and 1970s and is presently happening in the Maghreb countries (i.e., Morocco). On the other hand, developed countries such as France, Italy, and Spain have been able to significantly overcome problems of negative tourism externalities (i.e., cultural commodification, demonstration effects). Issues such as the rights of minorities and women, political instability, religious fundamentalism, and the preservation of historical resources and natural environments pose serious challenges for the future of the region's tourist industry – especially for the less developed states. Such problems will eventually need to be confronted and managed, if the tourist industry is to continue to be a major agent of social and economic development in these countries – but not only for tourism's sake.

Finally, although it may appear difficult to draw general conclusions, Mediterranean tourism holds immense potential for further growth and development. The promotion of natural, cultural, and historical resources as well as of alternative forms of tourism, which take advantage of regional peculiarities, can undoubtedly contribute toward this direction. Each country needs careful tourism planning to protect its physical and cultural environment and avoid monoculture. Planning decisions should respect the rights and nuances of different value-systems and cultures and reinforce and preserve special identity aspects. The commonality of problems reinforces the need for a unified effort to share resources, to undertake co-operative promotion, to exchange information, and to co-ordinate actions with regional co-operation. Tourism should not be regarded only as a business, but as interaction – a channel to understand other peoples, and develop and maintain positive national images.

In summary, the relationship between Mediterranean tourism, socioeconomic development, and cultural change is contingent. Above all, it depends on the structure of the tourist industry itself and the nature of the local, regional, and national economy. There exists, therefore, enormous diversity of socioeconomic and cultural experiences in the Mediterranean as the following chapters illustrate.

Organization and themes of the book

The unprecedented growth of tourism originating in the 'Old World', as a consequence of the post-World War Two emergence of mass travel, has resulted in a significant tourist influx to the 21 Mediterranean nations and a high ranking of tourism in their national economies. The tourism sector extensively affects policy-making, employment, culture, the physical environment, political stability, and the balance of payments. Further, tourism is continually emerging from the shadows of economic policy to take a centre-stage position. It is imperative that the Mediterranean tourist industry be evaluated and critically examined in terms of recent trends, economic organization, and contribution to both socioeconomic development and cultural change.

There have been previous attempts to investigate whether tourism is a 'passport to development' or whether it is a 'blessing or blight' and to broadly measure its economic, political, sociocultural, and environmental effects. From these efforts, several books on Caribbean, Pacific, Southeast Asian, and Western European tourism have been published. The Mediterranean, however, as a separate and touristically substantial region, is rather under-represented. This book, as the first of its kind in the international literature, aims to introduce the largest, oldest, and most developed touristic region in the world as a coherent unit of analysis in the international literature. The Mediterranean region is a system in itself and must be understood as such – therefore, its sociocultural and environmental well-being is a priority, first and foremost, for the region itself. Both its future and the shape and form of any partnerships it may have with outsiders, need to be determined in the first instance by insiders: regional issues need regional responses (Selwyn 1996). It is the purpose of this volume to set the stage for filling this significant gap in the international literature. It focuses on the identification and analysis of recent trends in Mediterranean tourism; it provides a wide-ranging synthesis of the progress made in related research; and it concludes with a discussion of the future prospects of this research.

The biggest challenge facing the editors of this volume was to locate and commission high-calibre scholars and practitioners of Mediterranean tourism. The search yielded a cadre of 21 academics and professionals from 12 countries around the world who have dealt with the unique challenges faced by and caused by Mediterranean tourism, in their roles as professors, researchers, consultants, managers, and advisors. Their experiences have afforded this volume meaningful and applicable insights on the complex phenomenon of Mediterranean tourism. Efforts have been made to include chapters that go beyond mere description of tourist destinations and to provide critical analysis of the dynamics of the region's tourism development. In doing so, it is hoped that this book will be utilized for its potential contribution not only to debates over Mediterranean tourism, but to tourism in the global context. It is also hoped that it will serve as a catalyst for further research and scholarship on Mediterranean tourism. With these in mind, the book is intended for three main audiences. First, it targets tourism specialists in the social sciences who may be interested in the organization, structure, and effects of Mediterranean tourism. Second, it aims to serve as a scholarly text for students of geography, sociology, tourism studies, anthropology, management, and planning and all those interested in enhancing their understanding of Mediterranean tourism. Finally, this text should be of interest to the various players representing the tourist industry, but also policy-makers who wish to enhance their cognizance of the sector's workings.

The book is divided into 19 chapters and organized along six broad themes. While each chapter is written so as to take into account the particular characteristics of tourism in each country, certain themes recur throughout the text. The most important of these are: changes in the organization of the tourist industry and its responses to demand shifts; tourism's contribution to socioeconomic development; tourism's effects on culture, heritage and traditions; inequalities in access to tourism; the political economy of tourism; tourism planning and policy; sustainability, mass tourism and alternative tourism forms; innovative forms of tourism management; tourist investment patterns and types; and the relationship between tourism, political instability, and religious fundamentalism. Further, although the volume includes chapters on the tourist industries of countries and topics for which little or no substantial work has been conducted to date (i.e., Algeria, Croatia, Egypt, Lebanon, Morocco, Syria, Tunisia, Turkey, residential tourism, marine tourism), there are some unintentional omissions, such as the couplet of metropolitan (urban) tourism and agrotourism (rural tourism), as well as several countries with Mediterranean coasts but less focus on the tourism industry. We hope to be able to rectify this in a subsequent edition.

This chapter of the *Introduction* explains the need for the existence of the book and provides a comparative overview of the dynamics and prospects of Mediterranean tourism as a whole. Apostolopoulos, Loukissas, and Leontidou set the stage for the book and focus on the history and geography of the region and further analyse the importance and contribution of tourism in the region's development. Part I comprises five chapters which focus on the tourist industries of five countries situated in the northern shores of the Mediterranean basin. The four main Mediterranean destinations of Spain, France, Italy, and Greece along with the newly established state of Croatia, which includes the famous Dalmatian shores and islands of the Adriatic Sea, are examined by Monfort and Ivars, Boyer, Montemagno, Apostolopoulos and Sonmez, and Vukonic respectively. Part II includes five chapters which centre on the eastern shores of the Mediterranean, dealing with the fast growing tourist markets of Turkey, Cyprus, Syria, Lebanon, and Israel. Var, Ioannides, Gray, Mallat, and Mansfeld respectively present the uniquenesses and peculiarities of the tourist sectors of the five aforementioned countries. Part III comprises four chapters, which concentrate on the southern shores of the Mediterranean, and that deal with the interdependencies of development, tourism, and Islamic fundamentalism. The chapters of Var and Imam, Poirier, Zoubir, and Barbier focus on Egypt, Tunisia, Algeria, and Morocco, and illustrate how political turmoil and instability along with religious fundamentalism can influence tourism development. Part IV comprises three additional chapters focusing on the spatial distribution of tourism and the emergence of new types of tourism in the Mediterranean. Loukissas and Skayannis, Leontidou and Marmaras, and Kradonellis examine the issues of sustainable development and environment, residential/second-home tourism, and marine tourism respectively. Finally, in the Epilogue, Apostolopoulos and Sonmez revisit the issues discussed and underline the overall prospects of the tourism sector in the Mediterranean, suggesting that globalization should be dealt with through innovative forms of restructuring and co-operative marketing.

Notes

1. The Mediterranean basin is an almost enclosed sea covering an area of 3,000,000 km². Twenty one countries share its coastline: the European countries of Spain, France, Monaco, Italy, Malta, Slovenia, Croatia, Bosnia-Herzegovina,

Yugoslavia, Albania, and Greece; the Asian countries of Turkey, Cyprus, Syria, Lebanon, and Israel; and the African counties of Egypt, Libya, Tunisia, Algeria, and Morocco.

2. The collapse and subsequent instability of Eastern Europe, the Gulf War, the civil wars and bombing in the former Yugoslavia and the Balkans, the financial turmoils in Southeast Asia, Japan, Russia, and Latin America and their subsequent sociopolitical impacts, and an overall fluid international state of affairs have been serious setbacks for the region's tourism.

3. The Blue Plan and The Economist Intelligence Unit have outlined several possibilities for the development of tourism between the years 2000 and 2025, according to varying hypotheses for economic development. By the end of the year 2000 it is estimated that arrivals in the 21 countries could vary from between 268 million and 409 million (Grenon and Batisse 1989) or 330 million (Jenner and Smith 1993) while by 2025 arrivals could reach between 379 and 758 million (Grenon and Batisse 1989). See also Williams (1997).

4. In North Africa, tourism accounted for 6.8 per cent of GDP in 1998 (annual growth of 6 per cent) and contributed 2.2 million jobs (7.4 per cent of total employment). On the other hand, in the Middle East, tourism contributed 2 million jobs (6.1 per cent of total employment) and accounted for 7.3 per cent of the GDP (annual growth of 5.2 per cent) (WTTC and TBR 1999).

Part I

Northern Mediterranean shores

Transformation of the mature tourist destinations

2 Towards a sustained competitiveness of Spanish tourism

Vicente M. Monfort Mir and Josep A. Ivars Baidal

Introduction

In this chapter we will analyse tourism in Spain (a country whose territory's surface is approximately half a million kilometres and which has nearly forty million inhabitants), one of the most important tourist destinations in the world, greatly benefiting from its Mediterranean location and features. After briefly referring to the historical evolution of tourism in this country, we will make a profile of the resulting tourist model, directly associated to the rise of mass tourism in Western Europe. The Spanish model highlights tourism's important implications in the economic as well as in the environmental and socio-demographic contexts. From the point of view of tourism, Spain is a very rich model to study in terms of the magnitude of its tourist establishment and its great tourist attraction capacity, within the framework of massive consumption patterns.

In a context of significant changes affecting demand and tourist-production schemes, it is interesting to do research on how a first-order traditional destination adapts to these dynamics. The Spanish case offers important teaching examples and highlights factors which need correcting. This is why the future of tourism in this part of the Mediterranean faces certain risks while simultaneously having interesting opportunities at its disposal. All these aspects are subject to analysis in order to detect which are the key factors that must be taken into account if we are to maintain Spain's competitiveness in the global-tourist market.

The genesis of tourism in Spain

If we briefly look over the evolution of tourism in Spain in recent years, various periods can be seen to follow one another, periods that are different in their duration and coincide with just as many ways of understanding and practising tourism-related activity, both from the perspective of administrative intervention (what is known as tourist policy) and in its practical aspect, as it has been developed by its social and managerial main agents.

In the late nineteenth and early twentieth centuries, the notion of tourism was linked to the recreation of a wealthy or well-to-do minority of the population – no matter if people were dealing with trips to summer resorts in the same country or if they were referring to the reception of foreigners who showed an interest in discovering the Spanish heritage with regard to art, monuments and landscape, the heritage of historical, enlightened and romantic travellers. They were minor movements of people which hardly affected the everyday life of the areas visited, and had no influence on the national economy. Nevertheless, their incidence was increasing during certain months of the year and grew incrementally (building of residential villas on the coast, development of recreational facilities: casinos, theatres, and so forth). At that time it was already becoming clear how important the contribution of foreign currencies could become in a country lacking these resources derived from foreign trade.

In the early 1950s, the boom of European tourism, associated with the recovery of economic growth, was spread all over the western Mediterranean, while the international isolation of Franco's regime was broken. In 1951, the number of foreigners that went through the border exceeded one million: the flow of foreign currencies, so necessary for the Spanish economy, grew to considerable proportions. The new situation led the government to reconsider the status of tourism: in July 1951 it created the Ministry of Information and Tourism. Private initiative became aware of the opportunities created by the beginning of the boom in Spain, due to its proximity to origin markets, the great appeal of its resources (climate and beaches), exotism and the low price level. In 1956, the volume of hotel supply surpassed 100,000 beds; in Benidorm, a Town Development Plan was approved which considered the expansion of tourist accommodation.

In 1959, the government launched what would be known as the 'Stabilization Plan', which laid the foundations of a certain opening-up towards the outside with the aim of modernizing the country's economy and achieving a limited recognition for its special political reality – in this way closing the autocratic stage that it had been dragging since the Allies' victory in World War Two. That plan allowed for the entrance of a new ruling class with other characteristics – the 'technocrats', whose knowledge and training turned out to be fundamental for Spain's tourist interests. Later development plans would mark the Spanish economic policy coming from the 1959 Plan, just as many Tourism Commissions would appear which would define the objectives and instruments of tourist policy (Bote and Marchena 1996: 303), obtaining, for this reason, an unprecedented recognition for tourist activity that did not continue later during the democratic period. In this context, the number of visitors increased several times, reaching more than 5 million in 1960.

The undeniable importance of tourism for the Spanish economy, particularly in regard to insular and coastal regions in the Mediterranean, was reflected in the government's policies. The access to the government of a technocrat ruling class, whose mission was the orderly management of economic development within the framework of the regime's political and social rigidity, gave rise to a reorganization of ministries in July 1962, as far as the Ministry of Information and Tourism was concerned. The incoming team were prepared to encourage the regulation of private tourist activity, inside the rigid control system, from the public sector. The fields in which they were going to work included the statute of private enterprises and activities, the encouragement of promotion initiatives (both private and institutional), the promotion of hotel credits to enterprises and Town Councils, the regulation of the tourist accommodation plant (hotels and campsites), the relaunching in the construction of 'Paradores Nacionales' (hotels resulting from the rehabilitation of buildings considered valuable as monuments or newly-built in places of interest for tourism, a concept created in the late 1920s), promotion abroad through direct participation in international fairs and professional forums, and so on.

Nevertheless, a lack of co-ordination existed between tourist policy measures and other branches of the general policy developed by the government and the local administrations. This co-ordination would have given the Spanish tourist offer in the most often demanded areas more efficiency and quality. Instead, areas such as land and air communication infrastructures, the care of the town and landscape environment and cleaning-up networks were neglected, and a long list of facilities which were not kept in step as they should have been, in order to achieve an optimal development of the Spanish tourist product. This brought about the configuration of the offer as a uniform, mass product and, therefore, scarcely differentiated among the various coastal destinations, with the exception of some municipalities.

Much more importance was assigned to the amount of tourists being received and their immediate profitability (the yearly maximization of foreign currencies coming in as a result of this activity), than to the option of gradual and conscious selection of different products. Products of pro-grammed quality, in accordance with the diverse carrying capacities of each tourist destination, in such a way that a diversified offer capable of meeting the needs of different types of demand with a varied spending capacity, could have been formed during the 1960s.

The Spanish tourist model

The contribution of tourism to the Spanish socioeconomic development: environmental and socio-demographic implications

The analysis of tourist magnitudes on a worldwide scale highlights the important role that Spain plays in this branch of productive activity. This situation is reconfirmed every year with the growth, both in arrival figures

Table 2.1 Main countries in terms of income and tourist arrivals

	Tourist receipts, 1999				Tourist arrivals, 1999			
	Rank 99 98	Millions of dollars	Rate of change 99/98	World share	Rank 99 98	Thousands of people	Rate of change 99/98	World share
USA	1 1	73.000	2.5	16.0	3 3	46.983	1.3	7.2
Spain	**2 4**	**25.179***	**9.5**	**7.2**	**2 2**	**51.958**	**8.8**	**7.9**
France	3 2	24.657*	6.3	7.0	1 1	71.400	2.0	10.9
Italy	4 3	31.000	4.0	6.8	4 4	35.839	2.9	5.5
UK	5 5	20.972	0.0	4.6	6 5	25.740	0.0	3.9
World's total		**445.000**	**3.2**	**100**		**657.000**	**3.2**	**100**

Source: World Tourism Organization.
*9 months estimation

for international tourist travellers and in their income replica, with advances above the world's average in both cases.

According to provisional data made public by the World Tourism Organization (WTO), corresponding to 1999, Spain possessed 7.2 per cent of the world's share of tourist income and 7.9 per cent of international traveller arrivals. These figures placed Spain in second place in the world for tourist income, simultaneously keeping its second place as a tourist destination, behind France.

After the extensive development of Spanish tourism was consolidated during the 1960s and 1970s, its growth has become more moderate, although it describes a tendency towards sustained advance. The most recent cause of instability was detected in the early 1990s, after the reduction in tourist arrivals as a result of the appreciation of the peseta in international markets. In spite of the loss of competitiveness-price that this meant, Spain has consolidated its position as a world-tourist destination, successfully competing against the more exotic destinations that have emerged in Asia and Central America and even other, nearer destinations such as the ones in the African Mediterranean and, above all, countries in Central Europe, which have vigorously incorporated themselves into the holiday panorama during the last decade. For some decades now, tourism has behaved as one of the most stable and booming activities in the Spanish economy. The assessment of the contribution of tourism to the Spanish productive system can be summarized in the analysis of the following macromagnitudes:

Table 2.2 Tourist balance of income and payments (thousand millions of pesetas)

	Year 1998	Year 1999	Rate of change 99/98 (%)
Receipts	4458	5085.1	14.1
Payments	747.2	868.9	16.3
Tourist balance	3710.8	4216.2	13.6

Source: Bank of Spain.

1   *Income, payments and balance from tourism.* Indicators in monetary terms indicate a clearly positive tourist balance, which has historically contributed, to a large extent, to the compensation of the structural imbalance existing in the Spanish commercial deficit.

In the context of the balance of payments of the Spanish economy the starring role played by tourist activity stands out, where the capacity to generate gradually-increasing income in the last 30 years has helped to compensate the balance on the current account and to succeed in rationalizing the Spanish trade balance.

The current account balance reflects the flow of goods and services which Spain maintains with the rest of the world. In 1996, this balance showed a surplus of US$1.77 billion, which explains the rationalization provoked by the income derived from the commercialization of the

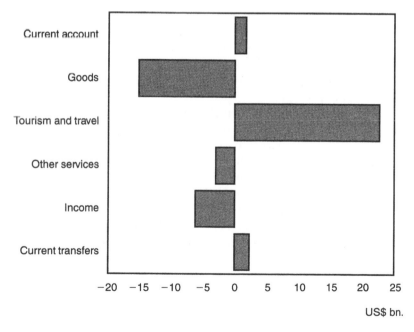

Figure 2.1 Current account balance 1996.
Source: Bank of Spain.

different tourist destinations in Spain. The tourist balance has the capacity to finance the loss-making parties in the Spanish balance of payments. Hence we can see the strategic value that tourism has established in the heart of the Spanish economy, since it has the capacity to make contributions of economic resources in an amount sufficient to channel the necessary imports in order to maintain the process of modernization and to gain in competitiveness for its productive system. This situation has been a constant feature in the Spanish economy since the Stabilization Plan of 1959, tourist income having the capacity to cover the deficit of Spain's foreign trade in recent years.

2   *Contribution of tourist activity to the GNP.* The economic effects generated by tourism directly affect the activities normally referred to as 'tourist sector', however, many other sectorial activities would be indirectly influenced by tourist expenditure. The intersectorial nature of tourist activity is reflected in the Tabla input–output de la economía la turística española (TIOT-92, Input–Output Table of Spain's Tourist Economy). The application of the table allows the accurate quantification of tourism's real economic impact, given that, after the calculation of various multiplying coefficients, we can obtain the effects induced by the total tourist demand in the economy as a whole. In general, the tourist intersectorial multiplier is estimated at 1.71, a figure that reflects how influential tourist activity is in the whole economic system.

After the application of this methodology, which takes into account the spreading phenomenon of tourism, the contribution of total tourist production (direct and indirect) to the Spanish GNP is estimated at about 18.55 per cent for 1995, which clarifies the specialization and how strongly dependent the Spanish economy is on the tourist sector at present.

3   *Employment-generating activity.* One of the most spectacular effects of tourist activity is, undoubtedly, the employment opportunities it generates. The difficulties mentioned previously regarding the assessment of tourist activity have not allowed us to carry out a census that is accurate enough to express the direct and indirect figures of tourist employment rigorously. The distribution of tourist employment in 1995 is situated at 1,147,000 jobs (9.52 per cent of the country's working population), whose structure responds to the following breakdown (data calculated from the TIOT'92 tourist intersectorial multiplier): direct employment: 671,000 workers; indirect employment: 476,000 workers.

It must be kept in mind that, whereas the country's working population only grew by 2.66 per cent in the 1995–96 period, tourist activity reached an interannual variation of 5.5 per cent. However, it must be remembered that employment in tourism suffers from two worrying effects: low salaries (among other reasons due to the low specialization) and, above all, a marked temporality as a result of the high seasonality the sector is characterized by.

From the social and demographic point of view, tourism has once again appeared as an agent of transformation in the structure of tourist areas, even with implications for the rest of the state. The clearest effects of tourism are the increase in the population of tourist areas, derived from a considerable immigration, coming both from inner areas of those tourist regions and from other Spanish regions with more limited possibilities for development. The immigration flow is made up of two different types of flows (Vera 1993: 496–9) with different effects on the demographic structure: on the one hand is the most important flow which is going to shape the work force in tourism and in the building sector – the young population with high growth rates which rejuvenates the age pyramid; and, on the other hand, a population contingent which fix their residence on the coast for various reasons, among which we can underline the almost permanent residence of foreign old-age pensioners in different sections of the littoral region, as is the case in the Costa Blanca or the Costa del Sol, whose effects are just the opposite to the ones mentioned previously, as they provoke the population's ageing. In several tourist towns, we can check high percentages of foreign population in the census, which can reach, in extreme cases, about 50 per cent of the total number of residents. As a result of the introduction of tourist activities, a tertiarization of the local economy takes place as well as a clear increase in the active population dedicated to the building sector in the provinces with the highest tourist development. In the same way, as a counterpoint to the deep social transformations, the 'quality of life' indicators in these tourist areas usually show registers above the country's average (Valenzuela 1988: 52–4). From the environmental point of view, the Spanish tourist development has a special significance, both for its undeniable effects on the environment and for being, at present, an essential way of increasing the sector's competitiveness and of guaranteeing its sustainability.

It is commonplace to criticize tourism as environmentally devastating, above all in a country that has experienced a high growth of a vast tourist plant; however, it is proved that the environmental impacts of tourism, like the sociocultural ones, cannot be attributed to the activity itself but rather to the way it has been developed. Environmental problems generated by tourism derive from the lack of planning and foresight of its impacts, in which we find the incidence of the permissiveness on the part of the Administrative Authorities (dazzled by the economic profits of tourism in the expanding period of this activity in Spain), the obsolescence of the previous legislation and the role of economic agents that want a short-term profitability for their investments (Vera 1993: 495), within a context of strong demand.

The concentration of tourism in narrow coastal fringes, very often in fragile and valuable ecosystems, accounts for a great pressure on scarce resources during the expansive stage in which the offer is created and in its later exploitation, with a high seasonal component which makes that pressure worse. Three factors are considered essential within the framework of the environmental problems generated by tourism: the land, the water supply and the landscape (Vera 1994: 137). The expansion of the tourist offer caused, and still maintains in many tourist municipalities, an evident competition with regard to the uses of the land. The traditional agricultural exploitation gave way to the new developments which made it possible to obtain quick profits in marginal agricultural lands, but also in fertile littoral plains, although the latter logically showed more resistance to the change in use, generating in many places an unadvisable tourist monoculture. These effects must be directly related to the huge amount of real-estate promotions received by Spanish tourist areas, which has made the building sector the one which has benefited most from the location of tourism on the Spanish coast. Already in 1987, it was estimated that at least 50 per cent of the seaboard was developed or classified for urban, port, industrial and, essentially, tourist uses (Secretaría General de Turismo 1994: 23), a percentage that increases considerably in the communities having a higher tourist specialization.

The different tourist-growth processes have caused evident transformations in the territory which have been felt in landscape modifications, very deep ones on some occasions. The use of tourism as an instrument for the preservation of the environmental patrimony, as well as of the built heritage, was left hidden behind the magnitude of tourism growth and its well-known effects. Nevertheless, it can be clearly seen that it is necessary to favour an environmental improvement in consolidated destinations and, likewise, to include the approach of sustainable tourist development in new products and emergent tourist areas.

The importance of foreign demand and the growing role of domestic tourism

Spain retains a huge volume of foreign demand, although with more moderate growth rates than in the years of development of mass tourism in the late 1950s and the 1960s. We can only underline two recession periods for arrivals of international tourists which coincided with the effects of the oil crisis (1974–76), and with the more recent negative upturn suffered between 1989 and 1991. Table 2.3, which refers to international visitors (not strictly tourists) illustrates this situation.

The concentration of motivation around sun and beaches enjoyment is a basic characteristic of the demand. In fact, the 'sun-and-beach' product accounts for a market share of 74 per cent of the trips to Spain made by Europeans, whereas products such as cultural, nature, business, rural, health or sports tourism, do not reach, in combination, 25 per cent

Table 2.3 Entrance of foreign visitors

Years	Visitors (thousands)
1959	4194.7
1960	6113.3
1961	7455.3
1962	8668.7
1963	10,931.6
1964	14,102.9
1965	14,251.7
1966	17,251.7
1967	17,858.6
1968	19,184.0
1969	21,682.1
1970	24,105.3
1971	26,758.2
1972	32,506.6
1973	34,558.9
1974	30,342.9
1975	30,122.5
1976	30,014.1
1977	34,266.8
1978	39,970.5
1979	38,902.5
1980	38,026.8
1981	40,129.3
1982	42,011.1
1983	41,263.3
1984	42,931.7
1985	43,235.4
1986	47,388.8
1987	50,544.9
1988	54,178.2
1989	54,057.6
1990	52,044.1
1991	53,495.0
1992	55,330.7
1993	57,263.4
1994	61,428.1
1995	58,359.9
1996	61,785.4
1997	64552.0
1998	70859.0
1999	72300.0

Source: General Secretariat of Tourism.
Note: From 1995 a new statistical system has been applied for measuring foreign visitors.

(Esteban 1996: 257). The predominance of this motivation logically explains the geographical concentration we are going to deal with later on in this chapter. As for the distribution of foreign demand according to nationalities, in Table 2.4 we can observe the preponderance of the

Table 2.4 Main origin markets for Spain in terms of overnight stays (%)

	1991	1992	1993	1994	1995	1996
Germany	34.9	34.1	33.8	33.5	34.4	34.2
United Kingdom	25.6	25.2	26.8	28.3	28.3	27.7
France	9.0	7.9	7.6	7.0	6.7	6.4
Belgium	4.3	4.5	5.1	5.0	4.9	4.8
Italy	6.2	6.8	6.9	5.8	5.2	4.2
Netherlands	3.3	3.2	3.6	4.4	4.0	3.9
Switzerland	3.0	2.7	2.6	1.5	1.5	2.2
USA	1.7	2.0	1.7	1.7	1.8	1.9
Sweden	1.4	2.3	1.7	1.6	1.2	1.3
Portugal	1.1	1.1	1.1	1	1.1	1.2
Other countries	9.4	10.4	9.2	10.1	10.9	12.1
Total	100.0	100.0	100.0	100.0	100.0	100.0

Source: Statistics National Institute (INE).

German- and British-origin markets, channelled towards the Spanish coast through tour operators located in the source markets, another of the elements inherent to the historical and current development of tourism in Spain. Indeed, tour operators exert a direct influence on the demand thanks to their negotiation power, derived, among other reasons, from the millions of tourists they carry every year, the marked processes of enterprise integration (travel agencies, airlines and, in many cases, hotels) and their aggressive communication and commercialization strategies. On the other hand, we can state that, as far as international tourism is concerned, the independent travel trend has not yet reached a significant level. Instead, we do consider the emergence of origin markets in some Eastern European countries like Russia, the Czech Republic or Poland.

Most tourists travel by car and plane. However, considering the entrance of tourists exclusively, the plane is the most commonly used means of transport. The demand's seasonal behaviour is another basic feature which is largely due to the type of tourist product in which Spain is specialized. This phenomenon has destructive effects in the sector's employment, the training of human capital, the enterprise profitability and the under-use of general infrastructures; whereas, on the other hand, it means a congestion during the summer months with impacts of different kinds (e.g. crowded tourist attractions, traffic congestion, noise pollution, natural resources under additional pressure, etc.) produced by the concentration of tourist flows, both in time and in space. However, seasonality varies depending on the tourist areas considered and it can already be seen how some receiving areas have succeeded in overcoming it, independently of the Canary Islands, which, due to the mildness of their climate, present a more uniform distribution of the yearly demand than the Spanish average. The case of Benidorm, located in the Valencian Autonomous Community, shows that it is possible to overcome seasonality in a 'sun-and-beach' destination. The destination has an occupancy rate of about 90 per cent throughout the year, thanks to a dynamic and renewed tourist supply, integrated in the urban environment, and with an intense promotion that enables it to have a good position both in the national and international markets.

The fragmentation of the main holiday trip into several trips throughout the year, a trend that is visible in the European demand, may be an opportunity to palliate the effects of seasonality, although this can only be efficient if products other than 'sun-and-beach' ones are developed, since the latter loses much of its appeal in short breaks outside the summer months.

The growth of domestic tourism is explained by taking into account the socioeconomic progress that has taken place in Spain as well as the importance assigned to tourist expense in the context of family expense structure. From the point of view of tourism, the recognition of the relevance of domestic tourism (that of Spanish people inside Spain) became clear in the recess of international demand which took place in the late 1980s and early 1990s, a moment in which national tourism helped to balance the loss of foreign tourism in many Spanish regions. In 1995, 58 per cent of the Spanish population, aged fifteen or older, went on some tourist trip with at least one stay overnight away from home, which means a significant leave rate. It shares the seasonal character with foreign tourism, although it offers some different characteristics (Instituto de Estudios Turísticos 1995: 16–20): the main accommodation used is 'free housing', either owned or transferred by relatives or friends (a circumstance that refers us directly to the importance of non-commercial supply in Spanish tourism and the evident weight of the second residence and the non-hotel offer in many tourist areas). Most trips are organized by the travellers themselves, thus the use of travel agencies represents a very small percentage; the most commonly used means of transport is their own vehicle. Only 8 per cent of the trips realized in 1995 had a foreign country as their destination.

The concentration of tourist supply in the continental Mediterranean and in the insular archipelagos

In accordance with what has been explained previously, it is easy to prove the marked concentration of the Spanish accommodation supply in the continental Mediterranean coast and in the archipelagos of the Balearics and the Canaries. Figure 2.2 shows the varied contribution of tourism to the GNP of the different Autonomous Communities. However, it is advisable to go beyond the analysis of the traditional supply based on the figures available for hotel beds, since they represent a modest percentage if compared to the total number of beds, although the remaining ones show great difficulties for their statistical treatment. Spain's receiving capacity is estimated at 11.5 million non-hotel beds (self-catering, second homes, including unregistered supply), which can have tourist use all year round or during some part of the year, whereas the hotel supply, of which

Figure 2.2 Contribution of tourism to the GNP of the Spanish autonomous
           communities.
Source: Leno (coord), 1995.

Table 2.5 Regional distribution of hotel and non-hotel places

Autonomous Community	Hotel beds	Non-hotel beds	% of non-hotel beds from the total number
Andalusia	135,319	1,816,000	93.07
Aragon	24,364	251,000	91.15
Asturias	12,901	199,000	93.91
Balearic Islands	260,247	544,000	67.64
Canary Islands	90,717	596,000	86.79
Cantabria	14,860	149,000	90.93
Castilla-La Mancha	15,731	558,000	97.26
Castilla and Leon	34,083	791,000	95.87
Catalonia	197,962	1,784,000	90.01
Extremadura	9999	239,000	95.98
Galicia	38,458	643,000	94.36
Madrid	49,490	521,000	91.33
Murcia	10,965	400,000	97.33
Navarre	6695	75,000	91.80
Basque Country	13,023	195,000	93.74
La Rioja	3327	59,000	94.66
Valencia	78,210	2,219,000	96.60
Total	996,351	11,039,000	91.72

Source: General Secretariat of Tourism.

Note: Data for non-hotel places correspond to an estimate carried out in 1992. With the aim
of being able to compare the data, we have equally used the data corresponding to 1992 as
regards the hotel offer.

a much more rigorous census has been taken, reaches a figure near one million (Instituto de Estudios Turísticos 1997: 35), which means 5.3 per cent of the world's hotel capacity according to the data provided by the Spanish Hotel Federation. The disparity between hotel and non-hotel places (whose regional breakdown can be seen in Table 2.5) is, to a great extent, explained by the real estate property implementation in Spanish tourism (Vera and Marchena 1996: 329–31), according to which tourist enjoyment of the coast was channelled, apart from hotel investments, through the building of tourist apartments or houses, with different typological results (blocks of apartments segregated from the traditional urban settlements in the seaside, complexes of tourist residential areas, and so forth). This is spurred on in many areas by an important foreign investment, and has sensitive economic, social and territorial impacts.

To make comparisons easier, we have consistently used 1992 data regarding the hotel supply. The regional distribution hides the existence of subregional spaces with more tourist intensity inside each Autonomous Community. Thus, the highest concentration in tourist supply, if we consider hotel places exclusively, is found in the provinces of Baleares, Santa Cruz de Tenerife and Las Palmas in the Balearic and Canary Islands, Girona, Barcelona, Tarragona, Alicante and Malaga on a Mediterranean coast with varied degrees of tourist specialization; and Madrid, taking into account its position as capital city of the state. The notorious concentration of the supply becomes evident when we see that only five Autonomous Communities (Andalusia, the Balearic Islands, the Canaries, Catalonia and Valencia) own 76 per cent of the whole hotel supply. From the point of view of the distribution of non-hotel offer, its spectacular development is clear in Andalusia, Catalonia, and Valencia, whereas in insular communities, the Balearics and the Canaries, the percentage of non-hotel offer is reduced in favour of hotel supply. Some of the most outstanding features in the structure of the hotel offer are the speed in its building, which goes parallel to international demand, and the encouragement of tour-operators who collaborated in financing the building of hotels in littoral destinations. In 1973, about 70 per cent of the present hotel capacity had already been built, this percentage being higher if we refer to specific coastal destinations. The dramatic growth of the offer has implied the obsolescence of a large part of the hotel plant which must receive renewal investments in order to keep its competitiveness in a process that has luckily been taking place for some years in the Spanish hotel industry. Likewise, we observe a substantive decrease in the lower accommodation categories (cheap hotels and boarding houses), the predominance of four- and, above all, three-star hotels in coastal destination areas, and the significant increase in high-quality hotels in urban contexts, mainly Madrid and Barcelona.

As far as the non-hotel offer is concerned, it is necessary to refer to its high illegal component, that is, unregistered offer on a tourist-exploitation basis. This offer, out of control by definition, means unfair competition for regulated establishments, escapes its legal treatment (administrative, fiscal, and so on), shows the problem of obsolescence of the offer which becomes worse in some areas, and its quality level (both in terms of physical parameters and as regards service quality) is, to say the least, uncertain, affecting the tourist's level of satisfaction and generating a negative image of some Spanish destinations. Without doubt, the existence of this huge illegal offer has to do with the exploitation of the important volume of houses that have appeared as a result of real estate property development, which, related to tourist enjoyment, has taken place in Spanish littoral areas. Both the Central Administration and the Autonomous Communities are making efforts, either through legislative developments or through an increased inspection in tourist areas, in order to avoid the pernicious effects generated by the commercialization of tourist places outside the current legislation.

The trend towards the diversification of tourist products

We have so far presented the basic structural features of the Spanish tourist model, that is, the demand's concentration in terms of motivation and its correlation with the location and the type of tourist offer developed. However, new elements must be incorporated in order to explain a more complex Spanish tourist model, which arises as an adaptation to the trends detected in the national and international demand: experienced tourists, fragmentation and multiplication of trips throughout the year, independent travel growth, greater environmental requirements, search for new leisure possibilities other than the traditional uses, an interest in authenticity, higher prestige for active and cultural tourism, a marked segmentation of tourist demand from different points of view (motivational, according to socioeconomic and demographic variables, geographic, psychographic or according to the consumer's behaviour).

These new patterns in tourist consumption have had two clear effects: the need to improve traditional littoral products in order to maintain their appeal and the creation of new products on the grounds of the appearance of the 'new tourist demand'. The implementation of new products, still the first and foremost ones demanded by the tourism practised by Spaniards themselves taking into account the international demand's preferent motivation, allows the spreading of tourism into inner spaces, which see in this activity a chance for regional and local development through the economic diversification of their productive structures. This is the case both in rural areas (largely dependent on inner Spain), on a declining primary sector, and in urban spaces (which obtain, by means of tourism, a new source of income and employment generation as well as an instrument for the tertiarization of their economies). In a very tight synthesis, we can illustrate the incipient Spanish tourist-diversification process paying attention to the current situation of nature and urban tourism, including, as a symptomatic example, the tourist development initiative for 'The Green Spain'. Nature tourism can be conceived in a wide sense, including different tourist products such as rural tourism, eco-tourism, certain sports and adventure tourism, and so forth. Despite the nonexistence of a specific definition, we can identify this type of tourism as the one having its main motivation in nature enjoyment. It goes without saying that as the ecological awareness grows in this end-of-the-century society, growth expectations for this kind of tourism are encouraging. Figures speak for themselves: in 1999, Spanish national parks received over nine million visitors; in 1995, the last research available for the whole of Spain, more than five million Spanish tourists made at least one trip into rural spaces, considering source towns with more than 100,000 inhabitants and excluding weekend-trips, which means that we could even talk about higher figures (CSIC 1995). The demand for rural or inland tourism in Spain thus constitutes a significant volume and an important economic flow, distributed all over the Spanish regions, since rural tourism is widespread, unlike littoral tourism, whose main feature is its concentration, although certain regions also exist with more reception of visitors in search of rural tourism. Two main motivations stand out: family ties (derived from the rural exodus which went with the emigration from the country towards the town during the country's industrialization and urbanization process, above all during the 1950s and the 1960s, which now generates contrary flows in the shape of 'return' tourist trips) and nature enjoyment. The source of this demand is mainly found in big cities; holidays in small towns or villages prevail, some of the most outstanding features being that holidays are organized by the tourists themselves and the abundant use of private accommodation (houses owned by relatives or friends, second home).

Parallel to the growth in the demand, a considerable increase has taken place in the rural-tourist offer, in the shape of hotels adapted to the demand's new tastes along with other types of rural accommodation among which stand out rural houses. Likewise, the number of enterprises offering recreational activities in the rural milieu (sports, educational stays, and so on) have increased several times over. Undoubtedly, rural tourism development is accounted for by the greater inclination on the part of the demand towards the consumption of these products, the decline in traditional rural activities and the important role played by tourism in current policies for rural development (Common Agricultural Policy, Structural Funds, Mountain Policy, European Community Initiatives like LEADER, and so forth). Within the framework of the LEADER I initiative (1991–93), 52 programmes were approved in Spain for the development of rural areas, in which investment in tourism represented 52.5 per cent of the total investment made coming from the contribution of the European Community structural funds, the national public administrations and the private sector (Beltrán 1995: 27–31). Co-financed activities offer very different characteristics as they range from the simple elaboration of a tourist brochure to the creation of an accommodation or recreational offer, without forgetting others such as regulation and management of rural tourism.

Urban tourism also shares good growth expectations from various factors: the cultural attraction of many Spanish cities; business tourism; fairs, congresses and conventions tourism, and the development of events with the capacity to attract tourism. As the tourist improves their formation and interest in culture, as cities have incorporated new cultural elements to their heritage, and as the tourist supply becomes more versatile and adapts itself better to a new profile of consumer (special prices for weekend visits, discounts for cultural events, and so on), we detect the appearance of tourist motivations in urban areas which go beyond the visits in the tourist circuits conventionally commercialized by wholesale agencies. A good example of this situation, still in its initial stages, is the joint promotion of the 'World's Heritage Cities' (Toledo, Salamanca, Santiago de Compostela, and so forth). Business tourism, derived from urban economic activity, as well as fairs, congresses and conventions, are very attractive due to their good expense capacity, and are present in all strategies for the development of urban tourism in Spain. However, these types of tourism are strongly dependent on the evolution of the number of fairs, congresses, and conventions as well as that of their characteristics. This implies extreme sensitivity on the part of the tourist offer to recessions in the congress and fair market and the need to join efforts on a town basis for the promotion of these types of events, a circumstance that has its representation in the almost generalized constitution of 'Convention Bureaus' in Spanish medium-sized and large cities.

Event tourism experienced an unusual boom in 1992, when the Olympic Games were held in Barcelona, when Madrid was declared Europe's cultural capital city, and the World Exhibition was in Seville. These events highlighted the importance of big events from the point of view of tourism and boosted, more or less successfully, the tourist renovation in the cities involved. At present, the development of events with tourist significance finds itself between the strong investments that great operations require, inaccessible for many cities, and the difficulty to develop minor events with enough importance as regards tourism.

Finally, the development of the project of 'The Green Spain' represents a very interesting experience from the point of view of Spanish tourist diversification. It deals with the implementation of a joint Action Plan for the tourist development of the Autonomous Communities on the Spanish northern coast – Galicia, the Principality of Asturias, Cantabria and the Basque Country – its vertebral axis being the Cantabrigian Sea and the existence of binding complements, namely, the landscape, nature, culture, history and gastronomy. There are three basic objectives to pursue (Blanco 1997: 10): creating a global image of tourist destination by taking advantage of the synergy of these four communities and constituting the most important supraregional brand in Spain; giving a boost to joint promotion and commercialization policies and creating the conditions suitable for business development and the creation of products under the same brand. This project was started in 1997, which is why its future development is still unknown, although it is illustrative that four autonomous communities should go for the development of tourism in their regions, when the current contribution of tourism to their Gross National Product is between four and five per cent and the travellers staying in hotels do not reach 10 per cent of the total number in the whole of Spain.

The evolution of tourist policy in Spain: a decentralized institutional structure

The 'tourist boom' that took place in Spain in the late 1950s justified the government's intervention in the promotion and regulation of the tourist activity. The magnitude of the tourist phenomenon and the repercussion of its economic effects on the country's economic development gave a boost to the promotion of Spain as a singular destination under the slogan 'Spain is different'. Tourism played a decisive role in the Francoist dictatorship's foreign policy, which did not obtain official recognition from European democracies.

Amongst the most important aspects with regard to the state's intervention in the search for the growth of tourist activity at any rate, the control over prices in the accommodation tourist plant stands out. This policy on the fringes of market rules stimulated Spain's comparative advantage as for prices, but had two negative consequences for the future tourist development: the deviation of investments towards nonregulated accommodation that escapes administrative control (holiday flats to rent and second homes), and the forced concentration of the offer on a low purchasing-power client, closing the possibility of reaching different types of potential demand. The development of this activity ended with an unbalanced tourism in the territory lacking its own structures and complementary infrastructures, disorganized with regard to the tourist sector and commercially dependent on tour-operators.

The model of spontaneous and accelerated development in Spanish tourism experienced a serious crisis in the 1970s, coinciding in time with the institutional change in the country. The establishment of the democratic state caused a strong administrative decentralization towards Autonomous Communities (ACs), which directly affected the tourist policy. According to what is stipulated in the Spanish Constitution of 1978, ACs acquired full capacity in terms of promotion and regulation of their territorial areas and although this exercise was not free from conflict, each one of them currently assumes its power share and tries to co-ordinate it with that of the others, communities being fully independent as far as the approach to their own regional tourist policy is concerned.

The process of administrative decentralization has distributed competence in tourist policy into three levels: the central government, the autonomous government and the municipality. However, this articulation has not meant an important variation in the way of dealing with tourism with regard to policy, since the inorganic staff of the Central Administration has been transferred to lower units and the budgets consigned to tourism have not considerably grown. Nevertheless, the competences assumed by the autonomous communities and municipalities with regard to tourism promotion and regulation, make it possible to develop action strategies nearer to regional characteristics, in order to create a really differentiated tourist policy, adapted to the characteristics of each product/destination, which does not limit the necessary real co-ordination among the different Administrations. At this point, it is advisable to stress that co-ordination is not only important among territorial Administrations (central, regional, and local), but also between sectorial administrations, since competences such as land planning, environment, education and training, transport, and so forth (which, strictly speaking, do not correspond to tourism), definitely affect this activity and handle enormous financial resources, much bigger than those consigned to it in tourist-specific administrations. For this reason, in 1994 the Interministerial Commission for Tourism was created, this being a board on which all the departments that carry out actions related to tourism on a state basis are represented, for the sake of improved co-ordination.

ACs have the faculty to freely legislate and regulate all that corresponds to tourist activity within reach of their jurisdiction, and fully assume the regulation of the tourist supply and the management of their own infrastructures, as well as the regulation, inspection, and control of tourist enterprises and activities which are developed in their physical territory. Regional tourist policies have contributed to a more rigorous and sophisticated planning, leaving behind the characteristic spontaneity in the development of Spanish tourism.

The state's territorial configuration grants the local Administration a high degree of autonomy and wide competences (land planning, law and order, preservation of the environment, traffic, economic promotion, tourism, and so on). Therefore, this level acquires unquestionable significance to satisfy the tourist demand. However, there are evident financial problems for the right development of local competences in a tourist municipality, since its budget is based on the population in accordance with the census, and it has to lend services to a much larger population derived from tourist flows, more or less seasonal. This situation leads to the demand for special funding for tourist towns, which has not yet been properly solved in any autonomous communities, and, at the same time, we must not forget that these municipalities show per capita income levels above the regional average and that many of them have considerably increased their incomes through the concession of building licences in the context of an indiscriminate real estate property development.

During the 1980s, a loss of dynamism was seen in Spain's tourist activity. A change of taste on the part of tourists, technological innovations, and the appearance of rival destinations for the Spanish 'sun-and-beach' tourist product led to the evolution of tourist policy towards a new conception of tourism. Quality and competitiveness are essential in this new conception.

The Central Tourist Administration reacted with the I Plan for Spanish Tourism's Competitiveness (1992–95), known as 'Plan FUTURES'. The original conception of the plan is based on stimulating the sector's competitive advantage in international markets, promoting the creation of innovative offers, complementary to the traditional 'sun-and-beach' product which will facilitate the adaptation to the new requirements in tourist markets, paving the way to the idea of a 'quality tourism' as opposed to 'quantity tourism'. The Plan admitted a deep imbalance between the existing demand and the tourist-product offer supplied by the Spanish market, above all from a qualitative point of view (MICyT 1992: 10).

The successful execution of the project for the improvement of industry, which counted on the collaboration of Public Administrations and private enterprise initiative, gave rise to the elaboration of the II Plan for Spanish Tourism's Competitiveness (1996–99), continuing the task of defining a new model of tourist policy to improve co-ordination, sensitivity, and co-responsibility exerted by the different agents in previous years.

In this context of growing protagonism of regional and local policies, we must stress the important influence of the European Union as far as tourism is concerned (Bote and Marchena 1996: 311–25), in spite of the residual role of tourism in the process of European construction. In addition to the European Community's lines of action (sustainable development, consumer's defence, air transport liberalization) and measures related to the Community's Plans in favour of tourism, the role of tourism in European regional policy is worthy of mention. This can be deduced from the European Community Support System and from the investment of structural funds in projects having a clear incidence with regard to tourism, ranging from road infrastructures to the training of human capital in the tourist sector.

Current strategies to maintain competitiveness

At present, the conservative Spanish tourist Administration coming from the general election held in March 1996, has elaborated a strategic plan which focuses its attention, on the co-operation of ACs, the support to enterprises and the projection of the Spanish tourist policy in the heart of the European Union.

This plan is the result of the maturity reached by the administrative structure of the territories arising from the Magna Carta of 1978. It aims at overcoming the effects of isolated actions which mortgage and scatter resources, of optimizing the budget funds available, and is provided with the purpose of ensuring that tourism will keep contributing to the creation of richness and employment in the whole of the Spanish economy. In short, the Plan elaborated by the state's Central Administration defines the following strategies (SECTyPYMEs 1997):

* Consolidating co-operation between the different Public Administrations.
* Reinforcing the role of TOURSPAIN as the axis of promotion and commercialization of Spanish tourist products.
* Improving the system of tourist statistics as well as the spread of the information and research coming from the Institute for Tourist Studies.
* Encouraging the presence and participation of Spanish tourism in the European Union's institutions.
* Establishing a Tourist Training system, on all educational levels, which improves the qualifications of professionals working in the sector.
* Improving Spanish tourism's competitiveness by correcting its structural deficiencies.
* Diversifying the Spanish tourist offer.
* Developing environmental sustainability in Spanish tourism as an essential condition for the survival of products and destinations. In this respect, the experience in Calvià, in the Balearic Islands, is an interesting step forward towards the achievement of sustainable tourism in consolidated 'sun-and-beach' destinations. The municipality of Calvià is applying a Local Agenda 21 which sets the pace for growth in terms of economy, town development and the environment, without giving up social, work and environmental sustainability. The elaboration of the Agenda has counted on the citizens' participation and has been helpful in order to carry out numerous actions: creating infrastructures, modernizing establishments, and an improved town planning based on sustainable development which envisages recovering the landscape, as well as creating public spaces and green-belt areas (which has already begun with the demolishing of obsolete hotel and tourism facilities), due to both private and public initiatives.
* Tourist Quality as a differential element in international markets. The hotel sector has recently created the Instituto para la Calidad Hotelera Española (Institute for Spanish Hotel Quality). During 1997, quality management models were being introduced in 10 destinations, along with a group of independent hotels, which meant that over 400 hotels and 150,000 beds were involved on a national scale. This initiative showed that quality improvement had stopped being a technical and political claim, and had become something real.
* Internationalization of the Spanish tourist enterprise.
* Improve actions in consolidated or emergent tourist destinations, with the joint participation of all the Public Administrations.

- Complementary support to enterprises and professionals.
- Reinforcing the function of 'Paradores de Turismo de España' within the framework of the tourist policy.

Through the lines constituting that proposal for a strategy in the tourism sector, the aim is to promote improved quality in tourist services and increase competitiveness in the industry of Spanish tourism. This will hopefully lead to the consolidation of a prosperous tourist sector which can remain as a leader in the generation of economic welfare in the Spanish state.

Conclusion

Spain is closely associated to mass tourism due to a great magnitude tourist development. Tourism has been an instrument of inestimable importance for the country's economic development and holds a significant place in the current economic structure, a circumstance for which it is necessary to continue to maintain Spanish tourism's competitiveness, despite the existence of structural deficits which bring up unknown factors for the future.

Spain's vertiginous tourist-development process, based on quantitative objectives of constant increase in the number of tourists, have had the following results:

- a high concentration in the 'sun-and-beach' product market, which is highly price-sensitive;
- a stereotyped international image of mass destination;
- dependence on European tour-operators;
- problems derived from seasonality;
- environmental degradation of many tourist spaces;
- the influence of exchange rates for the peseta with respect to the currency of source countries on the demand's behaviour.

If we consider the negative predictions about the future of mass-production tourism, or the growing competition of trips to long-haul destinations, the present situation of Spanish tourism could be interpreted as somewhat worrying. Instead, statistical indicators are still positive and interesting trends start to be noticed that aim at the qualification of traditional tourist spaces (modernization of the supply; incorporation of new attractions: theme parks, complementation of the littoral area with natural and cultural resources in inland areas, and so on; special emphasis on the environmental management of tourist enterprises and destinations) and the diversification of the supply with the introduction of new products and new geographic areas into the tourist market.

However, it becomes clear that we cannot continue with the exploitation of the traditional comparative advantages (proximity to major European markets, mildness of the climate, geo-tourist resources, and so on) in a scenario that is now different to the one that gave rise to mass tourism. For this reason, Spain must incorporate new competitive advantages from the continuous improvement in the quality of its products, relying on the experience curve acquired by tourist enterprises and destinations as well as taking advantage of its leadership in the international tourist market.

Finally, it is essential to modify the conception of tourism as a tool for economic development (compensation for the commercial deficit, employment generation, intersectorial effects, and so on) and to recognize, in practice, its strategic character through the implementation of actions allowing it to reach the sustainable development of the activity, taking into account its economic and social profitability.

3 Tourism in the French Mediterranean

History and transformation

Marc Boyer

The Mediterranean area of France

Only a small part of France is Mediterranean (10 per cent); France has fewer kilometres of Mediterranean coastline than Italy, Spain, Greece or Tunisia. However, it is where many innovations in tourism have taken place.

The French South is not a 'small deal' Mediterranean area. Provence, Languedoc and Corsica all have the characteristics of this original space: a dry and often hot summer, a mild winter, abundant rainfall, but irregular and falling in a few days; the mosaic of the country,[1] the vegetation which never falls asleep in the winter, but becomes in summer, a matting. The first winter tourists in the eighteenth century admired first of all around Nice the greenery, the flowers and fruits of winter.

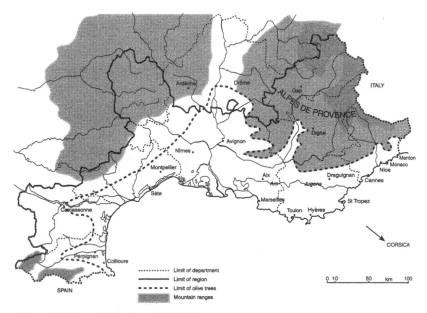

Figure 3.1 Map of the French Mediterranean area.

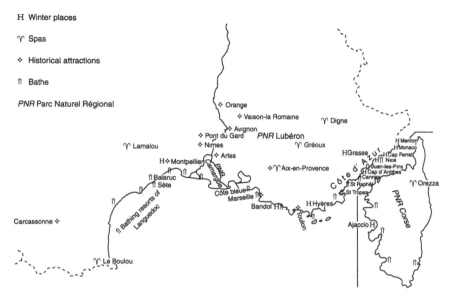

H   Winter places

ϒ   Spas

◇   Historical attractions

⇑   Bathe

PNR Parc Naturel Régional

Figure 3.2 Main resorts.

The traditional Mediterranean countryside brings together *ager, sylva* and *saltus*. The southern part of France has all three. Travellers admire the mixed cultures (*culturae mixtae*) climbing up walled terraces (in Provençal language: *restanques* or *bancàu*); tourists today are concerned about the results of forest fires set ablaze by the mistral: entire forests of various species have been destroyed. There are several sorts of pines, cedars, cypresses, all the varieties of oak (holm oak, cork oak, kermès and white oak), laurels, carob trees, and strawberry trees.

The *saltus* is the degraded forest, the maquis or garrigue according to the nature of the soil, it is an odour that the first literate travellers smelled. 'La Provence est une gueuse parfumée', (Provence is a perfumed rascally wench) wrote Des Brosses. Napoleon, on St Helena dreamed: 'I would recognize Corsica with my eyes closed – just by its fragrance.'

Another salient feature of the Mediterranean countries is its grouped habitations. To economize the arable soil and protect themselves from incursions (from the Barbary Coast), the large fortified burgs (boroughs) brought the population together; in the eighteenth century, the now useless ramparts were replaced by boulevards – the *Cours*. In the French south, tourists take pleasure in strolling along the narrow streets, wandering idly on the small squares, and the 'Cours'. Mediterranean coast is hospitable only in two small roadsteads. For a long time, the flat rectilinear coasts, those of Languedoc and the eastern plain of Corsica, attracted few inhabitants and no tourists.

The Mediterranean region – and the French part more than the others

– has a great power of attraction and integration. Examples are numerous: in Antiquity (8–6 BC), its natural inlets received colonies; the Phoenicians founded Sète, the Phoceans Massilia (Marseille). By sea, there arrived new religions, including Christianity; successive waves of immigrants were stranded in these ports. All were called for assimilation of which Rome was the first model. *Arma et togae romanae* subjected populations, and pacified the countries around the *Mare Nostrum,* while *Urbs sive Roma* dominated. The Imperator assured *panem et circenses* (food and fun).

The humanist travellers, first of all sought, around the Mediterranean, for vestiges of this civilization. The Romans built solid constructions: they left many monuments, some transformed, others restored. From the eighteenth century, the States of Languedoc, thinking of cultivated tourists, repaired the Pont du Gard, disengaged the Arenas of Nimes. Journeying to Rome, these distinguished voyagers, travelling in Provence or Languedoc, could tread along *viae romanae,* climb up along the aqueducts, go down into the amphitheatres (the Arenas of Nimes heralded the Coliseum), visit the Thermae, the Cryptoportiques (Aries), admire the theatres (of which the wall of Orange is the best preserved; in 1869, in Orange, the first Festival was created).

The great invasions gave rise, from the fourth century, to the fall of the Roman empire, but did not destroy everything. Nomadic hordes, over the centuries, were attracted by the Mediterranean, a lamplight of civilization. Coming from the North (Germans and Huns) or from the East (Arabs, Turks) they were civilized by the peoples they had subjected, and, in the North Mediterranean, christianized. In the Mediterranean area, the pattern is always *polis* or *civitas.*

The specific Mediterranean features are positive factors for tourism; which appears obvious today. However, it is not so obvious as that; it needed the discovery of these attractions. They were discovered one by one, at different times; the discoverers were almost all British aristocrats. The first sites were in Italy (especially Rome) and in South France (between Montpellier and the Alps).

Rome, for a long time, was the first attraction of a travel, especially cultural. 'The Grand Tour', in the eighteenth century, made the young British aristocrat into a gentleman. For them it was not a matter of discovering but of going to the sources, of reconnoitring. Along the way, these pretourists gave themselves over to sightseeing or *videnda:* Provence, *Provincia romana,* was a must: because it turned out to be Italy's antechamber.

Winter in the South of France

The second form of Mediterranean tourism was wintering in the South of France. Leaving the 'North' to sojourn for several months in the 'South' was the discovery of a few British aristocrats. This took place in Provence, from 1763 (treaty of Paris). Dr Tobias Smolett invented Nice and lauded it in *Letters to Nice from Nice*. Charlott Smith described Hyères where the English of her novel *Celestina* gathered together. At the end of the eighteenth century, Nice housed one hundred and fifty families each winter, most of them British; Hyères about fifty; Montpellier and Pise, only a few.

The three main factors attracting winter residents were the mildness of winter, the beauty of the vegetation and the hopes raised by therapies. The winter season lasted from October to the end of April, no longer, and sometimes less. The winter residents were not all sick, and there were residents of all ages. Most of them were persons of property or of independent means and the British, up to the mid-nineteenth century, were the most numerous. Thereafter, all the countries of Europe and North America sent their quota of winter residents. Incontestably, from the eighteenth century up to the 1929 crisis, the winter season was the most preferred. This mystical South comprised only a few resorts; the pioneers, Nice and Hyères were soon joined by Cannes, St Raphael, Menton and a few others. The greatest density was around Nice, the true winter capital. The 'Coast', beyond the Rhône, remained *terra incognita*. Curiously, a few winter resorts appeared on the Atlantic coast (Pau). In the eighteenth and nineteenth centuries, all winter resorts were completely separate. All had a triple function: therapeutic (a healthy climate), hedonic (paradise regained), ostentatious (these privileged persons led a worldly life while others shivered in the mists).

Hippocrates, the founder of medicine, had already advocated the benefits of a change of air. At the beginning of the nineteenth century, Laénnec built up a theory: 'There is no illness, said he, that cannot be relieved by a change of air.' During that time, the Romantic 'mal du siècle' was 'phtysis'. Laénnec who, around 1820, invented auscultation and the stethoscope, diagnosed tuberculosis but did not know how to cure it.[2] To stop or moderate its progression, the physicians recommended the South of France, provided the patients 'made good use of the climate of the south'. During the nineteenth century, many treaties and guides on the Mediterranean began in this way. The debate concerned the respective qualities of each winter place. The Englishman, Dr Bennett, published many works on this theme and came up with Menton. He settled down there and, around 1870 launched Menton. The former resorts still have their adepts; but the distinction still leads further on. Thus, the British, during the Romantic period, were attracted by the winter climate of the Atlantic. After 1815, English officers made of Pau an English winter town where it was possible to go hunting; in the hinterland, the English had their Arcachon and their Biarritz, different from that of the summer resorts of the Bordelais and the Spanish. Well situated on the route to India, Est and Madeira – then Alexandria, Cyprus and Malta became ports of call for long stays – so British!

Physicians, throughout the nineteenth century, talked over 'the good use of the South France' – a title given to many of their treaties, and of the respective virtues of the various resorts. All agreed that one should not arrive too late; to see the cemeteries, such as that of Menton, it was obvious that the advice was not followed. Towards the end of the nineteenth century, some physicians cast doubt on the Mediterranean – claiming the climate was too humid or too windy – and recommended the mountains oriented to the south. The Swiss set up high altitude resorts Leysin, Montana, Davos. The French Mediterranean was up against severe competition. But the reputation of wintry mildness remained; in the twentieth century, consumptive people continued to find refuge in Provence, such as Katherine Mansfield at Bandol. Retired people especially arrived in crowds to spend winter there.

There remained also the reputation of Eden. This coast that Stephen Liegeard called, in 1886, the Côte d'Azur, had become one garden, the Eden of Adam and Eve, the garden of the Hesperides, 'the country where the orange trees blossom' (A. Thomas, *Mignon*). By its inebriating flowers, its golden fruits, even its name, the orange tree has been the PARADIQUE since classical times. Every prince had his orange orchard. From the sixteenth century, the Garden of the King, in Hyères acclimatized new species. Generations of gardeners on the Côte d'Azur have acclimatized exotic plants, designed gardens with changing colours. Winter holiday-makers have been treated to mimosa and eucalyptus trees, the agaves, aloes and cactuses on the embankments, the pomegranate trees and Japanese medlars, the palms along the avenue, the hibiscus, bougainvillea, camelias decorating the gardens. Nothing is native, everything was imported. Not to make a commercial profit (sell a product), but through ostentation. The Mediterranean and particularly the Côte d'Azur invented and renewed the mover of distinction. Whatever the guides may say, it was not one degree more, less wind or the exceptional beauty of the site, which convinced the winter tourists to choose this or that resort; it was the quality of the gate-keepers who found the place, it was the arrival of an individual in high places. Nothing could beat Queen Victoria, who wintered at three different places: Grasse, Hyères and Cimiez. The discovery and launching of Cannes was exemplary. Prosper Mérimée, a regular winter tourist, wished to have the honour of the discovery; the inspector of historical monuments had no weight against Lord Brougham and Vaux, chancellor, leader of the Wights. In 1834, he was stopped on the Var by cholera, he retreated to the small port of Cannes, decided to construct a beautiful villa and to spend all his winter there. Mérimée was astonished: the thin, deaf old man was always alert and his 'salon' attracted great British names. They had invented another manner of holiday making; instead of renting furnished apartments, they built their own villa or mansion, always enclosed within a park. Everyone built according to his own taste, which could not be in the Provençal style. The pastiche was much admired; it was at Cannes that they began to juxtapose different styles: Palladian and gothic, Mooresque and byzantine.

And it was on the Côte d'Azur that the important acclimatizations occurred. Hyères and Cap d'Antibes were the first; all the coast from Hyères to Rapallo joined in. Alphonse Karr wished to be a gardener at Nice and St Raphael. The Handbury brothers at Vintimiglia, the baroness Ephrussi de Rothschild at Cap Ferrat, the Viscount of Noailles at Hyères and Grasse, and Ferdinand Bac at Menton, were great innovators. Lastly, it was also on the Côte d'Azur that, for the first time, the winter holiday-makers became interested in the sea; up to then, they had paid little attention to it. They were enraptured by the intense blue; hence the success of the concept invented by Liegard, a mediocre poet, and the arrangements made to admire it, such as the terraces of the Casino of Monte-Carlo and the Casino of La Jetée de Nice.

In the 1860s, the winter resorts of the French Mediterranean had just discovered a new attraction: gambling in a solemn and worldly setting. It was, up to that time, a custom of spas. During the Romantic period, gambling made the fortune of Baden; the minor Germanic princes found their profit. Adventurers like François Blanc and his brother-in-law, Benazet, became rich. But Bismark wanted to unify Germany. Fearing the Prussian discipline, Blanc and Benazet left Homburg and offered their services to the Prince of Monaco. Poor Grimaldi who, in 1848, lost Menton and Roquebrune! The principality had become the smallest state in the world, reduced to two rocks, one for the palace, the other, arid, became Monte-Carlo, named after Charles III. He decided to set up a resort with the most sumptuous of casinos, the most beautiful palace, the Hotel de Paris, both built by one of the greatest architects, Charles Garnier, who had just finished the Opera of Paris. The S.B.M., Société des Bains de Mer of Monaco, managed everything, designed wonderful gardens, set up Russian ballets and made the roulette spin. Luck was blind, like the prince Charles III, but brought in a fortune for the ones who 'held the bank' – the S.B.M., in other words the Grimaldi. This turned out to be a success because the rich flocked to the resort. As worded in the posters: 'there is nothing more "chic" in the world than Monaco.'

For a quarter of a century, Monte-Carlo was almost alone. Several European states, aware of what they were losing, legalized gambling provided it was worldly and situated in tourist resorts. Such was the spirit of the French law of 1919 regarding classified resorts. The winter resorts of the French Mediterranean, the spas, copied Monte-Carlo, with success in Cannes and Nice in particular.

Even up to the 1960s gambling maintained a character of luxury and ostentation in Europe. The worldly veneer hid the vice. What a difference with the vulgarity of the American Far West gambling, the sordid taste of Las Vegas! The epilogue is well known: over 20 years, the installation of money-making machines in the basements of the most renowned casinos, including Monte-Carlo, bring in more than the green tables. Such free entry places were a great success for a coarse public. The prince Rainier and the mayor of Nice, Jacques Medecin, both of whom had American wives, wished to transform the Côte d'Azur into a California, while preserving an elite character. Welcoming more and more people, laying cement everywhere, including the land gained from the sea (Monaco), 'concrete' unscrupulously, is this compatible with the preservation of exceptional quality? Some replied negatively: 'the Côte d'Azur has been assassinated', 'tourism has devastated the countryside' (J. Krippendorff).

Such a policy runs counter to the traditional segregation which was characteristic of the Mediterranean winter resorts. This segregation can be attributed to the British in Nice. They settled on the other bank of the Paillon, around the Croix de Marbre. Alexandre Dumas noted: *Nice New* upon the *Promenade des Anglais* is the opposite of *Nizza antica*, the old town, where one goes only for the Carnival.

Villas, mansions, palaces (after 1880) were not constructed according to local criteria. The medley of colours of the buildings and the exotic nature of the gardens are characteristic of all the winter resorts. The model invented in the South of France spread to all sides of the Mediterranean. The innovators were almost always British, especially in regard to the palaces, which are quite different from the existing grand hotels; the latter were alone, the main avenues, while the palaces were surrounded by parks. They were larger, with 200 to 600 rooms, with spacious suites. All the ground floor was devoted to worldly sociability. Steps, pergolas and broad staircases were signs of the ostentation of palaces – the word was perfectly demonstrated in this period – the rich winter holiday-makers, beginning with the kings and princes, Victoria or her son Edward who now had a residential palace, knew how to waste space, time and money.

Indeed, it is art. The American Thorstein Veblen who, in 1899, wrote the *Theory of the Leisure Class*, explained this. To be different with distinction is an explanation for the birth of new resorts. It was not the spread of the same phenomenon but rather of successive inventions.[3] The railway facilitated this selective diffusion; on the coast of the Esterel the station was there exclusively for the hotel.

The elitist segregation

The French Côte d'Azur also invented the appropriation, both physical and symbolic, of the finest spaces for the 'finest' of the 'world' – the princes and aristocrats who chose the shaded vales (Costebelle near Hyères or Valescure near St Raphael), the hills with panoramic views (Cimiez and Mont Boron overlooking Nice, the California above Cannes) and the protected peninsulas such as Cap Martin where the Empress

Eugenie took refuge, Cap Ferrat relished by Sissi and the King of Belgium Leopold, who found here 'the earthly part of paradise'. The Cap d'Antibes was the refuge of novelists like Jules Verne as was the peninsula of Portofino, near Rapallo. The present day Côte d'Azur no longer has this dominant winter resort character, and the colonization has moved on: the peninsula of St Tropez attracts writers (Colette), artists (Dunoyer de Segonzac), films stars (Brigitte Bardot). Cuelles constructed a false Venice at Port Grimaud and imitation provençal villages for the lonely rich (Castellaras). First seen on the French Côte d'Azur, the highly preserved residential area now characterizes the renowned Mediterranean tourist sites, in France, Italy, the South of Spain, and can be found in America on the coasts of Florida and California.

The colonizers were not always British, but the results were identical. The Austro-Hungarians wished to have their Mediterranean winter season: Opatija, amid the greenery of laurel trees, exhibits academic palaces; today, it is perhaps the place which the best evokes the luxury of the days before 1914, along with Yalta and the Crimea, the tsarist Côte d'Azur.

Florida is also a daughter of the Côte d'Azur, with the capitalist spice. Flagger, tycoon of the east American railroad, saw the profit that could be made from launching, in America itself, a winter season. The rich American clientele who used to cross the Atlantic bound for Nice, will now take the railway to Florida. So Flagger extended the railroad for Miami, then Key West, going to Miami Beach and West Palm Beach. During this season, which is always wintery, based on the amenity of winter, the attractions were different: the limpid green lagoon, the long beaches of fine sand, the swaying of the coconut palms were at the basis of a new happiness. The decor displayed emancipation: the hotels and palaces, in the Art Deco style, are painted in candy pink. The azurean model henceforth had a maritime rival, in the image of a tropical paradise. Its other rival, which appeared a little earlier, is white gold (snow and ice) or more precisely the invention by the British, again, (Sir Arnold Lunn at Davos) of a new winter pleasure: slipping with skis imported from Norway. Thus, around 1880–1900, winter sports began; during the twentieth century they would go on to supplant all that was representative of the Mediterranean paradises as a winter elitist resort. Thus, the process of elitist invention continued to give a sense to this chaotic history. The main hotel owners of the Côte d'Azur quickly became aware of what happening; between 1907 and 1913, they sent 'spies' to find out that some of their usual clients were at Davos, St Moritz or St Anton.

For these reasons, the Mediterranean, after 1920, lost some of its winter prestige. Does it still continue to be ignored in the summer? It is known that the first form of worldly summer season was invented by the British, again from the beginning of the eighteenth century and that the pretext was therapeutic: at first, people took the waters at Bath which, through the *savoir-faire* of Richard Nash, as soon as 1700, became a luxury spa in a Roman theatrical decor. Other mineral waters in England, then on the

Continent from the end of the eighteenth century, copied Bath and with great success. During the Romantic period, the fashion – the push – was directed to spa, to the Germanic Baden. The Mediterranean space was secondary in this attraction. The small Italian princes, like the Duke of Tuscany, following the example of what the German princes did, found the means to transform their Aquae and Bagni into important spas: Montecatini or Abano Terme. The Mediterranean slowly followed the example of Aix en Savoie. Although of Roman origin, the mineral springs of Languedoc-Rousillon (Le Boulou, Lamalou, Balaruc) and Provence (Aix, Digne, Gréoux) were of little renown.

The British had also invented a variation of sea bathing, once again the pretext was health. Immersion in the fresh waters of the ocean provokes a salutary suffocation. The English discovered 'the desire of the shores'[4] and transposed to Brighton the play decor of Bath. At the beginning of the nineteenth century, they took part in the launching of continental seaside health resorts on the other side, on the shores of the North Sea (Scheveningen and Ostend), and of the Channel (Dieppe). The Continentals in the nineteenth century launched other resorts: Trouville, Deauville (the Duke of Morny), Biarritz and San Sebastian. But there was nothing on the Mediterranean; a few attempts were made during the Second Empire as in Marseille but were failures. The reputation of heat was too strong, and the Mediterranean did not stir up any emotion. The summer Mediterranean is a tourist invention of the twentieth century. It was no longer of British origin; essentially it was North American. But it was also carried out in Provence, in other places than those that were frequented in the winter. What occurred in Juan-les Pins as from 1925, thanks to the Americans of the lost generation,[5] was not the mere extension of ocean bathing; it was even less a notoriety which might have happened to the indigenous dips of Marseille or Palavas les Flots – these places had, and still have, a local clientele and practices which recall those of the Parisians and the inhabitants of Lyon on the banks of the Seine or the Saône. No! The entry of the Mediterranean into summer tourism was sensational and localized. A seaside resort was established at the Pinède: this piece of land in the locality of Juan les Pins had been bought by the town of Antibes in 1915; planted with pines, bordering a sandy beach, it interested none until the arrival of a few Americans, extravagant gate-keepers. Some American authors such as Scott Fitzgerald[6] and Ernest Hemingway described the events between 1925 and 1930 at Juan les Pins where they resided. Together with Franck Jay Gould,[7] this society built a dozen hotels disseminated in the pinède, a palace with 250 rooms (le Provençal), beautiful villas, a flashy casino in the heart of the new resort. At Juan, nights were never ending, costumes were extravagant; people danced in the night clubs, wore pyjamas outdoors, passed by enticing girls, encountered the greatest American film stars – Charles Boyer – Mistinguett in the night-club called *The Hollywood Folies*.

The launching of Megève, as a ski resort and Juan les Pins for the summer was contemporary and parallel. Almost everywhere capitalists were the initiators (at Megève, the baroness de Rotschild), stars of literature or showbiz played an essential role in the celebrity and kings and princes in the consecration. At Megève, the king and the queen of Belgium; at Juan, the king of Greece, the Duke of Windsor.

Why had these Americans chosen the summer and Juan? Because, for them, the Mediterranean summer is bearable – in relation to the heat waves of North America, where there were no air conditioners yet; moreover, many American soldiers had appreciated the climate in 1918–19 in the hotels of the Côte d'Azur where they spent their convalescence. The lost generation liked, on the Côte, what looked rich and is false. After a few delirious nights on the Côte, the natural question was: 'Well, John, not too sorry not to have died during the war?'

Launched at Juan les Pins, the Mediterranean summer season quickly became a success, because it has a very favourable cultural climate. The path had been marked out by artists, painters who many years before had settled down in the villages perched out above the Côte (Cros de Cagnes, Vence, then Vallauris) or in the fishing ports (St Tropez), all off the beaten track. They had found the light – Colette, who had a house in St Tropez, described it well: 'There is no departure except towards the sun. There is no voyage except one towards an increased light.'[8]

Since the beginning of the twentieth century, men – and women especially – have discovered their bodies. In the water, one could take pleasure in swimming; the bathing costume was introduced; the warmer seas which allow one to stay longer in the water – the Mediterranean – had an advantage. The body became more exposed; the milky British white was no longer a criterion of beauty; a tan was appreciated. Also settled in St Tropez, Dunoyer de Segouzac painted the *Côte roties* (by the sun, the Mediterranean coast and the bodies of the women were also roasted) and Leon Paul Fargue commented: 'The human body has crossed centuries of clothes before becoming undressed again as from … 1914–18. What a number of hours spent angelically roasting under the scorching rays of the sun without a single back moving.'[9]

Mediterranean tourism today: France, a privileged place of observation.

An appraisal would be necessary at the end of the twentieth century; to do this, I have the elements. But my contribution must be short; I will simply present the main outline:
1 The French Mediterranean remains an observation laboratory: one can easily follow the evolution of contemporary tourism.

Mass tourism, at the end of the twentieth century, is especially a summer phenomenon; for half of the holiday-makers, their destination is the sea. The first touristic area in the world is Europe; the dominant flow is north–south and the Mediterranean space is the first beneficiary. The French part receives a large proportion of these tourists who – with difficulty – descend along the Rhône Valley.

2   The Mediterranean space is frequented almost exclusively on these shores, with three exceptions, and they carry weight: Italy, Greece, France. The three countries are of great historical and archaeological interest. Provence and Corsica combine the attraction of their shores and villages, the influence of a civilization.

3   In the French Mediterranean, space goodwill policies regarding tourism were elaborated and implemented. Originally, the idea was make use of tourism to develop the Mediterranean regions and to make the whole country benefit from it. The Spain of Franco began: the coastline tourism would be its Marshall plan; its Costas – Brava, del Sol – were developed, i.e 'concreted'. The Tunisia of Bouguiba imitated Spain. It was a strategy of 'bed production'. Then the problem was to find the clients, which put Spain and Tunisia at the mercy of tour operators. France, as from the 4th Plan (1960) desired to develop without deteriorating. It wanted to become the world leader of tourism through the creation of ski and seaside resorts; and, for this, it would master all the stages of construction. The snow plan gave rise to integrated resorts, functional ski factories implanted at high altitude; towards 1980, France possessed the largest skiable domain in the world. In parallel, France implemented plans for developing the coasts of Languedoc-Roussillon, Aquitaine and Corsica which are a model of concerted development; the State acquired the basic mastery, set up plans for developing the land where new resorts alternate with preserved and more wooded zones. Each resort is the work of an architect. In theory, these developments, so profitable to the regions, correspond to a sustainable development. Beyond an ambitious concept, the reality is less cheerful: these new resorts lack life, except in summer, and have artificial activities. The local inhabitants have scarcely adopted them.

4   France conducted, although rather late on, a policy of protection of the nature in two areas considered to be more fragile or more interesting – the mountain and the coasts. The national parks were created late by the law of 1963 and are all in the mountains except one, the park of Port Cros, a Mediterranean prototype. France created the National Conservatory of the Coast and Lakes which buys portions of the coast in order to 'freeze' them: 200 areas have thus been acquired, a third of which are in the Mediterranean region. Many regulations in France have stopped the 'cementization' of the coast; the Ute d'Azur is the most affected because the local collectivities are very sensitive to pressure. But France benefits from the old regulation of Colbert which makes the area between land

and sea (the area which can be covered by heavy seas) a public area of free access, whereas most of the other coastlines have been privatized. This liberty of access to the sea is a considerable asset for mass tourism.

5   Sociologists and anthropologists readily choose the Mediterranean area to study, in regard to their fashion, tourism, i.e., to describe a negative impact on the environment and the local destructured populations. And when they have finished studying the Mediterranean islands, from Ibiza to the Greek islands, our sociologists depart for Indonesia (Bali). Such a procedure does not seem to us pertinent; in fact these field surveys do not study tourism but traditional populations brutally confronted with modernity; without tourism, they would also have been destabilized, even if only by the television.

It would be better to look at the process of 'touristification'. Was it rapid or not? Conducted from outside? Imposed on the inhabitants? Because others changed their lives without asking them their opinion, the Mediterranean populations have rejected tourism. The same was observed in Languedoc where there is a refusal to become the 'ass tanner of Europe'. The inhabitants of the Cévennes did not wish to be disturbed by the neo-countryfolk, even less by the tourists. Claude Chabrol could say: 'If the tourists like us so much, let them send their money by the post.' The rejection was the strongest in Corsica where the villas of the continentals were blown up. On the other hand, there was no problem in Provence where tourism had existed for two centuries.

6   To study tourism is to revisit the question of Mediterranean integration. This sea, for centuries, facilitated the assimilation of populations; it sometimes allowed the cohabitation of communities (as in the Lebanon). It has never tolerated for a long time the 'plural societies' or a group with a much higher standard of living, grabbing the power. This 'plural society' was a failure in French Algeria after 1958 and in the Algerian Algeria after 1962. Less violently, the suddenly introduced tourism was at the origin of clashes between the Mediterranean societies – the local inhabitants, the tourists and the new inhabitants, arriving as retired people or as immigrant workers. Cohabitation was not easy, even in Provence today, an area nevertheless accustomed to being a tourism area for the last century or two. At the beginning of the twentieth century, one could have believed that on the Côte d'Azur there was a successful melting pot around the 'gentle way of life'.[10] It is less certain today.

7   The Mediterranean of the tourists is rich in myths and values. France is well placed to show it. Winter in the South was the realization on earth of a great expectation: regain the lost paradise which was a garden. After 1920, the exposure of a naked, or almost naked, body to the sun and the caress of the sea, was a new hedonism, begun in Juan les Pins, continued

in the Levant. For all sorts of Mediterranean peoples, and for the hunted down (Russians and Jews) some areas were a refuge: the Côte d'Azur as always, the Gulf of Naples, Capri. To finish, we can evoke the Mediterranean Club (Club Med). Gilbert Trigano and Gérard Blitz had invented a concept of happiness through the means of holidays where all is possible, 'If I like'. A name was lacking; they chose that of Mediterranean. The bankruptcies of the clubs should, then, have prompted them to call them Polynesia. During a third of a century, the club, which always had an idea before the others, brought to the holiday-makers a concrete Utopia under the banner Mediterranean. Today – the Trigano being replaced – the Club would become a 'Disney Product'.[11]

All is not said: one cannot do what one likes with a Mediterranean which is a contrasted space with a characteristic population, with a great strength of myth and dream. Better than any other, the French part of this Mediterranean is there to provide the proof.

Notes

1. For the tourists, the Mediterranean is composed of tiny ports, bourgs, perched on the hills, a partitioned landscape. The attraction-image does not include the marshes (those of the deltas), the steppes (the Crau), the monotonous plains (the wine-producing Languedoc), nor the deserts (the Agriates).
2. Tuberculosis ceased to be the great scourge in the middle of the twentieth century only thanks to penicillin and the BCG vaccine.
3. On the theme the Invention of Tourism, I have written a work in the collection *Découvertes* published by Gallimard, 1996, translated into Italian in 1997. My doctoral thesis, 2340 pages in 21 fascicles, concerns the *Invention of Tourism in the South-East of France from the XVIth century to the end of the XIXth century* (reproduction by the University of Lille III).
4. Alain Corbin-*La Tentation du vide. L'Occident et le désir du rivage*. 1750–1840. Paris-Cham. 1988.
5. Expression of Gertrude Stein: it designates Hemingway, Dos Passos, Scott and Zelda Fitzgerald, North American writers and artists who could not bear the aftermath of the war. They wander from Greenwich Village to Montparnasse and on the Côte.
6. F. Scott Fitzgerald, *Tender is the Night*.
7. F. J. Gould is a railway tycoon; Barbara Gould holds a highly-frequented salon in Montparnasse especially by the intellectual elite. The Société immobilière des Bains de Mer of Juan les Pins brings together Gould, Ernest Baudouin (the Casinos), Aletti (the palaces of Vichy and Algiers).
8. Colette, *Prisons et Paradis*.
9. L. P. Fargue and Dunoyer de Segonzac, *Côte Roties*, a magnificent album.
10. Title of a volume that Jules Romains devoted to Nice in *Les hommes de bonne volonté*.
11. Allusion to the fact that the former director of Disney has become the owner of Club Med, with a marketing logic.

4 Italian dualism and new tourism trends

The challenge of Mezzogiorno

Giovanni Montemagno

Italian dualism in economy, society and tourism

As has been well known for at least a century, one of the most serious economic and social problems in Italy is that of 'dualism' between the regions of 'Centro–Nord' and that of 'Mezzogiorno'. A dualism that has neither been removed nor significantly reduced by nearly fifty years of policies of special intervention by the state, the regional governments, and the European Union. Until now, in fact, the most relevant economic indicators show very important differences between the two parts of Italy, and almost all the predictions indicate a deepening of the phenomenon in next years.

Even if we think that what we mean with the expression 'Mezzogiorno' is already well known, it may be useful to remember that, just after the building of the 'Kingdom of Italy' (1861), the seven regions of the former 'Kingdom of Naples' had an economic and social development which was quite different from that of the other regions. Moreover, another region, previously belonging to the former 'Kingdom of Sardinia' (precisely, the island Sardinia), as it also had a weaker structure and a general backwardness, was included in this general framework of Mezzogiorno. At that time and in that way took shape, in the history and in the geography of Italy, what we call Mezzogiorno, and for 120 years a wide literature of essays and articles about the economic and social aspects of the 'questione meridionale' has been written by a large rank of authors, both Italian and foreign.

Having a more precise idea of the weight of Mezzogiorno, compared – from a quantitative point of view – with Italy as a whole, can be very useful for our discussion. We have just said that Mezzogiorno consists of eight regions over the 20 which make up Italy, and they are: Abruzzo, Molise, Campania, Puglia, Basilicata, Calabria, Sardinia and Sicily. There are 2556 cities and towns ('Comuni') in these regions, that is 31.5 per cent of the Italian total of 8101 'Comuni'. Their surface area is 123,059 km², that is 40.8 per cent of the total surface of 301,323 of Italy (Touring Club Italiano 1998: 66–8). Population, finally, in 1997 was in the Mezzogiorno of 20,959,000, that is 36.4 per cent of the Italian population of 57,567,000 (SVIMEZ 1998: 23), but we want to stress that at least another five million people born in southern regions live nowadays in the regions of central and northern Italy, because of the great flows of migrations that took place – direction south–north – in the 25 years from the end of the World War Two until the beginning of the 1970s. In conclusion, we are dealing with a significant part of a rich and important country like Italy. Moreover, we can also affirm that, in the economic and sociological literature of the last half century, the problem of the Italian Mezzogiorno has acquired a very high symbolic importance, perhaps higher than its real consistence. Therefore, we now have to take into consideration some economic data of Mezzogiorno, compared with the data of Italy as a whole, wherever this is possible.

The main economic comparative indicators of development or backwardness of a region are that of income and employment. We will start, in this short analysis of the economic conditions of the Mezzogiorno, from some data referring to these quantities or, better, of their variations in the last years. In 1995, the GNP (Gross National Product) has had a growth of 1.7 per cent in Mezzogiorno compared to 1994, whereas the rate of growth of the same datum has been more than double in Centro–Nord, that is 3.5 per cent (SVIMEZ 1998: 9). This is a very clear indication of a double trend that also existed in the previous four years, so that the risk of a deeper dualism between the two parts of Italy is more and more real. Unemployment data further illustrates this. In fact, in 1995 the number of employed increased in Centro–Nord by 0.1 per cent, whereas in the Mezzogiorno it decreased 2.2 per cent; in the same year, the rate of total unemployment in Mezzogiorno increased – with respect to 1994 – from 19.2 per cent to 21 per cent, and in Centro–Nord from 7.6 per cent to 7.8 per cent. These data give a general idea about the Italian dualism, which can be reinforced by some other data, like the ones about long-length unemployment. Respectively this is 14.9 per cent and 4.9 per cent of active population; and, above all, by unemployment of young people up to the age of 24 years, which has been – even in 1995 – in the Mezzogiorno 49.3 per cent for men and 54.6 per cent for women, against 18.9 per cent and 28.5 per cent in Centro–Nord. Finally, if we consider the trend of investments, we find that they increased 2.8 per cent in 1995 in Mezzogiorno – but they were 20 per cent less than the previous four year period – and in Centro–Nord they increased by 7.4 per cent, being only 4 per cent less than in 1992. The last reference we want to deal with, among many others which are helpful to explain the Italian dualism, is that of the demand for infrastructures and of the existing equipment of these infrastructures. Both these data – demand and equipment – demonstrate a very large difference between the provinces of Centro–Nord and those of Mezzogiorno. If we put the average datum for Italy equal to 0, the Province of Milan has an indicator of demand of 1.09 and the last one, that of Agrigento, is −1.48. As far as the equipment is concerned, the first Province, Trieste, has an indicator of 2.32 and the last one, Caltanissetta – also in Sicily – has an indicator of −2.55 (SVIMEZ 1998: 412–17). We wanted to stress these last data because we think that they are very important for our main subject, tourism in Mezzogiorno, which we will move on to now.

In regard to tourism, dualism is even stronger when compared to the global situation: Mezzogiorno, which represents 40 per cent of the national territory, has a very low share of the tourism sector. This depends upon many aspects affecting almost all the elements of tourist supply. Beginning with the territory, it is well known that, from one side, Mezzogiorno has resources with an enormous tourist value, but on the other side, in southern regions territory has been managed badly, so that nowadays, it is – mostly in the more valuable coastal areas – heavily defaced by structures and buildings, which are very often unauthorized holiday houses. In many areas, the landscape is so altered that it has lost its beauty, and this situation has very much reduced the competitiveness of Mezzogiorno compared to Centro–Nord. We will now turn to the subject of infrastructures, which means primarily transportation – entering the area and circulating within it. The situation, from this point of view, shows a very stressed dualism: airports are not well connected with the greatest continental and international hubs; their carrying capacity is very often exploited out of limits, and it is not sufficient to receive an increased traffic. In the southern part of the Italian peninsula, highways are practically limited to the 'Salerno–Reggio Calabria', a very old highway, in very bad condition, just like the general road network. In Sicily the road network situation is virtually the same: the few existing highways are kept in bad condition. We also observe a similar situation with regard to the ports system and, above all, marinas and landing places for pleasure boats. Other problems are connected to water resources supply and distribution, or to the disposal of solid and liquid waste. Another element of dualism, the most important perhaps, is that of lodging facilities, a subject where the gap of the Mezzogiorno is really relevant. In Mezzogiorno we find only 5240 hotels (15.0 per cent) with 377,452 beds (21.06 per cent), while the corresponding data for Centro–Nord are 29,309 (85.0 per cent) and 1,346,881 (78.4 per cent). (SVIMEZ 1998: 156–8). Also the composition of these two wholes is very unbalanced: if we consider three groups of hotels referring to their qualification (A = 4–5 stars; B = 3 stars; C = 1–2 stars) the respective ranking in Centro–Nord is 22 per cent, 55 per cent, 23 per cent, while in the Mezzogiorno it is 12 per cent, 30 per cent, 58 per cent. These qualitative and quantitative gaps in hotel endowments correspond, of course, to the data referred to tourist flows – always in 1995 – that is the main indicator of the tourist sector. The number of tourist nights in the Mezzogiorno was 52,685,000 – in the whole, both in hotels and in other structures – that is only 18.2 per cent, compared to the 237,073,000 of Centro–Nord (81.8 per cent), and 289,757,000 overnights in Italy as a whole. It is easy to understand that – apart from infrastructural and hotel problems – other difficulties have affected both the use of natural and cultural resources of Mezzogiorno and the supply of facilities; otherwise we couldn't explain such a strong dualism.

Nevertheless, we cannot reduce our analysis to these general considerations about dualism in the economic conditions of Italy and, consequently, in the tourist sector. We need to deepen and to extend our analysis, as our main goal is to indicate paths of development that could eliminate or reduce this dualism in the tourist sector and, therefore, in the economic system.

A SWOT analysis for the Mezzogiorno tourism system

In the previous section we analysed the problem of Italian dualism, and examined its presence and consistency with particular reference to the tourist sector. In this section, we will leave out the comparative analysis between the 'two Italies' in order to conduct a deeper survey of the tourist sector of Mezzogiorno, using a SWOT (Strengths, Weaknesses, Opportunities, Threats) analysis scheme. Our main purpose is to shape a concise but complete picture of the present situation in the tourist system of Mezzogiorno; as a consequence, the following sections will deal with the indication of the main goals of this system and the strategies which appear to be necessary to attain them. As a conclusion, we intend to underline the opportunity to transfer both the analysis schemes and the strategic models of development to the other regions of the Mediterranean which present the same characters of resources' richness on one side and of economic backwardness on the other.

We have already underlined the state of serious decay of wide areas of Mezzogiorno, especially where they have been exploited for the traditional tourist fruition, that is to say the bathing season in the coastal areas. On the other side, it is necessary to point out that a big part of the territory of Mezzogiorno, mainly in the minor urban centres and the internal areas, may constitute an important factor in the building of an innovative tourist supply that these regions can create. Large extents of protected areas in some great national parks (for instance, the Park of Pollino or Aspromonte in Calabria) or in regional parks (as the three existing in Sicily, the Park of Etna, the Park of Nebrodi and the Park of Madonie) or in nature reserves, with a wide variety of environmental habitats, or cultural and rural spaces, may represent strategic resources for an integrated supply of eco-tourism, environmental tourism or rural tourism. In the same way, hundreds of towns and villages may become the protagonists of a modern and effective cultural tourist supply, both urban and nonurban. In the last few years we have assisted the exemplary growth – as tourist destinations – of the two big capitals of Mezzogiorno, Naples and Palermo, and of other towns such as Catania and Bari. This trend may certainly be strengthened, while other destinations could also reaffirm their image of artistic, cultural towns and places to spend a pleasant and interesting stay.

The present situation of the two strategic elements of tourist supply of Mezzogiorno, that is to say, infrastructures and facilities, is not helpful to the development of tourism; we could even say that this situation is responsible for the gap existing between the southern regions' tourism and the rest of Italy. On the other hand, the regions of Mezzogiorno have big potentialities especially in this field. The whole of infrastructures related to transports, as far as space availability and inner shape are concerned, could be implemented in a medium term of two or three years, to let Mezzogiorno be competitive, while in a long term of five to ten years, they could make Mezzogiorno the leading region in the Mediterranean area. Therefore, a weak aspect is expected to become a strength factor in the short to medium term, provided that a new awareness of the situation is achieved, mainly to change course.

Tourist resources are all those elements, both natural and artificially created by people (cultural or historical resources) that constitute a motive for the moving of a tourist. The Mezzogiorno of Italy, with regard to both types of resources, has a real position of strength, in comparison with almost all the tourist regions of Europe, or of the Mediterranean basin: climate, beaches, landscapes, mountains and woods, parks and nature reserves are of a beauty so rare that they have a high capacity of attraction, which is for the most part unexploited. We can say the same about the inestimable treasure of archaeological and historical vestiges, of art or architectural masterpieces, of hundreds of villages, towns and cities, each one possessing something to show to visitors. Nature and culture, as integrated resources in a well-built tourist supply, have been until now *a* point of strength for the tourist sector of Mezzogiorno, but they are likely to become *the* most important or the only ones. It will depend upon the capacity to work them efficiently and to organize the whole tourist supply of Mezzogiorno in a better way.

We should analyse two other basic kinds of resources in order to assess if they are more likely to be points of strength or weakness. The first of these elements is represented by human resources. Once again, we are obliged to distinguish between the present state of the art and the future perspectives. We can immediately affirm that, at the moment, the level of employment and the qualification of employees in tourism are really low, compared to the dimension of the sector. In particular, the main gap in Mezzogiorno is related to a large number of hotel and restaurant staff, whose training level and hospitality culture is rather inadequate to the exigencies of a customers' demands, which are becoming more and more high and refined. Another significant deficiency that we need to point out concerns the field of global management of tourism planning and promotion, both in private and public sectors. Travel agencies and tour operators working in Mezzogiorno operate at 80–85 per cent in outgoing activities and very little in incoming ones. The first, immediate consequence is that the great majority of tourist flows are promoted and organized by agencies and tour operators of northern Italy or foreign regions. The second set of consequences is tied with sharing income and profits, creating employment and, generally, with the interest to maintain or not some destinations or products on the national or international markets. As far as public management of tourism is concerned, the situation is even worse, at every level, starting from the central government. As a matter of fact, a Department of Tourism now depending on the Ministry of Industry exists, but it has had neither a global policy of tourism development at national dimension nor any strategy or plan to develop the regions of Mezzogiorno, which are the most backward ones as far as the present results are concerned, but the richest for resources and perspectives. We can give the same judgement of inefficiency on the ENIT (Ente Nazionale Industrie Turistiche [National Agency for Tourist Industries]) as far as its task is concerned, that is the promotion of Italian tourism abroad. If the presence of central organisms is practically nonexistent, the governments of the regions are really active and endowed with substantial financial funding for promotion and management. However, as for quality and effectiveness, the results are not at all comparable to the level of expenditures, and the same objection can be raised with regard to the actions of all the other local bodies (Departments, Communes, etc.) or specialized organisms (A.P.T. = Aziende di Promozione Turistica [Agencies for tourist promotion]). So, in conclusion, even if the autonomous competence and the capacity of funding of so many organisms could appear a point of strength, it turns out to be a source of weakness, for the lack of skills and efficiency and the excessive bureaucratic and political ties.

A last matter that we want to emphasize is that of financial resources that are available for the tourist development of Mezzogiorno. Our statement is the following: if we consider them for their absolute, quantitative level, they are largely sufficient to fund an effective development policy. As far as the EU funds are concerned, we have to remember that tourism in Mezzogiorno can utilize many forms of financial flows: that is, all the structural funds (FESR, FSE, FEOAG) directed to the regions of objectives 1 and 5b, and furthermore, programmes LEADER, INTERREG, MEDA, etc. To have a more precise knowledge of the amount of flows directed to this aim, we will only remember the 'Programme for the development and valorization of sustainable tourism in the regions of objective 1 – Mezzogiorno', launched in 1995 and co-financed by the central government (Department of Tourism) and funded, for its three subprogrammes containing eight measures, with a total amount of 282 Mecu (Settimo Rapporto 1997: 707–12). Within the 'Global Grant Tourism', we must consider a grant of 59 Mecu from the FESR, managed by an Italian body named INSUD, and at least 200 other Mecu of indirect funding coming from URBAN, SME and LEADER. Fundings by BEI have had a lower importance in the last years (less than 30 Mecu in 1996), and furthermore, only 12 per cent are directed to the Mezzogiorno, while 75 per cent are given to the northern regions (Settimo Rapporto 1997: 713–17). The central government participates in global fundings together with EU, so that the other very important source of financial resources are, for the Mezzogiorno regions, regional governments, with their own finance. Once again, figures are very important: in the triennium 1994–96, the eight regions financed the tourist sector with a total amount of 2365 billions of lire (and Sicily is the first one with a budget of 806 billions of lire!) (Settimo Rapporto 1997: 718–30). Of course, a deeper analysis of flows, financing and efficacity of expenditure could be very interesting, but it is impossible, for the limits of this research: our aim is only that of indicating the case of Sicily, where the distribution of the budget between 'capital account' and 'current expenditures' is, respectively, 48 per cent and 52 per cent! The other Regions have a better subdivision, that of about 80 per cent versus 20 per cent. This indication clearly shows the reason why, in Mezzogiorno, the tourism sector has not had the right take-off and growth that it could have. A lack of efficacity in public planning and expenditures and a lack of capacity in investment and in management of private enterprises can easily explain our negative diagnosis about the tourism sector in Mezzogiorno.

This conclusion can well summarize the essential points of weakness of the Mezzogiorno tourism sector in Italy; and the threats to this sector are dependent on the continuity of this situation, both in the public and private sectors. Moreover, we have to add all the problems already stressed, concerning environment, lack of infrastructures (above all, airports, ports and marinas) and quantity and quality inadequacy in lodging facilities. But we cannot forget all the points of strength that we have also found in the tourism sector of Mezzogiorno. Wide areas in southern Regions are still in good condition, and they are very open and attractive for new tourist flows; with regard to infrastructures and structures, they have great prospects of development, in order to become adequate to tourist demand. But environmental and cultural resources are the most valuable areas that Mezzogiorno could exploit to become the first region in the Mediterranean basin and perhaps in the world, for quality more than for quantity. Human and financial resources are more than sufficient: they only need to be better organized and exploited. Therefore, the final diagnosis can be positive, but we have to indicate therapies, that is to say plans, strategies and policies to create better products and to get a real, widely-sustainable tourist development.

New tourism policies for a new model of development

After having underlined the economic and social problems of Mezzogiorno, and analysed the strength and weakness of tourism in that area, we can now pose the question of whether tourist development could be a right and effective solution for those problems. Our answer is affirmative, and for many reasons: first of all, from the forecast that, in the next ten years, about 1,850,000 jobs could be created, in the EU countries, in the T & T (Travel and Tourism) sector (WTTC [World Travel & Tourism Council] 1997: 3–13). They shall be direct and indirect jobs, well distributed in a large range of activities, both manual and intellectual, with higher wages compared to the average level, and particularly accessible to women, minorities and young people entering the labour market. Moreover, they will require high-level skills and shall be created in urban and rural areas, just where structural unemployment is stronger. Finally, the T & T sector shall produce significant, indirect effects in many sectors which may seem very far from it, like buildings, telecommunications, trade and manufacturing. This is the global scenario in which we want to set our statement: the T & T sector can be a right and very effective answer to the backwardness problems of Mezzogiorno, because of its general conditions based upon significant resources allocations, geographical and physical characteristics and so on, which can easily implement this growth path.

As a logical consequence of our previous statement, there are new questions to answer; precisely, how much tourism for Mezzogiorno, and what kind of tourism? The first answer may appear obvious, but this isn't true: the instinctive reply is 'more, much more than today, possibly double than the actual flows'. We can agree with this statement but only if we consider a set of conditions and restrictions: the first one is the relation between tourism and the environment; in reference to this area, we have to stress that new tourist flows shall never exceed the carrying capacity of our territory. A very important concept is that new flows have not to be concentrated, in time and space. Until now we have always had a very strong concentration in three to four months of the year, in terms of time; in coastal areas in terms of space. A correct development could be accepted only if it could solve this problem. This statement leads us to the second question we raised: what kind of tourism? We will find the definitive answer at the end of the chapter but we intend to express now some statements about tourism development in Mezzogiorno. First of all, it has to be an *endogenous* development: finance, planning and management should be the expression of local potentials, in order to avoid decisions and investments that would be bound to a short-term exploitation of existing resources. It has to be a *sustainable* development, respectful to the environment, both cultural and natural, just to avoid strong concentrations of buildings, structures and people; definitely, a non-polluting development. Finally, it has to be an *employment-oriented* development: since unemployment is the main socioeconomic character of Mezzogiorno, and because of the abundance of qualified human resources, tourism development models have to be largely labour-using; this is possible in many ways, which will be better pursued by an endogenous management of tourism. Then, how can we conciliate the fulfilment of these goals with such general conditions and particular restrictions? We will soon face this matter.

We have just stated that tourism can be a real answer to the problems of Mezzogiorno, not only through the conditions we underlined but, above all, with an immediate and strong change in policies and actions. The first responsibility is at the central level, both government and parliament, to whom all the operators of the tourist sector ask for a reform of the present legislation (State Law for Tourism, L. 17 May 1983 n. 217); a much more effective reform, not like the one the government has submitted to the parliament during these days. Recognizing the role of regions, like the proposed bill intends to do, is undoubtedly very important, but what is much more important is to settle an efficient network of local bodies (Aziende), in the whole Mezzogiorno not corresponding to the departments, having a perfect knowledge of their territory (with its structures, facilities, resources, people) and charged to create the local tourist products that will be promoted and sold in national and international markets by specialized regional or national agencies (Montemagno 1987: 52–60; 1998: 1–13). We do not approve of the actual policy of the Italian government that prefers to entrust the tourist development of Mezzogiorno to a special agency (INSUD); in our opinion, the position of this organisation is too far from local realities, and the level of its management is inadequate. Coming back to the task of local bodies (Aziende) to create innovative and competitive tourist products, we think that first of all they have to recruit teams of skilled people, able to understand that a good tourist supply is not made of expensive summer seasons of various shows, but, instead, has to comprehend a lot of components. Therefore, they have to plan their products by having a good knowledge of markets and demand, and then by operating with a precise model. With regard to this statement, we want to refer to a model of I.T.P. (Integrated Tourist Product), that we have elaborated during these years and which can be a very useful reference for every local management (Montemagno 1989: 823–37).

We can write our I.T.P. model with the following function:

$$i(T, I, S, R - A, G - F)$$

where T is Territory (the primary content of a tourist demand, in both its aspects of extension and quality); I is Infrastructures (above all for transportation, to enter the territory and circulate within it); S is Structures (that is lodging facilities of all kinds, but also catering and recreation facilities); $R - A$ are Resources and Activities (all those elements, both natural or created by men, constituting a motive for tourists' move); $G - F$ are Goods and Facilities (habitually demanded from tourists, apart from lodging, transportation and so on).

This very simplified presentation of our model is, in a mathematic form, a function of minimum value; it implies that all the elements have to be represented simultaneously in a tourist supply and that the consistency of the whole supply is determined by the lowest element, where for consistency we mean either the carrying capacity or the level of tourist flows we want to obtain in our planning. In other words, even if it might be obvious, our proposal for a tourism development in Mezzogiorno is that of a co-ordinated planning of all the elements, which must have a precise correlation, both in dimension and quality; this means to stop planning big events without structures and facilities in the area, or to create large resorts without existing and attractive tourist resources and activities (environmental or cultural). We must add that the model of I.T.P. we have presented here can be much more complex, if we just consider other elements and insert them in a general model of I.T.S. (Integrated Tourist Supply, that includes, besides the product, conditions, prices and promotion), considering every element, in turn, as a function (of minimum value or of summation) of a set of subelements. However, what we want to state is the necessity and urgence to plan the new, sustainable tourism development of Mezzogiorno on a correct and effective technical and scientific basis.

A new tourism, planned by new subjects using new models, can be a solution for the unemployment problems of Mezzogiorno, but this needs, above all, new typologies of tourism. As a matter of fact, up to this moment, the main tourist flows in Mezzogiorno belong to the classic typology of the '3S' tourism, that is 'Sun, Sea, Sand' tourism; that means summer, seaside tourism, concentrated in time (three months, June–August) and in space (narrow coastal belts), with saturated beaches and big buildings, and with wide resorts, very often controlled by national or international chains of hotels and Tour Operators. In conclusion, this is a hard typology of tourism, with a very heavy environmental impact and a comparative low level of yield and employment. The new typology we want to propose can also be indicated by a similar expression, that of '3C' tourism, that is 'Culture, Cities, Countryside' tourism (or, in Italian, 'Cultura, Città, Campagna'). With reference to well known typologies of tourism literature, it means 'cultural tourism', 'urban tourism', and 'rural tourism', but we want to stress that we give to these expressions a broader sense than is normal. So, 'urban tourism' is not restricted only to big cities or to the so-called 'art cities', but also to the hundreds of middle cities and little towns of Mezzogiorno that have in their landscapes, churches, buildings, treasures of art and culture, almost always unheard of by most of the people in Italy and Europe in general (Montemagno 1985: 77–87). Similarly, 'rural tourism' consists of all those forms of environmental tourism which have become so fashionable recently, and not only of tourism in rural spaces or of farm holidays (Montemagno 1994: 115–35). Both these forms can be described in the more general expression of 'cultural tourism', where 'culture' is not restricted to arts, architecture, archaeology and so on, but extended to traditions, folklore, popular and religious feasts; we refer, for instance, to the hundreds of Easter celebrations, to craftsmanship, gastronomy, etc. (Montemagno 1986: 99–128).

The products we think about, and that have to be created and sold by the local bodies we have just outlined, could be *day trips* (for residents in the region, schools and tourists staying in other sites), *stayings* (of seven, ten, fifteen, thirty, or sixty days) and *itineraries* (of three, five, seven or ten days). They all require that all the cultural and environmental resources could be visited, frequented and exploited by visitors, but it is also necessary to check very strictly that the flow of visitors does not exceed the site's carrying capacity. Above all it is necessary, however, in order to have a successful supply, that all these products could be managed and guided by well-skilled people. These could include specialists in economics, management and marketing for the first function, and, for the second one, a lot of people licensed or educated in the fields of history, arts, languages, literature and so on. This role of animation is, in our opinion, the most important to satisfy the exigencies of the visitors, and it has to be organized for small groups of five–seven people, ten as a maximum. In this way, we will get two results: the first is to develop a very attractive tourist activity, and the second is to create a lot of jobs corresponding to the characters of Mezzogiorno unemployed people: young people, women, graduates. If we also consider the jobs that might be created in hotel and restaurants, in transportation and other sectors, we can affirm that really tourism, precisely this kind of tourism, can be an effective solution for the economic and social problems of Mezzogiorno. The last point we want to make is in reference to seasonality: in our opinion, this model of tourism – and two others, not dealt with in our research, that is congress and religious tourism – can be offered for ten months out of twelve; and with this last observation, we think that our demonstration could be considered complete.

The simplified model of an I.T.P., based on a '3C' tourism, can be considered effective not only for all the regions of Italy's Mezzogiorno but, in its broader sense, to almost all the regions of the Mediterranean basin. In the whole, these regions belong to 20 countries, and in some of them (by instance, Côte d'Azur and Languedoc-Roussillon in France, Catalonia, Valencia, Andalucia, etc. in Spain, Greek Islands and so on) tourism is already very developed, but this development is – for the 90 per cent – based on the old model of '3S' tourism, with all the connected problems of saturation, pollution and so on. A new development of tourism is highly suitable for this area in order to reduce the strong pushes to migration caused by demographic growth and general backwardness (above all in North Africa). However, it can be the traditional seaside tourism only in a very lower percentage, because of its concentration – in space and time – and its position in international markets, for the competition of other areas with similar attractions (Indian Ocean, America, etc.). Many Mediterranean regions can adopt our model of '3C' tourism, because they have all the elements we have specified for it, including a treasure of cultural heritage lost by so many civilizations, from the ancient Egyptians or

Phoenicians until today. Of course, the old model can't be effaced or reduced, but it has to be combined with the new one, to increase its attraction and competition capacity. In our opinion, from a European point of view, we must add another aim to the ambition of overcoming unemployment – that is, a better demographic equilibrium. Thus, a strong intervention of the EU in the development of the tourism sector in the whole Mediterranean basin can be justified. In the spirit of the Conferences of Barcelona (1995) and Malta (1997), this intervention or euro-Mediterranean co-operation can be realized through the activity of institutional subjects, funding of initiatives (see the MEDA Programme), planning, protection of cultural heritage, research and exchange of data, information and models, formation and training of human resources, realization of inter-regional tourist products (Settimo Rapporto 1997: 573–6). This short sketch, of course, can be enlarged and deepened, and, to conclude, we want only to stress that a real development in this direction could not only promote economic growth, but also reinforce the peace processes, so important for many areas of the Mediterranean basin.

5 The 'new old' tourist destination

Croatia

Boris Vukonić

Introduction

Croatia is a middle-sized European country consisting of two regions quite different in their natural and geographic characteristics. Both geographically and culturally the continental Croatia belongs to central Europe, whereas costal Croatia lies on the Mediterranean. Hence, Croatia is both a central European and a Mediterranean country. Croatia, a country with a thousand years of history, first appeared as an independent state on the political map of Europe in 1990. It was the result of the disintegration process, ended by war against Serbia and Montenegro, two former Yugoslav Republics, which remain within the borders of the former state (Yugoslavia). For the last 76 years of its history Croatia was a part of Yugoslavia. The Republic of Croatia become an independent and sovereign state by the Constitution proclaimed by the Croatian Parliament in 1990 and was internationally recognized as such in 1992.

Croatia covers an area of 56,610 km². The Mediterranean part of Croatian territory covers 31 per cent of the total state surface. The coastline, including the islands, is 5789 km long. Croatia's territorial waters cover about 138,000 km², i.e. close to 5 per cent of the total surface of the Mediterranean sea. Almost half (48 per cent) of the total Adriatic coastline belongs to Croatia. Croatia has 1185 islands, rocks and reefs. Only 66 islands are inhabited. Krk, Cres, Brač, Hvar, Pag and Korčula are the biggest Adriatic islands. According to the 1991 census figures, 4.8 million people live in Croatia. The most inhabited region is Zagreb (308 inhabitants per square kilometre).

The highest authority on tourism is the Committee on Tourism of the Croatian Parliament, and the highest executive body of the Croatian government is the Ministry of Tourism, particularly active in the legislative area. At the national level, tourism is promoted by the Croatian Tourist Association and the Croatian Chamber of Commerce, including hotel and catering associations, travel agencies and organizations promoting nautical tourism. At the legislative level, basic legal acts in the field of tourism were created and enacted.

The system of schooling and education for tourism is based on the need for lifelong training and education for tourist purposes. Training and educational institutions are predominantly state-owned. Tourism is taught in a great number of secondary schools all around the country and at the university level, i.e. at the faculties in Zagreb, Opatija, Pula, Zadar and Dubrovnik.

An outline of Croatian tourism history

Croatia has a 150-year-long tradition of tourism, but it is only since the 1960s that it has assumed the shape of mass tourism. There are various underlying reasons which have contributed to that, all of them relative to the favourable characteristics of Croatian tourist potentials. Everything had started back in the nineteenth century. Only 22 years after Thomas Cook had organized his first tourist trip, Brothers Mihaljević did the same in Croatia, in 1863: they organized a rail trip (package tour) from Zagreb to Vienna and Graz. It was the beginning of organized tourism in Croatia.

When the Austrians and Hungarians saw the benefits of the mild climate and beautiful landscape, the development of tourism in Croatia began. The Viennese Society of Southern Railways built the *Kvarner Hotel* in Opatija in 1884. Their aim was to turn it into an organized seaside health resort. The *Therapia Hotel* in Crikvenica was built in 1894 and three years later the *Imperial Hotel* in Dubrovnik was opened. And so hotel building has continued up to the present.

The beginnings of tourism are also linked with the foundation and activities of tourist boards, especially those on the island of Krk (in 1866) and Hvar (Hygienic Board in 1868). Poreč and Pula had tour guides as early as 1845, while in Zagreb the first guidebook, called 'A Guide for Natives and Foreigners,' was published in 1892.

Tourism potentials

Croatia lies at the meeting point of the Mediterranean, the Alps and the Pannonian plains. The characteristics of each have merged into a unique and charming harmony of opposites which became the tourist product of Croatia. Although its main tourist attraction is the Adriatic coast and islands, Croatia also has some attractive tourist potentials in the interior of the country. In addition to its variety and natural beauty, another point in Croatia's favour is its preserved environment.

Croatia is undoubtedly one of the least saturated tourist countries in the upper Mediterranean. Today some 7.5 per cent of Croatian territory is protected within the national park system or under some other regional protection. Of the seven national parks, probably the best known are Plitvice lakes and Krka river in the mountainous regions, and islands Kornati, Brijuni and Mljet in the Adriatic region. Most touristically-notable are the monument cities of Dubrovnik, Split, Trogir, Šibenik, Poreč, Rovinj and the capital city of Zagreb; the islands of Hvar, Mljet, Lošinj, Brijuni, Rab, Brač; and the Baroque castles and thermal spas in the hinterland, of the region called Hrvatsko Zagorje.

Tourism is characterized by leisure and recreation, which require good resources and a favourable climate, along with tourist facilities and their distribution. Particularly important for Croatian tourism are all forms of maritime activities because of specific natural features and the indented Adriatic coast.

Favourable characteristics of the Adriatic coast tourist offer stem from the characteristics of the climate (warm Mediterranean climate), physical geographic characteristics (beauty of the landscape and wealth of cultural and historical monuments) and ecological status (cleanliness of the sea and absence of pollution, especially on the east Adriatic coast) all on one hand, and a very favourable geographic position with regard to the communications with major tourist-generating markets and well-appointed tourist facilities on the other hand. The Adriatic belongs to the warm seas, the sea temperature never below 11 degrees centigrade, and during the summer the sea temperature never drops below 22 degrees centigrade. Whilst the western coast is low-lying, shallow and without islands, the eastern coast is very indented, featuring numerous islands and bays which contribute to a unique experience for a seaside holiday. Biological quality gives the eastern coast an incontestable advantage over the western Adriatic coast: the population is half of that on the western Adriatic coast, which results in a lower level of littoralization and impact of other human activities, especially industry, on the environment. Therefore, the total level of anthropogenic impact on the sea is much lower on the eastern than on the western Adriatic coast.

Air and sea temperatures are the crucial climate factors contributing to the appeal of a seaside tourist resort, bearing in mind that out of a long stream of tourists, over 50 per cent is made up of children and elderly people. The average annual air and sea temperature on the Adriatic reads around 15°C and 17°C respectively. Most of the visitors to the Adriatic, both domestic and foreign, tend to stay in the region during the four summer months (June–September), when nearly 90 per cent of tourist flow is realized.

Tourist flow and other tourist data

Relatively poor organization of various tourist services is the main reason for some natural and other attractions being under-used for tourist purposes. This particularly refers to continental Croatia and its cultural landmarks, spas, hunting and fishing grounds and other possibilities. Today, the most touristically-developed part of Croatia is semi-island Istria with more than 60,000 beds in various accommodation facilities. In total,

Croatia has about 800,000 beds, but only about 20 per cent in hotels and similar accommodation units. In 1950, 84,000 nights of foreign tourists were registered in Croatia while in 1987 this number reached 59 million. The largest number of bednights was recorded in the 1986–87 period, 68 million, of which foreign tourists accounted for 86 per cent. Traditionally, most of them were Germans (40 per cent), while visitors from the United Kingdom, Austria and Italy altogether accounted for 30–35 per cent of all bednights.

In 1985, Croatia had at its disposal 5.4 per cent of total accommodation facilities, had registered about 7.5 per cent of all tourist arrivals and about 4 per cent of total receipts of international tourism in southern Europe. Today, as a consequence of war and a total erosion of quality, Croatian tourism has lost its market position, losing the international market share gained from the late 1980s. Despite a continual, although slight, increase in accommodation, the tourist flow has been on the decrease since 1987.

There were a thousand catering firms in the tourist sector in 1989. Croatia held 4 per cent of the total accommodation facilities in the Mediterranean countries; it realized 15 per cent of the total number of bednights, but earned less than 4 per cent of the total tourist revenue. Domestic and foreign tourists in 1991 were offered 570,000 beds in 479 hotels, 48 motels, 78 tourist complexes and 175 camps. Three years later, in the middle of the war, the picture was less favourable. Croatia's share in the total accommodation facilities dropped to 3 per cent and its share in the total number of bednights was reduced seven times, barely exceeding 2 per cent. Its share in the total tourist revenue made up a symbolic 1.2 per cent.

After Croatia gained its independence, tourism was facing many serious problems, the biggest are those caused by war, waged from 1990 until 1995. Although the maelstrom of war virtually brought the development of the tourist industry to a halt, it did not completely deter tourists from coming, particularly to areas unaffected by the fighting. Besides a steep fall in the volume of trade, many tourist attractions experienced mass-scale destruction, which in turn defined the scope of post-war reconstruction. After a one-year break, tourism revived in 1992 but did not really recover until the next year when 2,300,000 tourists and 13,000,000 bednights were registered.

The war's impacts, direct and indirect, made Croatia an undesirable tourist destination, in spite of the fact that in former Yugoslavia more then 80 per cent of foreign tourist traffic was registered in Croatian territory. Direct war impacts could be seen in destroyed infrastructure and tourist facilities and, even more, in partly or totally destroyed historical monuments and natural beauties (over 300 cultural monuments were razed to the ground, some of which were of inestimable value).

Damage to hotel and catering facilities is estimated at DEM 230 million. Furthermore, many refugees and displaced people were given accommodation in hotels and similar facilities, and the resulting damage to such facilities, which was considerable, needs to be taken into account. The cost of repairs to these facilities has been estimated to DEM 170 million (Ivandić and Radnić 1996).

The middle and southern part of the Dalmatian coast, the vital part of Croatian tourism destinations, were among the most badly affected regions in Croatia by war. International tourism flows to Croatia were formally stopped, except, in small numbers, in the northern part of Croatian Adriatic coast (Istria and northern islands), not affected by war.

The indirect effect of the war was to cause the drop in the physical volume of tourism flow and, consequently, in total tourist expenditure. It was estimated that in the 5-year period of conflict, Croatian tourism lost 304 million bednights due to the military aggression against Croatia. It was also estimated that the Croatian tourist sector lost about US$6.8 to US$10 billion of tourism revenue.

In 1996 Croatia was visited by 2,700,000 foreign tourists, which generated 16,600,000 bednights. The overall number of tourists was 3,900,000, with 21,500,000 bednights, a rise of 67 per cent from 1995. Domestic visitor nights increased by 12 per cent and foreign by 94 per cent. It is a good sign of recovery for Croatian tourism.

Today, the tourism industry in Croatia generates an additional market of 60,000,000 consumers who spent an average of 140–50 days in tourist facilities, of whom 40 per cent come from foreign countries. In the pre-war years, the additional market involved 70 per cent of foreign tourists who spent nearly $2,300,000,000. The volume of tourist consumption accounted for 10–12 per cent of gross domestic product, creating direct and indirect employment for 200,000 people and generating 80 per cent of the overall commodity exports and 61.2 per cent of the service exports.

The share of basic accommodation in the overall accommodation potential of Croatia was 23 per cent. About 20 per cent of basic accommodation was high category and 30 per cent medium category. The largest number of foreign bednights was registered in hotels (38 per cent), camps (27 per cent) and rented rooms (21 per cent), while more than 70 per cent of domestic tourists sought cheaper accommodation.

One of the significant characteristics of Croatian tourism is the marked concentration of tourist facilities along the Adriatic coast, which indicates that the tourism industry is more developed along the coast than in the interior of the country. Another prominent feature of Croatian tourism is its seasonal nature. The basic accommodation facilities are used 104 days by foreign tourists and 36 days by domestic tourists. Complementary accommodation is occupied 31 days by foreign and 27 days by domestic guests, 58 days of full occupancy. The average length of stay in accommodation varies from 4.73 nights in 1991, 5.46 in 1993, 5.87 in 1994 and 5.29 nights in 1995.

Research results show that most foreign tourists come to Croatia individually, mainly by passenger car (more than 90 per cent).

Table 5.1 Arrivals by mode of transport (thousands)

	1993	1994	1995
Air	249	408	419
Rail	379	370	365
Road	15,387	17,700	15,237
Sea	23	50	78

International tourism receipts totalled US$832 million in 1993, increased to over US$1400 million in 1994 and reached in excess of US$1500 million in 1995. This was not more than 0.7 per cent of total tourism receipts in Europe or 7.4 per cent of tourism receipts in Central and Eastern Europe. International tourism expenditure in 1995 was about US$600 million.

According to the results of the research on *Competition of Croatia's Tourism in the European Markets*, conducted by German company Steigenberger Consulting (1997), the present image of the Croatian tourist product in the international market has some explicit characteristics (Table 5.2).

Table 5.2 SWOT analysis

Strengths	Weaknesses	Opportunities	Threats
Attractive coastlines	Car accessibility	Incline of demand	War image
Variety of islands	Building obstructions	Trend to sun/sand/sea	Image as mass destination
Established tourism sector	War image problems	High degree of fame	Regional instability
Low crime rates	Lack of packages	Special interest facilities	Re-engineering process
People's hospitality	Tourist information	Weekend destination	Infrastructure
City of Dubrovnik	Nightlife facilities	Convention destination	
Historic towns	Accommodation quality	Governmental activities	
Sport infrastructure	Variety of restaurants		
Hinterland attractions	Public opinion		
Spa resorts	Seasonality Price levels		

New tourism marketing strategy

Tourism was always among sectors Croatia expected the most from economically. After gaining its independence, Croatia's expectations of tourism as one of the most important pillars of national economy became a part of the contemporary state economic policy. That is why Croatia elaborated The Master Plan of Croatian Tourism, aiming to establish at least the overall goals for future tourism development and its contribution to the national economic recovery after the war was over. Even during the war, Croatia's government had never given up the belief that tourism could and should be one of the most prosperous economic sectors. In conditions of war it was not easy to conduct the proper tourism policy; but it was not easy to even believe in any tourist development.

The roots of today's national tourism policy were established during the civil war, when Croatia had conducted its marketing activities on the international tourism market, mostly of a promotional nature. At that time, the main difficulty resulted from the fact that there had been very little similar experience in the world: war was a 'market obstacle' which is not very common in both the theory and practice of world tourism development. Evaluation of the former Croatian tourist product shows that the former hectic and passive policy led to a loss of the position in the market, to a mostly negative image, and finally, to unfavourable terms of placement.

So, the first goal of contemporary Croatian tourism policy should be concentrated in changing the tourist policy from passive to active, including the repositioning of Croatian tourist products in the international market. For achieving such a goal, many modifications of products, markets and elements of the marketing mix should substitute the former tourism policy. In such circumstances, the application of marketing seemed to be a very questionable one. But, if we accept thinking about war as any other market obstacle than, at the same time, it seemed that marketing would also be the only possible way to overcome war difficulties in the post-war period.

In the opinion of many Croatian tourist experts, there was another vital question: how to implement marketing in the transition period of Croatian economy on the one hand, and in the war and post-war period of 'no peace and no war', on the other. Croatian tourist potentials in the past had been presented to the world market as a part of former Yugoslavia. Thus Croatia, as a tourist destination, had generally been put on the international market bearing the name of Yugoslavia, rather than its own. So, one of the first goals of marketing policy was to launch on the market 'a new old product', now under its own name, the product with a new identity and image.

The national tourist promotional campaign was concentrated to support these goals, to minimize war damages and to regain the confidence of the international tourism market. It was obvious that usual 'promotional invitations' could not be successful, and Croatia decided to lead the promotional campaign by saying, 'We are here to stay'; and, for those who had already visited Croatia in pre-war time by saying, 'Remember the nice holiday you once spent in Croatia?' Increased efforts in recent years to systematically analyse the international tourist market resulted in a new sign and logo for Croatian tourism, and a slogan 'Small Country for a Great Holiday.' Promotional events on the international tourism market have been organized under this initiative almost since Croatia became an independent state. Simultaneously, posters and two basic tourist brochures were prepared. Hundreds of regional and local brochures followed, attempting to invoke a new visual identity for Croatian tourist destinations, trying to meet the basic promotional message: that there were a wide range of tourist attractions and numerous tourist activities in almost all parts of Croatia.

It is already obvious that, through certain promotional activities, the conditions will have to be created which will enable the return of tourists from countries who regularly visited Croatia before the war. Likewise, the strategy of the promotion has to consider a very large number of the Croats in diaspora, living practically on all the continents. The patriotism Croatian diaspora has shown during the war, has been considered a strong additional element in their decision to spend holidays in Croatia. So, some specific promotional material was produced to attract them to Croatian tourist destinations.

The future prospects

The results Croatia already achieved from tourism in the past confirm the hope that tourism could be one of the main pillars for Croatian economy in the future. Such a trend has already been observed during the last two years of Croatian tourism development. In the 1997 tourist year, many Croatian tourist destinations reached 60–70 per cent of results achieved before the war. Most of the damage caused to the tourist facilities has been repaired, as have most of the roads. More than that, about 50 km of new highways were constructed, especially on the way from Croatian borders to the Adriatic coast. It is a good sign of recovery, as roads were in the past (and still are) the worst part of Croatian tourist infrastructure.

For future tourism development, Croatia is concentrating mostly on the European tourism market, bearing in mind that some other markets like USA, China, and Japan, would be very prosperous over the coming years. Croatia was expected to reach the pre-war numbers in international tourist arrivals, bednights and foreign tourist receipts by the end of 2000.

6 Greek tourism on the brink

Restructuring or stagnation and decline?

Yorghos Apostolopoulos and Sevil Sonmez

Introduction

During the Greek classical period, visitors and foreigners were considered sacred and were believed to be protected by Zeus. In the spirit of this ancient tradition of hospitality, contemporary Greece has instituted a tourist industry of considerable magnitude, and the country has continually been classified among the world's most popular tourist destinations (*Condé Nast Traveller* 1998). Greece's touristic gamut[1] comprises an array of spectacular attractions encompassing natural beauty and rich history and culture. A plethora of mainland destinations along with 15,000 km of coastline and over 2500 islands (spread out into the Ionian and Aegean Archipelagos), which add to the country's 'mass charter attractiveness', have been the catalyst for transforming the country into a vacation spot of international calibre. The phenomenal expansion of tourism development since the 1960s has transformed the basis of socioeconomic structure, altering the country's life chances and welfare, accompanied by adverse sociocultural and ecological ramifications (Kassimati *et al.* 1994; Loukissas 1977; Loukissas (co-ord) 2000).

This chapter considers the foremost economic sector of the country (i.e. tourism), which ironically continues to constitute a 'statistically invisible industry' (Leontidou 1999), in the midst of international geopolitical and economic reshuffling. Within this international framework, and without overlooking the crisis phase of traditional, mass charter, Greek tourism model, this chapter critically reviews the multifaceted dimensions of Greek tourism. We will focus, among other things, on production and consumption patterns, public policy, spatial polarization, and structural deficiencies. Furthermore, the chapter provides policy guidelines for the restructuring of the Greek tourism system in the context of a sustainable development strategy.

Tourism evolution, reshuffling and globalization

Travel and tourism can be traced from the classical Mediterranean civilizations of Greece, Rome and Egypt, through the medieval pilgrims, to the

Grand Tourist of the eighteenth and nineteenth centuries. Global tourism, as the product of the post-World War Two era, skyrocketed to 625 million in 1998 from 14 million tourists in 1948 (WTO 1999),[2] mainly as a consequence of a growing middle class and the technological revolution. Tourism holds a leading position in the world economy and is considered to be one of the economic sectors offering immense potential for long-term growth, especially underscored by the continuing worldwide expansion of the services sector. The tourist industry – despite setbacks such as the collapse and subsequent instability of eastern Europe, the Gulf war, the civil war in the former Yugoslavia, the financial turmoils in Southeast Asia, Japan, Russia, and Latin America and their subsequent sociopolitical impacts, and a fluid international state of affairs – has become the world's pre-eminent industry, producing $3.6 trillion of gross output and employing 225 million people worldwide (WTTC 1997).

One of the most notable developments in the fast evolving post-World War Two era has been the domination of the tourism sector by transnational corporations. While this domination has, on the one hand, resulted in an unprecedented expansion of the industry, with immense profits for the tourist enterprises of core nations, on the other, it has seriously weakened the negotiating power of developing countries. The trade practices of tourist transnationals have often had long-term detrimental impacts on peripheral and semiperipheral nations that are heavily dependent upon tourism (i.e., Greece). Such practices have brought about socioeconomic and spatial polarization and various types of economic dependency (Brohman 1996). The tour-operator oligopoly, in particular, literally controls the market, going so far as defining the means by which tourists consume the tourist product of developing countries (Timothy and Ioannides, in press). Moreover, as the entire society subsidizes infrastructure and often pays a high price in sociocultural and environmental costs and conflicts, tourism benefits are enjoyed by only a few groups – mainly local elites and outsiders.

The international patterns of tourist mobility have been spatially polarized with an overall dominance of Europe (with a marked decline in recent years), a continuing increase in Southeast Asia's and the Pacific area's share, and a decline in the Mediterranean (WTO 1999). In the European context (from where Greece draws over 90 per cent of its visitors), tourist flows and their spatial distribution reflect a differentiated process of production and consumption attributed to factors related to accessibility, supply of attractions, and inter-country differences (Komilis 1987). At the other end of the spectrum, the Mediterranean region constitutes the world's prevailing tourist destination, representing almost 40 per cent of the world's total tourist arrivals and 30 per cent of the world's revenues respectively (WTO 1998), of which the European Union countries of the Mediterranean dominate 80 per cent (Montanari and Williams 1995).

Within this sociopolitical and economic milieu in the world arena, tourism development in Greece, and its spatial distribution in particular, can be attributed to a series of exogenous and endogenous determinants (Komilis 1987). The former encompass geographical location and accessibility within the European transport landscape and the organization of the transnational tourist industry – as the networking power of airlines and tour operators can affect those European markets which are drawn to Greek destinations as well as into components of the Greek tourist product (Komilis 1987). The latter mainly contain Greece's stage in the economic development process, the central, regional and local governments' tourism-related policies, the tourist product's articulation and attraction (quality and price competitiveness) as well as the marketing policies geared toward international tourist markets and the industry's main agents (i.e., tour operators, travel agents, the state) (Komilis 1987). In this context, the Mediterranean's non-monopolistic tourism, with the existence of a plethora of alternative sunlust destinations, increases competition and reinforces the likelihood of substituting destinations under certain circumstances (e.g., price changes, tour operators' practices, etc.).

Spatial aspects of tourism development

Uneven demand and supply

Since the formation of the modern Greek state in 1832, the country has demonstrated several forms of underdevelopment (Mouzelis 1978). Under the so-called 'spheres of influence' of the post-World War Two era, Greece has been a satellite in international geopolitics with subsequent impacts on its socioeconomic structure and development. 'The various forms of economic aid coming from the British and primarily the Americans had as their declared objective, avoidance of the restructuring of the Greek economy and preservation of Greece as an exporter of raw materials and importer of manufactured goods' (Kofas 1989), casting the country into a long phase of 'dependent development'. Both the political and economic structures have experienced numerous problems,[3] and various economic indicators worsened as the unfavourable trends of the late 1970s continued into the 1980s and 1990s, causing further deterioration of macroeconomic imbalances[4] (Central Bank of Greece 1992). During the past five years, due to the 1994–99 convergence programme in the context of the economic and monetary unification of the European Union, there has been a marked progress in all Greece's macro- and micro-economic indicators (Central Bank of Greece 1998). Still, for sustainable stabilization trends to emerge, fiscal imbalances and low productivity have to be resolved through the use of radical reforms that will affect the size and functioning of the public sector and the organization of markets and production (Central Bank of Greece 1998).

Within these sociopolitical and economic contexts, Greece appeared in the contemporary touristic map only in the early 1960s. The subsequent emphasis on 'outward-oriented development strategies', for the purpose of stimulating rapid growth, transformed the country's economy to a monocrop. Further, the vulnerable tourism sector has been treated as a panacea despite inadequate infrastructural and superstructural preparation (Apostolopoulos 1996). In fact, as recently as 1985, when Greece was on (yet another) corrective fiscal programme, it was 'the bankers and the transnational capitalists in the OECD who had prescribed wage austerity (1985–86) as part of a new development strategy based on the lowering of labour costs to promote low-grade exports and *cheap mass tourism...*'. (Petras 1992, emphasis added). With this exogenous-based rationale along with the endogenous quest of boosting employment and balance of payments, Greek tourism (especially since the 1970s) has demonstrated vertical growth, becoming the 'national industry'.

Production and consumption of tourism

Since its genesis, tourism development in Greece has been symptomatic of intense spatial polarization (Leontidou 1998). Until the early 1970s, the predominant model of tourism development centred around a dominant central destination (Greater Athens) and several peripheral regions of two types: (a) small resort islands which were vitally dependent upon Athens for the transportation of their few seasonal visitors as well as for supplies and infrastructural needs and (b) major peripheral destinations which, while they were away from the centre, were linked via an international air-based transport system to facilitate their considerably greater seasonal tourist demand. While the foregoing tourism development model still exists, since the early 1970s tourism development has gradually moved toward the creation of a few 'autonomous' and touristically peripheral regions as well as a continually weaker and declining centre (Komilis 1987). In the context of these two distinct tourism development models, Greek destinations may be further classified into: (a) predominantly urban destinations where accommodations were developed via primarily legal procedures; (b) peripheral destinations where illegality and arbitrariness pertained to tourist establishments;[5] and (c) resort destinations with mainly transnational hotel establishments where predominantly foreign tourists consume the tourist product and international and indigenous labour produce the services (Komilis 1994).

International tourist arrivals during the 1950–90 period have recorded a massive increase, mostly due to the promotion of package holidays (approximately 70 per cent of arrivals are charter dependent),[6] although the growth rate is slower than that at other Mediterranean and European destinations (especially since the early 1980s). While in 1938 (before World War Two), 90,400 tourists were recorded, in 1950 (in the aftermath of two devastating wars[7]) arrivals dropped to 33,400 (GNTO 1997). Thereafter, in the three-decade period between 1960–90, a quantitative explosion took place: 400,000 tourists visited Greece in 1960, 1.6 million in 1970, 5.6 million in 1980, and 9.3 million in 1990 (GNTO 1998) establishing the country as an international destination. During the 1990s, spasmodic patterns of arrival were recorded, not only confirming the vulnerability but also the low competitiveness of Greece's tourism sector. Thus, in 1992 – following changes in broader geopolitical and national developments – tourist arrivals reached 9.7 million, 11.4 million in 1994, 10.6 million in 1996 and 11.1 million in 1998. The average of these aggregates for the 1990s accounted for approximately a 2 per cent share of world arrivals, 4.5 per cent of European arrivals, and 5.5 per cent of Mediterranean arrivals (GNTO 1998; WTO 1998), with declining trends.

The typology of inbound travellers and tourists includes approximately 70 per cent package vacation groups, 17 per cent business travellers, 7 per cent individual tourists, 4 per cent conference participants, and 2 per cent government officials (Agelis and Falirea 1996). In the 1990s, incoming tourists were primarily of Eurocentric origin (approximately 90 per cent originate from European countries), 86 per cent of whom originated from European Union countries, and over 60 per cent originated from the UK, Scandinavia, and Germany (GNTO 1998). At the other end of the spectrum, outbound tourism has been rising rapidly mainly due to the lifting of foreign exchange restrictions by the central government. While, in 1985, only 1.5 million Greeks travelled abroad, the current number exceeds 5 million (the equivalent of approximately half of all incoming tourists)[8] with rising trends.

The revenues from this tourist influx have compensated for a considerable portion of the merchandise trade loss as the Greek economy has been marked by a chronic trade deficit. The contribution of tourism to the Greek economy in both the 1980s and 1990s has been phenomenal; in 1980 tourist revenues contributed $1.7 billion to the national economy, $1.9 billion in 1985, $4.6 billion in 1990, and $6.8 billion in 1995 (GNTO 1998). For the 1996–98 period, tourist revenues have shown uneven patterns (similar to those of arrivals) with 'stagnating' trends when compared to competing markets in the region and Europe. In 1996, tourist revenues reached $6.9 billion, $7.2 billion in 1997, while in 1998 they exceeded $8 billion (GNTO 1999)[9] (Figure 6.1). Furthermore, for 1996 alone, tourism contributed 14.7 per cent to the GDP,[10] constituted 18.7 per cent of total exports, and generated 13.5 per cent of total jobs, remaining the country's pre-eminent industry (WTTC 1997).

During the past half century, the transformation of the infrastructure has marked all facets of tourist facilities and services. Conventional types of lodging were predominant until the mid-1960s, yachting appeared in the late 1960s, bungalows and apartment hotels emerged in the early

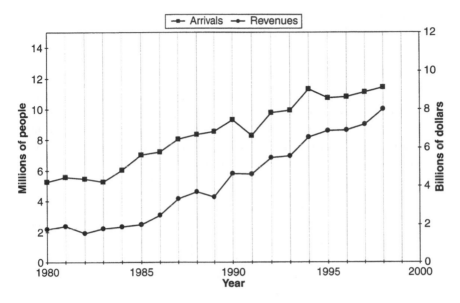

Figure 6.1 Tourist industry's growth in Greece 1980–98.

1970s, and organized cruises and summer villas along with foreigners' home purchases appeared in the mid-1970s (Zacharatos 1986). At present, 7500 official accommodation establishments with a total of half a million bed capacity have been recorded (GNTO 1997),[11] over a million beds were offered by unregistered and unregulated accommodation establishments in 1992 (GATE 1993), and approximately 23,000 beds were provided by cruise ships and yachts in 1994 (Eurostat 1995). There has been an uneven regional concentration of hotel capacity: 14.2 per cent of the total number of beds are concentrated in the Greater Athens area, 17.4 per cent in Rhodes, and 18.5 per cent in Crete (GNTO 1997), that is, more than half of the available accommodation facilities are concentrated in only three destinations. Further, the second-home phenomenon reflects a spectacular increase and as domestic tourism originates mainly from Greater Athens and Thessaloniki, second homes predominate on the fringes of these metropolitan areas (Leontidou 1998).

Finally, both the tourist product and the country overall as a destination remain very popular, despite fierce regional competition and the plethora of structural problems of the sector that have significantly impacted its competitiveness. According to *Condé Nast Traveller's* (1998) annual survey, Greece still holds a remarkable standing worldwide. In 1998, Greece ranked seventh in the world and the second most highly-rated destination in the Mediterranean after Italy. It was rated higher than the USA, Canada, the UK and France and much higher than its major regional competitors Spain, Turkey, and Morocco. Furthermore, the Greek islands (together with Cyprus) were classified as the favourite Mediterranean vacation islands while they ranked eleventh in the world, after 'exotic' destinations such as Bali, Hawaii, Barbados and Fiji.

Public policy trends in tourism

The Marshall-Plan-initiated infrastructural advancements, along with other types of foreign aid, contributed immensely to Greece's post-war reconstruction and subsequently to the gradual emergence of its foreign tourist flows. The Greek National Tourism Organization, which was established in 1929 but became an autonomous agency in 1950, played a crucial role in the planning, promotion, education, construction and management of accommodations and infrastructure as well as in financing private enterprises (Kassimati *et al.* 1994). Tourism's potential for economic development was realized relatively late when the industry was extensively promoted as a development tool by international financial organizations. A review of tourism policies of the 1950–98 period clearly shows the lack of a comprehensive state strategy for tourism policy and development (with a plethora of often disjointed and contradictory measures and laws), accompanied by state intervention in the market and a lack of effective legislative backing for properly directing or regulating the growth of tourism (Apostolopoulos 1994, 1996; Buhalis 1998; Komilis 1987, 1995a,b). The implementation of most plans and policies have been unsuccessful due mainly to the limitations of the administrative apparatus in establishing and promoting procedures for implementation, and due to the intensive resistance of interest groups against planned interventions (Agricultural University of Athens and University of Thessaly 1998; CPER 1994; CPER and University of Thessaly 1998) bringing to surface 'value-conflictive' situations between planning agencies and private interests (Komilis 1994). The tourism policies of the past 50 years distinctly illustrate the impact of the national sociopolitical history and sectoral developments.

The 'post-war reconstruction' period (1950–67): tourism as a foreign exchange earner

The fact that tourism did not emerge as a public policy concern until the 1960s, although foreign tourist flows were recorded a decade earlier, might be explained in terms of Greece's urgent foreign exchange needs – in the aftermath of two costly (in both physical and human capital) wars (Tsartas 1989). Tourism began to appear in state policies, and incentives for tourism development included direct or indirect subsidization of private tourist enterprises by loans for the construction of accommodations (56 per cent of all tourism investments during this period were allocated for hotels, see Leontidou 1998). The subsidization of the hotel sector involved – among other things – financial and fiscal incentives, interest rates subsidies, tax reductions, and operating cost reductions (Leontidou

1998). In addition, infrastructural improvements were initiated to facilitate tourism development, mainly in urban areas, without taking into account the developmental needs and problems of the country's peripheral regions (CPER 1994; Katohianou 1995; Komilis 1994), thus reinforcing the phenomena of spatial polarization and inequality.

The 'dictatorship' period (1967–74): tourism as a means for 'heating up the economy.'

Public tourism policy (laws 147/1967 and 1313/1972) during the dictatorship years also focused on the urgency for foreign exchange inflows with an even greater emphasis on construction, financing, and promotion. Special concessions (in the form of tax and depreciation allowances) were granted to private enterprises to build hotels in urban, semi-urban, and rural regions which were completely unprepared infrastructurally to receive the tourist influx. These concessions were adopted as a means of 'heating up the economy' (Leontidou 1998). Furthermore, this period was marked by an upsurge of private tourism investment with an overall decrease in public capital, while foreign investment (especially in hotels) was substantial, with controversial effects, including the loss of local control and identity among host communities, failure to create skilled jobs for the locals, intensification of labour exploitation, and inequitable distribution of the costs and benefits of growth (Apostolopoulos 1996). In 1968 in particular, 66 per cent of all tourism-related investment was of foreign origin and was concentrated in coastal localities (Leontidou 1998). In addition, mass-scale tourism was extensively promoted in infrastructurally-unprepared rural regions (Katohianou 1995) with intense regional investment imbalances and adverse ecological consequences (CPER 1994; Loukissas 1977; Loukissas and Triantaphylopoulos 2000; Tsartas 1989).

The 'return to democracy' period (1974–81): tourism as a tool of regional development

Tourism increasingly became the means for alleviating the chronic trade deficit (via the inflow of foreign exchange), as well as an effective strategy for economic development. The tourism development policies of the two consecutive conservative governments of this period were characterized by boosting regional development, by supporting the problematic regions of the country, while simultaneously discouraging further development in touristically-saturated regions and localities. Despite various governmental incentives, these policies were unable to halt a further development in already saturated areas, resulting, among other things, in the emergence of the para-hotel sector and an upsurge in unauthorized, illegitimate construction with subsequent ecological ramifications and market distortions (CPER 1994; Katohianou 1995).

The 'socialist experiment' period (1981–89): tourism as a form of social welfare

The socialists' move into power and the official accession of Greece into the European Economic Community (as it was then called) marked the economic development policies of this period. Tourism policies centred around the 1262/82 Law which provided investment incentives by subsidizing the development of small- and medium-sized tourist enterprises, as well as around the input of the 'Integrated Mediterranean Programmes' funds. These incentives focused, as previously, on already touristically-developed and saturated regions as well as on medium-sized and quality enterprises which ultimately did not contribute to the upgrading of the supply of tourist facilities (CPER 1994; Katohianou 1995). In addition, the illegitimate and often illegal construction of rental establishments reached a peak. Extra emphasis was placed on the qualitative upgrading of the tourist product along with the gradual development of new and alternative forms of tourism (i.e., programmes of 'social tourism', agro-tourism, etc.). Still, tourism regulation was difficult due to conflicting or non-explicit measures (i.e., the enactment of 'tourist saturated areas' through the Ministerial Decision 2647/86).[12]

The 'European unification' period (1989–98): tourism in the midst of regional competition

Worldwide geopolitical and economic reshuffling, the upcoming economic and monetary unification of the European Union, the emergence of a new, more pragmatic leadership in the Greek national politics, and less state intervention accompanied by stronger support for market policies, along with their impending impact, constitute the skeleton of this period. Within this context, the unco-ordinated and often conflicting policies of the past continue and the absence of a tourism strategy with clearly-defined objectives is apparent. Three additional laws dealing with economic development and tourism (2160/93, 2206/94 and 2234/94),[13] tourism programmes funded by the European Union, along with the 'Integrated Mediterranean Programmes' provide the direction and the funds for tourism policies for regional development with a focus on the diversification and upgrading of the tourist product.

Tourist transnationals have played a critical role in forming Greece's tourist product by exerting immense power and influence, and thus crucially affecting the sector's development and growth (Apostolopoulos and Ioannides 1999). Although nominal foreign investments in Greece do not constitute the majority of investment, the international tour-operator hegemony – through their control of incoming flows, managerial and marketing expertise, and financial resources – is overwhelming. Foreign interests in Greece do not exert control through direct ownership or shareholding (sufficient enough to maintain a predominant managerial influence), but rather they prefer making contractual arrangements with indigenous parties (i.e., hotels) offering them effective corporate control and profits without the risk of committing large sums of capital to build a resort complex. Finally, the phenomenon of absolute power exerted by international agents – regardless of the objections of Greek tourist enterprises – have been met with a *laissez-faire* attitude by the National Greek Tourism Organization.

Structural deficiencies of the tourism system

Despite tourism's impressive aggregates and its undoubted contribution to regional development and welfare, a systematic analysis of Greek tourism sheds light on a series of long-term structural problems and inadequacies. These are mainly the result of the way tourism has been developed since the 1960s and are associated with the lack of a strategic plan for the development of the most vital sector of contemporary Greek economy (Komilis 1995a). These problems have already hindered a faster growth and a higher regional share and have affected the sector's regional competitiveness. A comprehensive review of Greek tourism in the post-World War Two era reveals a series of critical areas classifiable into five categories.

Ineffective public policies

Public tourism policies have lacked vision and have subsequently impeded the growth and prospects of the sector. The lack of a comprehensive long-term tourism policy, inconsistent and often 'irrational' political intervention in the sector (with intense phenomena of clientilism) resulting in policies without continuity, the lack of co-ordination among different tourism-related agencies, an incrementalist and fragmented approach to emerging problems, policies based on insufficient scientific documentation and analysis (often based on subjective, personal and even conflicting judgements and paradigms), a lack of strategic marketing policy (ad hoc annual advertising campaigns undertaken by private companies), spatial and socioeconomic inequality and unevenness, and even the domination of the national political economy by foreign interests[14] constitute the major stumbling blocks in the context of public policy (Apostolopoulos 1996; Buhalis 1998; CPER 1994; Komilis 1995a).

Diminishing competitiveness

Endogenous and exogenous problems, associated with issues related both to the Greek and international tourism sectors, have significantly influenced the competitiveness of the tourist product, as well as the place of the sector within the national economy. As a result, the Greek tourist product has not developed a reputation as being of high quality, diversified and unique, analogous to its potential and its competitive advantages (Kalogeropoulou 1993; Komilis 1994, 1995a; Touloupas 1996).

The endogenous problems include infrastructural shortcomings, lack of articulation of tourism with other economic sectors (withholding development and growth of other sectors and even hindering attempts for 'endogenous' development), one of the highest seasonality patterns in Europe (over 80 per cent of international arrivals take place in a five-month period), an overall poorly organized and managed sector (especially in saturated regions) and an uneven spatial distribution of tourist supply (Agelis and Falirea 1996; Apostolopoulos 1996; Buhalis 1998; Komilis 1994, 1995a,b). The exogenous problems include shifts of the international tourist market to new destinations, increasing competitiveness of regional destinations (i.e., Turkey, Cyprus, Morocco), the entry of the former communist bloc countries into the regional tourist market (i.e., Romania, Bulgaria, Ukraine) and widely fluctuating tourism revenues due to the uncontrolled nature of geopolitics (Apostolopoulos 1996; Komilis 1995a).

Market distortions

The problematic organization and structure of the sector has resulted in an inefficient industry. These problems include an expanding underground hotel sector (unofficial reports claim that two out of three beds/night-stays belong to the para-hotel sector), widespread tax-evasion (estimated evasion exceeds 60 per cent of the total gross output of the sector), over-supply of diminished quality accommodations, management repatriation, foreign exchange leakages (substantial leakage of tourism earnings transferred abroad), an overdependence on the German, British, and Scandinavian markets (eight out of ten tourists come from Europe and 46 per cent of arrivals come from Great Britain and Germany), and unprecedented tour-operator intervention (Apostolopoulos 1996; Buhalis 1998; CPER 1994).

Adverse sociocultural ramifications

The lack of a master tourism plan has exacerbated the various effects of tourism on the sociocultural spheres. These include a loss of control over local resources, reinforcement of patterns of socioeconomic inequality and spatial unevenness, increasing crime, phenomena of staged authenticity, commercialization of traditions and customs, moral permissiveness, unequal relationships between tourists and locals, international demonstration effects, rising alienation among locals, and the emergence of 'parasitic strata' of tourist entrepreneurs becoming role models for youth (Apostolopoulos 1996; Briassoulis 1993; Kassimati *et al.* 1994; Tsartas 1989).

Marked alterations in fragile marine and terrestrial ecosystems

The haphazard development and expansion of tourist infrastructure has posed problems of considerable magnitude to environmental resources. These critical problems of varying intensity relate to: (a) the impact on the coastal open (natural and agricultural) and built (rural and urban) landscapes and (b) the demands and pressures on the local and regional (land and marine) resource base (Briassoulis 1993; Komilis 1987, 1994; Loukissas and Skayannis 1999). Specifically, land and water resources are abused due to the increasing use of coastal (especially) land and sharp seasonal demand for water. Acute problems of land-based and marine pollution due to inefficient systems of solid waste discharge and management are exacerbated and over-concentration of activities and infrastructure create diverse physical pressures and corresponding problems (Komilis 1995a,b).

Strategic planning for sustainable tourism development

The development discourse – with tourism as an integral part – has embodied the debate over the most appropriate or alternative forms of development *vis-à-vis* mainstream development processes.[15] This discussion has brought to the fore the concept of sustainable development as a positive socioeconomic change that does not undermine the ecological and social systems upon which communities and society depend. The successful implementation of sustainable development requires policy, planning, and social learning processes; its political viability depends on the full support of the people it affects through their governments, their social institutions, and their private activities. Thus, in the context of sustainable regional development, strategic tourism planning and policy acquire even more meaning within developing regions where globalization, uneven and unequal development, disorganization, corruption and governmental ineffectiveness, debt, poverty, marginalization of people, dependency forms, and so forth are in the foreground (Baker *et al.* 1997; Mawforth and Munt 1998).

The structural deficiencies of the Greek tourism sector and the subsequent ramifications on both the sociocultural and physical spheres have definitely placed the tourism system in a prolonged phase of flux. The pre-eminent strategic tool for sustainable tourism development is coterminous with restructuring and modernizing the entire tourism system, that is, 're-planning' and 're-developing' the tourist market, the tourist industry, and the tourist product by taking drastic corrective and adjustive measures (Apostolopoulos 1996; Komilis 1995a,b). The process of restructuring and reorganization involves deep economic, social, and cultural changes in the geography of production and consumption, influencing regions and localities. Thus, in the context of current trends towards 'postmodern' practices, the process of restructuring is associated with Post-Fordism and is characterized by qualitative shifts from mass production and consumption to more flexible systems of production and organization, as well as changes in the way that goods and services are consumed, with rapidly changing consumer tastes and the emergence of niche and segmented markets (Ioannides and Debbage 1998; Mowforth and Munt 1998). In tourism-specific contexts, the Post-Fordism paradigm captures the processes of global restructuring and the qualitative changes in the organization of both production and consumption, by sponsoring diversity in traveller preferences and proliferation of alternative destinations, multiplication of vacation types based on lifestyle research, development of new destinations and experiences due to fashion changes, growth of 'green tourism' and other vacation forms tailored to travellers' needs, 'de-differentiation' of tourism from leisure, culture, retailing, education, sport, and hobbies (consumption less 'functional' and increasingly aestheticized), and growth of a travellers' movement with more information provided through the media regarding alternative vacations and attractions (Lash and Urry 1994; Mowforth and Munt 1998).

Drawing on the main issues and points from the sustainability and Post-Fordism modes of production and consumption, there are certain parameters functioning as common denominators which underlie regional development policies and strategies. These include or relate to: (a) the inter-regional differentiation and diversity of regional tourism production; (b) the maximization of those economic benefits of tourism (both the formal and informal tourism sectors) over an entire region by providing the best interlinkages of tourism to other sectors of the regional economy; (c) equity and local involvement conditions that should prevail when tourism's growth is related to sustainable regional development; (d) the environmental considerations taken into account in policy-making and tourist product development; and (e) the continuity and adjustability of a region's tourism development within its wider environment (Komilis 1994; Mowforth and Munt 1998).

In the context of Greek tourism in particular, there is an imperative need for restructuring, reorganization, innovation and modernization of the system as a whole (Apostolopoulos 1996; Komilis 1994, 1995a,b). Such processes pertain to structural changes in the composition, articulation and spatial distribution of tourist consumption patterns *vis-à-vis* the tourist markets and products, while they further involve neoteristic approaches in the search for new tourist markets, development of new tourist products, as well as for improved management and marketing approaches (Apostolopoulos 1996; Komilis 1995a,b). Specifically, the processes of restructuring, reorganization, innovation and modernization may include: '(a) the diversification and qualitative upgrading of tourist supply and demand (articulation or mix of different segments of international tourists); (b) the interweaving and interlinking of tourism and other tourism-related sectoral activities into regionally identifiable and differentiated tourist products; and (c) the increase of the competitive advantage of the Greek tourist industry' (Komilis 1995a, p. 71).

In support of the restructuring process of the Greek tourist market and product, the promotion of certain new or less developed (non-mainstream by either Greek or regional standards) types of tourism is expected to utilize and mobilize existing idle and the regionally diverse tourist resources and, thus, conduce to increasing the region's competitiveness and share in the international tourist market. The development of a more diversified tourist product seems to be favoured by present conditions of a highly competitive tourist market. Past attempts to diversify the tourist product had been inhibited by particularly favourable demand conditions and by the rapid expansion with strong state support of one type of accommodation (hotels) and by private or state policy orientation to mass and resort tourism (Komilis 1995a,b).

Furthermore, the whole process of planning demands a drastic reorientation of the country's policies and practices in order to achieve its foregoing objectives in the framework of sustainable tourism development. Thus, planning policies should preserve and enhance: (a) diversity and unique regional physical and sociocultural elements by stressing local identity; (b) complexity or multiplicity of urban and rural functions as well as intersectoral and interfunctional (between land uses) complementarities and coexistence; and (c) balance between regional self-sufficiency, dependency on external inputs, and integration with the European Union's competitive environment (Komilis 1987, 1994).

Finally, in the context of restructuring the Greek tourism system, and especially when developing international tourism, tourist regions should be prepared to face the highly competitive international market that leaves them susceptible to exogenous economic forces. A successful national or regional tourism development strategy would require a careful consideration of exogenous and endogenous forces and concerns (i.e., scale of development, range of regional and community benefits, resource availability, and investment allocation). It is possible that through the restructuring of the Greek tourist product and through appropriate marketing policies (Apostolopoulos and Sonmez 2000), the main exogenous determinants of tourist flows could be influenced. This restructuring should, firstly:

> liaise with the major processes of economic restructuring and the heightened levels of territorial competition, which develop among European countries and regions today and exhibit certain contrasting spatial characteristics: increasing globalization and centralization of the production system, but also flexibility regarding specialization or localization in particular industries and spatial contexts
>
> (Komilis 1994, p. 71)

Secondly, restructuring should

> contribute to the maximization of regional development benefits in a way that utilizes and mobilizes the regional resource base, realizes regional intersectoral linkages and is compatible with local–regional economic interests, societal values and environmental assets
>
> (Komilis 1994, p. 71).

Concluding remarks

The end of the millennium and the era of economic globalization and transnationalization of tourism find Greek tourism in a state of flux. While the country is heavily dependent on tourism revenues, hesitant measures by both the state and the private sector during the past half-century have kept the tourism sector in a prolonged stagnation phase. Uneven supply and demand, a stubborn persistence in the traditional resort tourism model, slow and unco-ordinated emergence of new tourism forms, 'irrational' political intervention in the market, lack of strategic marketing policy, and fragmented and often absurd public policies have not contributed to the long overdue recovery. Although tourism has been viewed as the sector to 'revitalize' Greek periphery and 'equalize' various regional inequities in the country, its spasmodic developmental process and lack of articulation with other productive sectors have hindered and even withheld their development.

The sustained growth of the Greek tourism sector in the next century will be coterminous with the restructuring, reorganization, and modernization of the entire tourism system by re-planning and re-developing the tourist market, the tourist industry, and the tourist product. Linked with broader sustainable development principles, state and private sector policies should stress 'inter-regional differentiation and diversity of tourism production', linking tourism to other economic sectors, environmental considerations (capacity, impacts), 'flexibility and responsiveness' to evolving conditions of the international environment, equitable distribution of tourism benefits, as well as minimization of social conflicts by avoiding adverse ramifications of tourism growth.

Notes

1. Greece's touristic potential – due mainly to its cultural heritage and geomorphology – is immense, but still untapped to a large extent (i.e., lack of a diversified tourist product, poorly developed forms of environmentally-sensitive tourism, or tourism focusing on culture and archaeological and historical treasures).
2. Forecasts are even more optimistic and international arrivals are expected to reach 700 million by 2000, 1 billion by 2010, and 1.6 billion by 2020 (WTO 1998).

3. On the political level, this period has been marked by a 'peculiar' type of political elite, characterized by a voter–politician relationship in the form of clientilism (Petras 1992). Besides its subsequent ramifications, military interventions in the republic have become a common phenomenon, similar to other Balkan and Latin American nations (Mouzelis 1986). On the economic side, factors such as the skyrocketing increase in the amount of imported goods, a dramatic decline of investments, the intrusion and control of the economy by foreign capital, the unequal distribution of income and wealth, organizational problems, severe emigration and immigration problems, and critical regional problems have placed Greece in the category of countries with serious micro- and macroeconomic deficiencies (Central Bank of Greece 1992; CPER 1990; Karayorgas *et al.* 1990) transforming the country into a satellite of core powers.

4. According to 1994 data of the statistical service of the European Union, Greece is the poorest country of the European Union (Hellenic Resources Network, www.hri.org 1997a). Dominant features include an extremely high public deficit combined with an upsurge in inflation, a further widening of the current-account deficit, and a decline in GDP (Central Bank of Greece 1995). These problems, in the main, were due to serious structural weaknesses, inflationary pressure built up over a number of years as a result of large public deficits, and the increase in liquidity and nominal income, as well as substantial deterioration of competitiveness.

5. It is estimated that 50 per cent of rooms operate without regulations and license as part of a powerful and expanded para-hotel sector of the Greek tourist industry.

6. In 1994, 6.7 million international tourists arrived via charters, 6.2 million in 1995, and 6.4 in 1996 (GNTO 1997).

7. Greece was involved in World War Two between 1940–44 and soon after, in a civil war from 1947–49 (as a consequence of the power games of the two blocks) which were devastating in both human and physical capital.

8. Their proportion in the total population has increased sharply and while in 1980 it was only 14.5 per cent of the population, by 1993 had risen to 45.7 per cent (Leontidou 1998).

9. The international problem surrounding tourism statistics is very evident in Greece. There are no consistent reports of various relevant figures, and agencies such as the Greek National Statistical Service, National Greek Tourism Organization, and Central Bank of Greece report different figures as total tourism revenues. The present figures include travelling receipts, prepurchases of drachmas by tour operators, cruises, credit card purchases, and other invisibles.

10. While the official statistics report a 7 per cent contribution of tourism to GDP (9.6 per cent including multiplier effects), Arthur Andersen's report claims that tourism's share is much higher and close to 15 per cent of GDP (Piou 1995).

11. Accommodation capacity for 1995 was approximately 60 per cent (Agelis and Falirea 1996).

12. The 'tourist saturated areas' policy was not a response to ecological or social concerns but a response to the hoteliers' problems of excess capacity in certain regions. Ultimately, while prohibitive measures to hotel construction were successful, they have compelled investors and landowners to recourse to unauthorized construction practices, thus contributing to further growth of the informal accommodation (para-hotel) sector (Komilis 1992).

13. This law, which has utilized the concept of 'areas of integrated tourism development' as one of its major incentives for private investment and an ultimate

upgrading of the tourist product, has been severely criticized for its complete ignorance of 'tourism's socioeconomic and spatial development processes at either a national/regional or international level and scale' and the 'particular nature and idiomorphy of the tourist product (related to the fact that it is not a single physical entity) together with the organizational structure of the travel industry, particularly with regard to transport systems and tourist enterprises; two main factors instrumental of the tourism activities location at different spatial levels' (Komilis 1995a,b).

14. Phenomena of this type have resulted in forms of 'neocolonialism', where dependency between 'receiving' Greece and the 'generating' core replicates colonial, imperialist, or satellite forms of domination and structural underdevelopment of the past.

15. The failure of conventional – neoclassical, Marxist, and structuralist – schools of thought to provide overall satisfactory explanations to both academics and practitioners has led to an even stronger push to the centre of the public stage and to a consideration of alternatives.

16. The Secretary General of the National Greek Tourism Organization, Nikos Skoulas, in July 1997 reiterated that the 'Greek light, sun, and sea model of tourism is no longer adequate for foreign tourists' (Hellenic Resources Network, www.hri.org 1997b).

Part II

Eastern Mediterranean shores

A fast-growing tourist market

7 The state, the private sector, and tourism policies in Turkey

Turgut Var

Introduction

The lands of Turkey are located at a point where the three continents making up the old world, Asia, Africa and Europe are closest to each other, and straddle the point where Europe and Asia meet. Geographically, the country is located in the northern half of the hemisphere at a point that is about halfway between the equator and the north pole, at a longitude of 36°N to 42°N and a latitude of 26°E to 45°E. Turkey is roughly rectangular in shape and is 1660 km long and 550 km wide.

The actual surface area of Turkey inclusive of its lakes and rivers, is 814,578 km², of which 790,200 are in Asia and 24,378 are located in Europe. The land borders of Turkey are 2753 km in total, and coastlines (including islands) are another 8333 km. Turkey has two European and six Asian countries for neighbours along its land borders.

Because of its geographical location, the mainland of Anatolia has always found favour throughout history, and is the birthplace of many great civilizations including Hittites, Greeks, Romans and Ottoman. It has also been prominent as a centre of commerce because of its land connections to three continents and the sea surrounding it on three sides.

History of Turkish tourism

Turkey has always been perceived as an exotic destination. Construction of the railroad that served the Oriental Express through Wagon-Lits brought the heart of Europe to Istanbul where foreigners enjoyed a melting pot of East and West. Several foreign authors, like Pierre Loti and Agatha Christie, contributed to the charm and the mystique of Istanbul as an oriental destination through their writings.

Turkey had several domestic and foreign conflicts during the first part of the twentieth century. The domestic unrest starting with Young Turks ended the reign of Abdul Hamid the Second after thirty-three years and gave way to a more democratic government. However, the war with Italy over Libya in 1911, followed by the terrible consequences of The Balkan

Wars, left Turkey economically and militarily drained. Then, worst of all, Turkey found herself in the First World War followed by the war with Greece. The latter conflict lasted until 1922 and Turkey became a republic in 1923.

The first economic conference that was held in the early years of the Republic recognized the importance of economic development. The Izmir Economic Congress, which was convened in February 1923, marks the beginning of the active role of the new Kemalist state in the formulation of economic policy – a policy which was to guide the post-war economic development of Turkey along a capitalist path (Berberoglu 1981).

The Izmir Economic Congress ended with the reaffirmation by the state of its commitment to protect the national capitalist class from the metropolitan bourgeoisie and to reconcile the diverse interests of the dominant national classes, in order to rally them behind its programme to modernize and develop the Turkish along capitalist lines. Due to a very low level of capital, the state had to undertake many of the activities traditionally assumed elsewhere by the national industrial bourgeoisie. The state role in the economy began to expand as it entered various branches of local industry to develop the infrastructure, establish banks, and regulate commerce – all co-ordinated within the broader framework of the national economy. In addition to the direct role played by various state banks and credit institutions in encouraging the expansion of the national industry, many special laws were passed granting major concessions to private capital. The sugar industry provides one of the best examples of the nature and scope of the concessions granted to private capital during this early period. No tourism development could be discussed in a country that was devastated by long wars since 1911.

According to Berberoglu (1981) there were three main obstacles that contributed to the failure of the policy envisaged. The first was the resistance of landlords in areas where their interests were threatened by industrial expansion; secondly the failure of the expected transformation of the merchant class into industrial capitalists, and finally the unfavourable terms of the Treaty of Lausanne, in force from 1925 until 1929.

Under the provisions of the Treaty of Lausanne, Turkey had to recognize the economic concessions granted to foreign firms by the Ottoman Empire prior to 1914. Moreover it was further agreed that Turkey would keep her customs duties to the level specified by the Ottoman customs tariff of September 1, 1916. The state was unable, until 1929, to develop a customs policy that would offset Turkey's trade deficit (Berberoglu 1981). The lack of significant revenues from customs, coupled with the abolition of rural agricultural tax, worked against full scale participation of the state in the industrialization process during this period. The expiration of the Treaty in 1929 thus marked an important turning point for increased state intervention in the Turkish economy.

Although serious efforts were made by the Nationalist leadership to accelerate industrialization by providing domestic finance and a guarded acceptance of foreign capital in various branches of the national economy, Turkey remained a part of the metropolitan-controlled capitalist world economy until the end of the 1920s. Since Turkey was basically a supplier of agricultural and raw materials and importer of finished manufactured goods, she was subject to global forces. It is for this reason that the Great Depression of 1929–30 had a devastating impact on Turkish economy through the foreign trade. With the coming of the Depression, Turkey began to experience considerable difficulty in finding foreign markets for her products. The resulting drop in the value of Turkish lira led to a major decline in agricultural products prices and thus Turkey lacked the foreign exchange necessary to continue importing capital equipment necessary for development process.

The wide-ranging expansion of the State into all branches of the national economy during the 1930s was, in large measure, a direct response to the unfavourable conditions created by the financial collapse in the capitalist countries. These were enough to convince the Kemalist government to take decisive steps by assuming more forcefully the role of entrepreneur, taking on many of the tasks traditionally performed by the private enterprises. With the state beginning to assume the commanding heights of the economy in the early 1930s, Turkey thus entered a unique period of capitalist development which was later to be called 'etatism' and which marked the period of consolidation in later years of the decade.

In an effort to resolve Turkey's balance of payment crisis, the state launched an all-out campaign of nationalization of foreign firms throughout the 1930s. Almost all the railroads, utilities, transportation facilities, mining, new railroads, postal and telephone services, insurance, and some banking companies were nationalized. The last nationalization was the water system of the city of Izmir in 1944. This policy of the 1930s and early 1940s played an important role in strengthening the Turkish economy. It ended the outflow of capital, hence improving the country's balance of payment. Within this context the state assumed the role of planner and came out with two five-year development plans. During the First Five Year Development Plan, banks that would provide credit to various industries and mining were created. Several textile mills, shoe factories, paper and steel mills, and similar industrial and consumer goods factories were constructed and the state gave emphasis to education, and research in key areas.

Turkish tourism development cannot be separated from the industrialization policies of the various governments since the foundation of the Republic. These policies can be divided into the following periods:

1 Strong etatist policy (1924–50)
2 Guarded liberalization (1950–60)

3   Planned economy (1960–80)
4   Transition to private entrepreneurship (1980–90)
5   Private sector dominance and liberal economic policy (1990–Present)

Although Turkey was a neutral country during the Second World War, it was affected by the scarcities of the war. The economy remained stagnant until the Democratic Party won the elections in 1950 under the leadership of Celal Bayar and his Prime Minister Adnan Menderes who opposed the etatist policy of the People's Republican Party.

Strong etatist policy and tourism (1923–50)

The 1923–50 period concentrated on infrastructure and supra-structure investment with very limited emphasis on tourism as a tool for economic development. It suffered the difficulties of post-depression times and the subsequent Second World War. However, very important achievements were made during this period and the country became more reliant on its own resources and production capacity.

The First Tourism Congress was held in 1931 and this was followed by the convening of the International Women's Congress of 1935 in Istanbul, which gathered together many famous women leaders who were struggling for universal suffrage all around the world. During this period several archaeological sites were opened and awareness of tourism began to flourish in the rapidly progressing middle class. The International Izmir Trade Fair was established as a touristic and commercial attraction during this period. In short, due to the emphasis on primary industries and lack of capital in the tourism sector, any major development in tourism was almost impossible. Another important development was convening the First Tourism Advisory Conference in 1949 which prepared a document to set the foundations of a national tourism policy. It also determined the role of the state and private enterprise (Olali 1984).

In general there was very limited infrastructure and a very poor transportation, communication, health and other service sector industries.

Guarded liberalization (1950–60)

With the coming to power of the Democratic Party in 1950, Turkey embarked on a qualitatively different path of development from the one it had followed earlier under the state capitalist policy. One of the objectives of the Democratic Party was to transfer state enterprises to the private sector, and in this way, to return to the liberal economic policies of the 1920s. The DP government drew up a foreign investment law so favourable to overseas capital that it was later to throw many of the key branches of Turkish industry into the hands of multinational companies. The Law of Encouragement of Foreign Capital of 1954 stated that:

a   All areas of the economy open to Turkish private initiative are also open to foreign capital;

b   Foreign capital is not obliged to go into partnership with local capital;

c   Foreign corporations operating into Turkey may (if they wish) repatriate all of their profits to their home country, add to their principal investment, or invest in another corporation of their own choosing

As a result of this law a huge amount of foreign capital poured into Turkey especially in petroleum, pharmaceuticals, tires, fertilizers, food processing and other industries. Since there were several problems of developing the necessary infrastructure, tourism had to wait several more years.

One of the major achievements of the Democratic Party was its recognition of tourism as a major economic development tool. In 1953 the Bank for Municipalities and Foundations was given the responsibility to administer the credit mechanism for touristic enterprises. (Olali 1984) Later on this function was transferred to a new bank, called Tourism Bank. This bank basically provided credit to state-owned touristic enterprises. However, other independent measures were also used. For example, the municipality of Izmir, encouraged by tourism developments in other countries, attempted to build a new hotel along the Bay of Izmir. However, due to housing demand for the newly formed NATO, this building was converted to South-East Regional command post for NATO and Izmir had to wait several more years to have tourist accommodations. On the other hand, Conrad Hilton was able to get permission to set up a hotel in Istanbul that was financed by the retirement funds of Turkish public employees. This was the beginning of inflow of foreign capital into tourism-related facilities.

The first road that connected a major tourist centre, Ephesus, to Izmir was constructed in 1951, and limited bus tours to Ephesus and other historical locations and beaches began. It was during this period that tourism began to make itself known. One of the important foreign exchange tools to attract tourists was to give them a special exchange rate, which was much higher than the current rate existing at that time. It was also domestic tourism that attracted several bus companies to carry day-trippers to touristic sites. The expansion of bus services gave more flexibility to Turkish people to travel wherever and whenever they wanted. The Turkish Airlines, the flagship of the country, began converting its DC-3 fleet into modern and faster airplanes. New airports were planned and opened for domestic and international travel. The Turkish Maritime Lines began to operate modern ships between the domestic and foreign ports.

After two years of liberal policies and abundance in agricultural production, the weaknesses in the system started to show themselves. The Turkish exports and imports showed enormous increases coupled with very large trade deficits. Continued foreign borrowing and expansion of credit during this period meant ever-increasing foreign debt. Towards the middle of the period the liberalization policy led to inflation and the prices almost tripled in ten years.

The economic policy which started by following a very liberal course in both internal and external markets was drawn towards more and more controls as shortages and difficulties were encountered. By the end of the period, the economy was in a state of complete imbalance and impotence and could only survive through day-to-day measures. As a result of inflation and uneven income distribution caused by the liberalization policy, social unrest began. In 1958 a stabilization policy of considerable scope was adopted in both domestic and foreign trade. Turkish currency was devalued, foreign trade difficulties were eased with the help of substantial foreign aid and a credit squeeze was implemented.

The revolution of 1960 took place before the results of the stabilization measures could be fully known. Although no significant effort to promote tourism was made during the 1950–60 period, the annual number of foreign tourists increased about twelve per cent. In 1961 there were 258 hotels with 15,685 beds. (State Planning Organization 1964). Only thirty per cent of these accommodations could satisfy the requirements of foreign tourism. During the same year 129,000 tourists visited Turkey and spent $7.5 million (State Planning Organization 1964).

Planned economy (1960–80)

When the military commanders took power in 1960, the Turkish economy was in dire straits. Output was stagnant and there were high levels of inflation and unemployment; a sizeable trade deficit, a large and ever-increasing foreign debt and an associated balance of payment crisis. Shortly after the coup the generals formed a provisional government and the extremist elements were eliminated: General Cemal Gursel became the head of the government and the country went into a period of political reconstruction. A constitution giving new and expanded rights to labour, universities, political parties, press and intelligentsia was prepared and submitted to a national referendum. Despite its flaws, the new constitution seemed sufficiently attractive on paper to win approval by a large margin. Elections were scheduled for October 1961 and the resumption of multi-party activity was again authorized. Several new political parties were formed and Turkey went into a period of coups and coalition governments which made it difficult to follow a stable economic and tourism policy. However, one of the important outcomes of the coup was the creation of the State Planning Organization which was given the duty of preparing a 15-year economic development plan divided into five-year segments containing objectives and targets for production, exports, income distribution, wages, capital formation, balance of payments and other key elements in the national economy. SPO also created various committees dealing with uniform accounting principles and practices, price level adjustments or re-evaluation of state economic enterprises, creation of capital markets and recommendations for legislation concerning various obstacles for development. It should be emphasized that for the first time in Turkish economic history tourism was recognized formally as an important sector. However, in the First Five-Year Economic Development Plan, tourism had only four pages out of a total of 472 pages. Nevertheless this was a good start for Turkish tourism. The Plan gave the objectives of tourism development as follows:

> In order to close the gap in Turkey's balance of payments, full advant-age should be taken of the development of international tourism which expands with rising incomes. . . . The main principle is to attract this tourist potential to Turkey in as short a time as possible by making the necessary investments. As much attention should therefore be given to publicity and promotion, provision of services and the sou-venir trade, as to overnight accommodation.

> Demand is estimated on the assumption that past trends in the number of tourists to Turkey can be improved.

In SPO's studies it is estimated that the average length of stay of tourist could be increased from four days to six days and daily expenditures from $60 to $100 between the 1963 and 1967 planning period. Additionally, in order to eliminate waste of scarce resources, efforts had to be concen-trated on regions most likely to attract tourists and where these efforts would yield results in a short time. It was also recommended that certain developing or already-developed areas should be supported.

It is also interesting to see that tourism was included in the foreign rela-tions section of the Plan as a separate item under invisible category of the balance of payments. It was also stated that 'The most important invisible export item is that of tourism'. In the Plan, tourism and travel revenues were estimated to rise from $20 million to $28 million in 1967 (State Planning Organization 1964). During this period The Ministry of Tourism contracted with the Middle East Technical University to conduct a study for optimal dis-tribution of touristic investments, especially state supported, over the com-peting project. Charles Gearing, William Swart, and Turgut Var developed an investment decision model based on maximization of touristic attractive-ness. (Gearing, Swart and Var 1976) This model made extensive use of com-puters and application of operation research. The results of this study has helped Turkish tourism and gave the Ministry of Tourism an invaluable tool. Later on, the Soviet Socialist Republic applied the same methodology in the development of the Crimean Peninsula (Scherbina 1980).

Between 1963 and 1974, the Turkish economy grew, in real terms, at an average annual rate of between 6 and 7 per cent. This was one of the highest continuous growth rates with the OECD. The economy in these years was characterized by relative price stability, manageable budget deficits, high levels of consumption and investment in the public sectors, modest reliance on external borrowing and a sound balance of payments. The economic policy that characterized the 1963–74 period can be termed a 'mixed economy'. The Plan stated:

> The principal objective of the development policy is to attain within a free system a balanced rate of growth of 7 per cent and to achieve a just distribution of the benefits resulting from the sacrifices required for development. In establishing this policy, the public and private sectors have been taken into consideration as the two component parts of a whole, and not as two separate sectors with conflicting interest.

In order to have a fair participation of the private sector in economic development, the following points were emphasized:

a   Economic policy to be followed will be clearly defined in order to give both public and private sectors an opportunity to plan ahead and act in confidence.
b   The state will direct investments according to the plan targets and will inform the private sector on economic development.
c   The state-owned enterprises will follow a price policy designed to yield maximum profits and no unfair competition will be allowed for the state sector.
d   Interest rates and terms of incentives to be provided for the repayment of interest and principal will be determined by sectors and in accordance with the plan targets.
e   Import programmes will be determined by sectors, no discrimination will be made between public and private sectors (DPT 1963).

During this period, the state engaged in projects to improve roads, airports, marine facilities and electric power. It is also interesting to note that the first Turkish automobile was manufactured during this period, although the major components were imported. Similarly, household appliances like refrigerators, washing machines and vacuum cleaners, together with many other consumer goods, were poured into the market by the newly-formed Turkish private companies.

The Ministry of Tourism was established in 1965 but its impact was relatively minor until the passage of the Tourism Encouragement Law of 1982. Tourism did show spectacular results in terms of growth of foreign and domestic tourism. As shown in Table 7.1, between 1963 and 1974 the tourist revenues went from around US$7.7 million to US$193.7 million. More interesting, the number of foreign visitors surpassed the one million mark in 1972. And the tourism balance showed a surplus in 1970 for the first time during this period. The growth rates are also shown in Table 7.1. It is important to point out that the expenditures of Turkish citizens in foreign countries also reached in excess of US$151.8 million. This was mainly due to the increased travel generated by guest workers going to Germany and other countries in Western Europe.

From 1974 to 1977, sharp increases in the price of petroleum and the ensuing inflation, recession and high unemployment affected the Turkish economy. By 1977, Turkey's problems included a severe shortage of foreign exchange, a high rate of inflation, a large public sector deficit, a slowdown in growth and increasing unemployment.

The government announced a series of strong economic policy measures in January 1980 aimed at controlling high inflation, reducing the balance of payments deficit and stimulating economic growth. These policies demonstrated a basic departure from direct government regulation and control toward a greater reliance on market forces. The measures did not show their impacts soon enough to thwart another military intervention headed by General Kenan Evren. These series of measures, contained the following major elements in relation to the tourist sector:

a   After an initial 33 per cent devaluation of the Turkish lira against the US dollar, the adoption of a flexible exchange rate policy designed to maintain international competitiveness to attract workers' remittances from Europe;

b   strengthening public finances through substantial price increases and elimination of price controls;

Table 7.1 Tourism revenues and the number of foreign visitors, and average expenditures per person 1963–74

Year	Revenue (000 US$)	Number of visitors	Revenue per person (US$)	Growth of revenues	Tourist expenditures of Turkish citizens (000 US$)	Surplus or deficit (000 US$)
1963	7659	198,841	39	–	20,511	−12,852
1964	8318	229,347	36	0.09	21,807	−13,489
1965	13,758	361,758	38	0.65	24,310	−10,552
1966	12,134	440,534	28	−0.12	26,329	−14,195
1967	13,219	574,055	23	0.09	26,813	−13,594
1968	24,082	602,996	40	0.82	33,409	−9327
1969	36,573	694,229	53	0.52	42,231	−5658
1970	51,597	724,784	71	0.41	47,738	3859
1971	62,857	926,019	68	0.22	42,192	20,665
1972	103,731	1,034,955	100	0.65	59,320	44,411
1973	171,477	1,341,527	128	0.65	93,013	78,464
1974	193,684	1,110,298	174	0.13	151,797	41,887

Source: Hasan Olali, *Turism Dersleri*, Izmir: Istiklal Matbaasi, 1984, p. 40.

c   a strict observance of a non-inflationary monetary policy;

d   export promotion, including an increase in the scope of exporter's foreign currency retention and abolition of duties on imports used in the manufacture of export goods (this principle later on applied to tourism);

e   greater liberalization of trade and external payments;

f   a realistic interest rate policy that would increase domestic savings;

g   the consolidation of Turkey's private commercial debts which were not guaranteed;

h   institutional changes to expedite the formulation, co-ordination and implementation of economic policy;

i   implementation of a new policy regarding foreign investments. (Ministry of Tourism and Culture 1986).

However, from 1975 to 1980, as shown in Table 7.2, Turkish tourism was more or less stagnant. It was a period of great uncertainty due to extremist activities and a series of weak coalition governments.

Transition to private entrepreneurship (1980–90)

There is no doubt that the 1980–90 period was the most significant in the development of the Turkish tourism industry. The following legislations were passed by the Grand National Assembly:

1   Tourism Encouragement Law 1982;
2   Law of Protection of Environment 1983;
3   National Parks Law 1983;
4   several regulations concerning the quality of tourism facilities (1983),

Table 7.2 Tourism revenues and the number of foreign visitors, and average expenditures per person 1975–83

Year	Revenue (000 US$)	Number of visitors	Revenue per person (US$)	Growth of revenues	Tourist expenditures of Turkish citizens (000 US$)	Surplus or deficit (000 US$)
1975	200,861	1,540,904	130	0.04	154,954	45,907
1976	180,456	1,675,846	108	−0.10	207,893	−27,437
1977	204,877	1,661,416	123	0.14	268,528	−63,651
1978	230,397	1,644,177	140	0.12	102,476	127,921
1979	280,727	1,523,658	184	0.22	95,070	185,657
1980	326,000	1,300,000	251	0.16	111,100	214,900
1981	381,300	1,405,300	271	0.17	103,300	278,000
1982	370,300	1,390,500	266	−0.03	108,900	261,400
1983	408,400	1,623,000	252	0.10	127,300	281,100

Source: Hasan Olali, *Turism Dersleri, Izmir: Istiklal Matbaasi*, 1984, p. 40.

tourism investments and their auditing (1983), establishment of the Higher Tourism Co-ordination Committee (1982), and finally the establishment of an agency for tourism promotion (1982).

During the early 1980s, General Kenan Evren, the architect of the 1980 military intervention, became President. The constitution was amended and general elections were held in 1983 which gave a majority to the Motherland Party headed by Turgut Ozal. The new government announced sweeping regulations further freeing the government from the etatist bureaucracy.

The most important of these laws was the Tourism Encouragement Law which had the following purpose and scope:

> The purpose of the present law is to ensure that necessary arrangements are made and necessary measures are taken in order to regulate, develop and provide for a dynamic structure and operation of the tourism sector.
>
> The present law comprises provisions governing the tourism sector, including definitions of tourist regions, areas and centers; establishment and development of such regions, areas and centers; and encouragement, regulation and inspection of touristic investments and facilities.
>
> (Ministry of Tourism and Culture 1987)

The law gave certain criteria for determining tourist regions, areas and centres. These included such particulars as natural, historical, archaeological and sociocultural assets of the tourist trade; potentials for winter tourism, hunting and water sports; health resorts; and other existing forms of tourist potential. It also brought about the necessity of procurement of certificates for tourism investments. The law stated that:

a   it will be compulsory to procure either a tourist investment certificate or a tourist facility certificate in order to benefit from incentive measures, exceptions, exemptions and rights prescribed in the present law and other legislation.
b   certified investments will have to be started, completed and rendered operational within periods prescribed by the Ministry.

The prospective investor is first required to obtain an Encouragement Certificate from the State Planning Organization before an investment certificate is issued by the Ministry of Culture and Tourism. Encouragement certificates may be obtained under the following conditions:

a   a prescribed minimum investment;

b   accommodation facilities with a minimum 100-bed capacity; restaurants, a minimum chain of three with a minimum 30-person capacity;
c   yachting establishments with a 45-bed capacity fleet;
d   other facilities certified by the Ministry of Culture and Tourism in terms of the conditions of Regulation for Specification of Tourist Investments and Facilities.

With sweeping changes towards liberalization, Turkey has provided several incentives to prospective investors. The incentives provided by the Tourism Encouragement Law included the following:

1   *Land allocation.* In tourist areas and at tourist centres where infrastructure is provided by the state, state-owned lands may be allocated to investors on long-term lease or easement basis (extendible 49 years).
2   *Loans.*
    a   Low interest, long-term loans from the Tourism Bank (up to 60 per cent of the total investment cost).
    b   Low interest, long-term loans from the Tourism Development Fund (up to 15 per cent of the total investment cost for capacities over 300 beds).
3   *Exporter rights.* All tourist facilities are considered as exporters and granted all the benefits of exporters.
4   *Forestry fund instalment.* Forestry fund to be collected from investments within the state-owned forest areas (it was 3 per cent of the total investment), and is paid in five equal instalments.
5   *Preferential tariff rates.* All licensed tourist facilities in tourist areas and centres are granted reduced tariff rates for electricity, water, and gas.
6   Telex and telephone needs are met by the state.
7   *Foreign personnel employment.* Granting licensed tourist facilities may employ foreign personnel (up to 20 per cent of the total personnel).
8   *Alcoholic beverages sales.* License for local and international alcoholic beverages sales.
9   *Casino operation.* Casino establishment and operation license may be granted to those facilities earning a certain amount of foreign exchange (Ministry of Tourism and Culture 1987).

In addition to the incentives provided by the Tourism Encouragement Law, the SPO also gave more incentives which included a very generous investment allowance for income and corporate income tax write-offs for acquisition of fixed assets like building, heating, air conditioning, kitchen equipment, elevators, and other fixtures. For example, for those investments in the first priority developing regions were granted 100 per cent tax write-off. There were also exemptions from customs taxes, property taxes for five years, reduced equity requirement as low as 25 per cent for large investments, the right of foreign partners to transfer their profits outside the country, retention of certain foreign exchange generated by

Table 7.3 Figures of licensed accommodation facilities by types and categories

| Type | | Tourism Investment Licenses | | | | | | Tourism operation | | |
| Licenses | Category | At project state | | | Under construction | | | | | |
		(A)	(B)	(C)	(A)	(B)	(C)	(A)	(B)	(C)
Hotels	5 Star/Lux class	11	3754	7793	18	4879	10,414	15	3550	6824
	4 Star/1. class	15	2718	5810	20	3633	7402	21	3050	6077
	3 Star/2. class	65	5668	11,524	57	4284	8475	69	5919	11,699
	2 Star/3. class	63	2831	6183	106	5615	11,317	192	10,176	19,166
	1 Star/4. class	22	696	1360	72	2659	5099	236	8770	16,737
Motels	/1. class	2	34	68	16	1294	2555	25	1745	3750
	/2. class	14	316	612	21	656	1261	31	805	1656
Holiday Villages	/2. class	12	2744	6040	10	3013	6893	14	4448	9186
	/2. class	3	394	832	2	535	1122	5	629	1437
Boarding Houses	/unclassified	61	1062	2084	41	716	1329	72	1162	1401
Campings	/unclassified	2	79	237	8	903	2695	22	2195	7928
Inns		–	–	–	–	–	–	14	846	2957
Thermal Resorts		–	–	–	–	–	–	1	150	264
Special Licenses		2	69	128	1	63	150	1	897	1992
Apart hotels		–	–	–	–	–	–	1	–	55
Total		113	20,365	42,671	366	28,250	58,712	731	44,342	92,129

Note: Total bed capacity, together with those fulfilling formalities to receive operation licenses, will reach to 101,658 by the end of January 1987.
(A) Number of facilities.
(B) Number of rooms.
(C) Number of beds.

the company, cash rebates, mid- and long-term loans, and finally income tax exemptions for revenues generated in foreign currency. A maximum 20 per cent of the foreign exchange earnings of tourist establishments is deducted from the total gross profit subject to taxation.

As soon as the incentives were announced, a flood of investors began to knock on the doors of the State Planning Organization and the Ministry of Culture and Tourism. In 1984 more than 36,470 bed capacity was under construction and there were over 68,300 bed capacity granted tourism operation licenses. As a result of these efforts the number of foreign arrivals increased by 30 per cent between 1983 and 1984 and reached a record number of 2.1 million.

In line with the conclusions of planning researches within the guidance of Five Year Development Plans, physical planning studies were carried out by the Ministry in order to define the areas where infra- and super-structure were to be realized by public and private sectors. The state provided land use plans 1/25,000 scale for the tourism planning regions that would create approximately 600,000 additional bed capacity (Ministry of Tourism and Culture 1987).

In short, tourism became a buzz word for this period of development as it was for manufacturing in the previous periods. During this period TUGEV (Tourism Development and Education Foundation) was founded and an increasing emphasis was put on tourism education. Tour operators and travel agencies also became more organized and more open to the external markets. The promotion expenditures were increased through the participation of an emerging private sector. The state slowly abandoned its direct participation in tourism projects and concentrated on developing a better infrastructure and co-ordinating investments.

Table 7.4 Turkey's tourism receipts as a percentage of GNP and export earnings 1985–95

	Tourism receipts (million US$)	Tourism receipts as a percentage of	
		GNP	Export earnings
1985	1482.0	2.8	18.6
1986	1215.0	2.1	16.3
1987	1721.1	2.0	16.9
1988	2355.3	2.6	20.2
1989	2556.5	2.3	22.0
1990	3225.0	2.3	22.0
1991	2654.0	1.8	19.5
1992	3639.0	2.4	24.7
1993	3959.0	2.2	25.7
1994	4321.0	3.3	23.9
1995	4957.0	3.0	22.9

Source: Turkish State Institute of Statistics.

Private sector dominance and liberal economic policy (1990–Present)

Turkey entered the 1990s with impressive tourism statistics. The number of foreign arrivals reached 5.4 million in 1990. This was almost double the number of arrivals in 1987. Tourism revenue also increased from US$1.7 billion in 1987 to US$3.3 billion in 1990. The Gulf War was a tremendous blow to Turkish tourism in 1991. However, the impact was temporary due to shortness of the war. By 1995 the number of arrivals was over 7.7 million and the foreign exchange receipts were over US$5 billion. The number of bed capacity climbed to 276,000 beds in 1995, an increase of over 90,000 beds within five-year period between 1991 and 1995.

However, the economic benefits derived from this impressive growth in tourism activity have not been achieved without a series of associated costs. Unfortunately, much of this rapid growth was achieved in a spatially concentrated and uncontrolled manner. This resulted in a concentration of development in traditional coastal locations and the creation of significant degree of environmental disruption (Travel and Tourism Intelligence 1997).

Although the problems of rapid expansion are now recognized by the Turkish Government and major tourism operators alike, and more selective, and sensitive tourism policies have been developed as a consequence, this legacy still makes it difficult for Turkey to reverse the negative effects of those policies. Nevertheless, the Turkish Government remains firmly committed to developing a modern tourism industry to ensure it is able to effectively exploit the country's variety of tourism resources with due regard to its environmental impact and consequences. (Travel and Tourism Intelligence 1997)

Due to its relative political stability, Turkey is experiencing a tremendous increase in tourism revenues. In 1996 Turkey had 12 per cent growth in tourist arrivals, which was almost twice that of the OECD growth over 1995. The number of arrivals was almost eight million. (Table 7.1) and more importantly the growth in revenues was over 31.9 per cent, a record

Table 7.5 Turkey's tourism balance of payments 1985 and 1990–95 (million US$)

	Tourism receipts	Tourism expenditure abroad	Balance
1985	1482.0	323.6	1158.4
1990	3225.0	520.0	2705.0
1991	2654.0	592.0	2062.0
1992	3639.0	776.0	2863.0
1993	3959.0	934.0	3025.0
1994	4321.0	866.0	3455.0
1995	4957.0	912.0	4045.0

Sources: Departing Visitor Survey, Ministry of Tourism 1995; Turkish Central Bank.

Table 7.6 Tourism ministry-licensed accommodation establishments in Turkey 1995

Type	Existing			Under construction		
	No. of establishments	No. of rooms	No. of beds	No. of establishments	No. of rooms	No. of beds
Hotels						
5-star	82	23,089	48,615	33	8380	18,092
4-star	131	18,167	37,049	93	14,284	29,960
3-star	399	32,699	66,618	398	31,257	64,384
2-star	550	24,186	47,379	380	15,486	31,203
1-star	220	7673	14,679	97	2627	5191
Total	1382	105,814	214,340	1001	72,034	148,830
Motels						
1st class	11	639	1254	11	384	790
2nd class	27	675	7345	27	598	1184
Total	38	1314	8599	38	982	1974
Holiday villages						
1st class	59	18,501	40,186	55	13,555	28,868
2nd class	11	1505	3278	20	3555	7645
Total	70	20,006	43,464	75	17,110	36,513
Boarding houses	185	3155	6175	177	2946	5791
Campsites	22	2090	6110	14	1493	4371
Inns	7	441	1274	3	158	382
Aparthotels	24	652	1926	0	0	0
Thermal resort hotels	1	20	40	0	0	0
Special license[a]	60	1727	4066	17	536	1118
Training and practice facilities	2	81	163	3	266	528
Golf resorts	0	0	0	2	462	1839
Service stations	0	0	0	1	27	54
Hostels	0	0	0	1	40	137
Other	2	136	306	2	463	946
Total	**1793**	**135,436**	**286,463**	**1334**	**96,517**	**202,483**

[a] Restored historical buildings.

Source: *Bulletin of Accommodation Statistics*, Ministry of Tourism (General Directorate of Investments), Department of Research and Evaluation 1996.

for the OECD countries. In 1996, Turkey received over US$6.5 billion. The 1997 results give an even better picture. Between January and December of 1997, over 9.7 million foreigners visited Turkey, which represents over 13.8 per cent growth over 1996 (State Institute of Statistics 1997). The preliminary results indicated that revenue from tourism in 1997 was close to 10 billion US dollars and Turkey expected to receive over 20 million visitors in 2000. It was also expected that receipts would top 15 billion US dollars before the end of the century.

There are several factors that make these expectations realistic. First of all Turkey is no longer a low quality and cheap destination. It has a good quality and reasonably priced tourism product. Its tour industry has good guides and travel agents and most of the promotion is done professionally, in most cases, without the assistance of the State. In order to achieve a sustainable growth, both the Turkish government and the tourism industry are shifting their emphasis from volume to a more selective development strategy targeted to improve the quality of a broader portfolio of tourism experiences.

The most important factor is the realization by both the tourism industry and the government of the importance of tourism for Turkish balance of payments and the economy overall. As the tourism receipts approach 40 per cent of the total export earnings, more attention is given to sustainability. Several new airports are currently being planned or contracted. A bridge across Dardanelles will make auto traffic to the favourite western and southern destinations more convenient and will lighten congestion in metropolitan Istanbul. Another major project is to build a bridge on the eastern Sea of Marmara that would cut the travel time from Istanbul to Izmir drastically by eliminating the current ferryboats. Also, recent liberalization of private school regulations is expected to fill the qualified-personnel gap.

Another factor that is having positive affects is the change in the composition of tourists flows. In 1997, ten countries accounted for 66.9 per cent of the total tourist arrivals. Over 2.3 million Germans visited Turkey. This represents approximately 24 per cent of all visitors to Turkey. The Russian Federation contributed over one million, and accounted for 11 per cent. The United Kingdom, United States, Romania, France, Iran, Austria, Israel and the Netherlands each contributed over 2 per cent. Specifically, the United Kingdom had a share of 9.4 per cent followed by the United States with 4 per cent (State Institute of Statistics 1997).

Turkey suffers from a high degree of seasonality, especially for those coastal facilities. As is seen in Table 7.7, over 50 per cent of foreign tourist

Table 7.7 Foreign tourist arrivals in Turkey by month 1995

Month	Arrivals	% of total
January	274,680	3.6
February	302,407	3.9
March	368,195	4.8
April	535,462	6.9
May	732,394	9.5
June	810,419	10.5
July	1,008,709	13.1
August	1,070,234	13.9
September	1,054,871	13.7
October	836,025	10.8
November	393,023	5.1
December	340,467	4.4
Total	7,726,886	100.0

Source: Turkish General Directorate of Security.

arrivals are between June and September, and over 70 per cent between May and October. Facilities are under-utilized for the rest of the year. This is a typical pattern not only for Turkey but also other Mediterranean tourist destinations and has been consistent over many years (Travel and Tourism Intelligence 1997). The low level of off-peak travel to Turkey is likely to remain in the foreseeable future despite the serious efforts of the government and the travel industry. Recent development of modern conference and convention facilities and active marketing for international conferences, together with the promotion of city and religious tourism have, however, been very effective in generating more tourists.

Certain problems have been surfacing in tourism policy that cannot be handled in a fragmented way, where responsibilities for policy determination, implementation, and control are spread too wide. Although the Ministry of Tourism is the prime agency involved in determining tourism development policy, a plethora of other organizations and interest groups in both the private and public sectors, have various degrees of involvement in this process. The frequent changes of coalition governments and turnover of the top-level administrators have resulted in discontinuity. Political stability is essential for achieving a sustainable national tourism policy. During the past two years there has been an important change in general political life of Turkey. Although no party gained a clear majority, the Refah (now called Fazilet), probably the most conservative party, was able to form a coalition government with a party headed by Ciller, committed to a centrist policy.

The Ciller and Erbakan coalition lasted less than a year, and the country was thrown into a debate on the limits of religious activities in a modern European society. The new coalition headed by Mesut Yilmaz, the former Minister of Tourism and supported by socialist parties, seems to be following a middle of the road policy in order to end this debate. However, with the resurrection of Refah under the new name Fazilet, could soften the approach taken during the Erbakan coalition, because of the power vested in the Turkish Constitutional Court. The parties are closely monitored in terms of their compliance with the Constitution. The closing down of over US$450 million worth of casino gaming is a clear contradiction of policy. Turkey paid millions of dollars as incentives and allowances in order to attract these firms during 1980s, but was compelled to close them down because of corruption and religious objections. There is a clear need for the government to recognize that the past strategies cannot sustain for the longer term and require a fundamental rethinking.

The second important problem for Turkish tourism is the redistribution impact of hyperinflation. Turkey has been experiencing very high inflation, running between 57.6 per cent in 1990 to 107.3 per cent in 1994 and currently close to 100 per cent, and is the highest among the OECD countries. This high inflation is reflected in the value of Turkish currency leading to increased tourist exports and decreasing imports and travel abroad. The loss of confidence in Turkish lira is also pushing the whole country towards a US dollar-based economy. Inflation also produces very high interest rates, of nearly 90 per cent or more. In an economy like this even the entrepreneurs are tempted to benefit from higher and secure interest rates rather than investing in long-term projects including tourism development opportunities. The huge government budget deficit is also fuelling the existing inflationary tendencies. Larger fortunes are being made by a smaller group of people at the expense of the middle class, including fixed income groups like government employees and retired persons. The end result is a fragmented search for a party that would bring an end to this erosion of savings and income. The new elections are scheduled in March 1999.

The question of privatization and government subsidies to agricultural products need to be settled. Low agricultural prices and decline in production also show themselves in Turkish balance of payments by the increased need for further imports, of which some are essential for the tourist industry.

So far, the impact of terrorism has not been drastic. Turkey has been engaged in a civil disturbance for almost ten years and witnessed a number of terrorist activities during this period. Some of these activities were directed towards tourists but have not slowed down the inflow of tourists.

In spite of all these problems Turkey managed quite well between 1985 and 1995. Tables 7.8 and 7.9 clearly show the phenomenal changes in tourist receipts. 1991 was an exception because of the Gulf War.

On the bright side, new developments are taking place. Some of these

Table 7.8 Turkey's tourism development programme targets and actual tourism receipts 1985–95

	Target receipts (million US$)	Actual receipts (million US$)	Percentage change on previous year	Actual/ target (%)
1985	600.0	1482.0	76.4	247.0
1986	1150.0	1215.0	−18.0	105.7
1987	1225.0	1721.1	41.7	140.5
1988	1660.0	2355.3	36.8	141.9
1989	2675.0	2556.0	8.5	95.6
1990	3030.0	3225.0	26.1	106.4
1991	3300.0	2654.0	−17.7	80.4
1992	4000.0	3639.0	37.1	91.0
1993	3900.0	3959.0	8.8	101.5
1994	4700.0	4321.0	9.1	91.9
1995	4500.0	4957.0	14.7	110.2

Sources: *Departing Visitor Surveys*, Turkish Ministry of Tourism (1985–90); Turkish Central Bank (1991–95).

Table 7.9 Tourism development plans of Turkey versus actual receipts

Tourism development plan	Target receipts (million US$)	Actual receipts (million US$)	Actual/target (%)
First five-year plan (1963–7)	139.0	55.1	39.6
Second five-year plan (1969–72)	422.0	278.9	66.1
Third five-year plan (1973–7)	669.9	951.5	142.0
Fourth five-year plan (1978–83)	2418.1	1770.1	73.2
Fifth five-year plan (1984–9)	3388.2	9329.9	275.4
Sixth five-year plan (1990–4)	21,702.3	17,798.0	82.0
1995	4500.0	4957.0	110.2
Total	33,239.5	35,141.5	105.7

Sources: Turkish Ministry of Finance (1963–78); Turkish Central Bank (1979–95).

developments, like airports, bridges, dams, and energy projects are already in a construction phase and more are planned. Being a geographical bridge between Europe and Asia has created new opportunities for Turkey. A number of Turkish-speaking countries became reality after the dissolution of Soviet Socialist Republics. Turkey has very good relations with these countries in areas like tourism, energy, banking, manufacturing, and development of infrastructure. Through the scholarships offered by the Turkish government and private organizations, large numbers of students come to Turkey to enrol in various programmes, including tourism education and training. Turkish exporters provide much-needed credits and entrepreneurs go into joint venture projects in these emerging countries. In fact the largest department store in Moscow has recently been constructed by the Turkish private sector which is also taking part in running it. There are plans to expand this to other countries in Central Asia. Turkey is also a strong supporter of the Silk Road project supported by the World Tourism Organization. This project aims to complete a continuous net of highways that would connect Europe to the shores of China. It is an important project; when completed it would change the tourist flows as well as the economies of those countries along the Silk Road.

In short, Turkey has vast touristic resources and skilled labour. The number of universities in Turkey increased from less than ten to over 80 in the past decade and there is more opportunity to attend college without travelling to large urban centres. What Turkey needs is a stable economy with lower interest rates and prices, a better, fairer income distribution, and better planning on a national level with clear and realistic objectives that would not change from party to party and would be implemented without watering down. It is time to ask, 'Where were we, where we are now, and where would we like to be in 20 years?' These questions are related to the whole economy, and not just the tourism sector.

8   The dynamics and effects of tourism evolution in Cyprus

Dimitri Ioannides

Tourism development in island microstates

In numerous island microstates, particularly those situated in middle latitudes, tourism has become synonymous with economic growth. For the last four decades, policy-makers in these small countries have emphasized the development of tourism as a means of attracting the foreign exchange necessary for inducing economic diversification and rapid modernization. Indeed, the considerable obstacles associated with these microstates' insularity and relative isolation – namely the lack of natural, human, and institutional resources, declining agricultural sectors, and their extremely small internal markets – have oftentimes made the adoption of tourism an 'inevitable' development option (Wilkinson 1989).

Unsurprisingly, the phenomenal growth of tourism and its resulting impacts in so many small island states, have generated considerable interest in academe. During the last two decades, there have been studies of the dynamics of tourism's evolution in island microstates (Butler 1993; Debbage 1990; Ioannides 1992) and the sector's spatial characteristics (Pearce 1989; 1995). Moreover, there now exists a considerable body of literature investigating tourism's impacts on the economic, sociocultural, and physical environments of island nations (Britton 1978; Britton and Clark 1987; Bryden 1973). In recent years, researchers have paid increasing attention to the planning and policy implications of tourism development in island economies (Inskeep 1991; Ioannides 1992, 1994; Wilkinson 1989). Finally, a handful of authors have begun exploring the political dimensions of tourism development in island microstates, including the negative effects brought on by ethnic strife, sociopolitical tension, and war (Bastin 1984; Burns 1995; Ioannides and Apostolopoulos 1999; Lockhart 1993; Richter 1989).

As one of only two island-states in the Mediterranean, and a country bearing the legacy of intercommunal conflict, war, and partition, Cyprus comprises an excellent 'laboratory' for studying the dynamics and effects of tourism development (Figure 8.1). Since 1974, the two separate communities on the *de facto* divided island have witnessed widely divergent trajectories of tourism development revealing that they have reached dif-

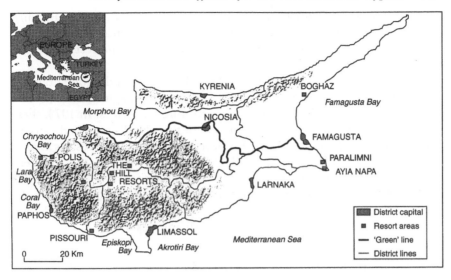

Figure 8.1 Cyprus – divided island.

ferent points of their respective resort cycles (Butler 1980). The Greek-Cypriot part of the island (The Republic of Cyprus) has used an aggressive tourism development strategy to overcome seemingly impossible obstacles and witness an 'economic miracle'. Conversely, political isolation of the self-proclaimed Turkish Cypriot state (the Turkish Republic of Northern Cyprus) has served to undermine efforts to promote this region as an international tourist destination and has, therefore, limited the region's aspirations for economic growth and diversification.

Tourism development in the post-colonial era

Cyprus only embraced tourism formally as an economic growth strategy after its independence from Britain in 1960. Until then, the industry had been dominated by small-scale, family-run businesses and there were only a handful of recognized 'high class' hotels (Christodoulou 1992; United Nations Development Programme 1988). Approximately 45 per cent of the island's total hotel bed spaces were located in the Troodos mountain resorts while the five coastal towns accounted for less than a third of the total room capacity. In 1960, the majority of the 25,700 overseas tourists who visited the island were from nearby Middle Eastern countries (Andronikou 1979; Cyprus Tourism Organization [CTO] 1970–95).

Following the recommendations of various international organizations (e.g., the United Nations Programme of Technical Assistance), a series of Five-Year plans stressed the promotion of the tourist industry as a means of encouraging economic diversification (United Nations Development Programme [UNDP] 1988). Concurrently, the growing demand for sun-lust tourism in northern markets (e.g., Scandinavia and Britain), fuelled largely by the increasing affordability of air travel through the introduction of inclusive tour packages, promoted interest in coastal resort development.

By the early 1970s, the island's tourist industry had undergone dramatic spatial restructuring with the coastal towns of Famagusta and Kyrenia emerging as the dominant destinations for foreign visitors. In 1973, these two resorts controlled almost 58 per cent of the island's total bed capacity (Ioannides 1994). Importantly, the development of these resorts caused Cyprus to shake off its peripheral image and break into the northern European market as a destination for sun-seeking tourists. Between 1971 and 1973 the number of arrivals to the island more than doubled, with the United Kingdom, the Federal Republic of Germany, and Scandinavia emerging as the three dominant markets (Ioannides 1994).

War and partition: tourism at a standstill

During the summer of 1974, an unsuccessful coup d'etat against President Makarios led to the Turkish invasion, a short-term war, and ultimately the occupation of 40 per cent of the island. The result of these events was a large-scale population exchange with approximately 200,000 Greek Cypriots moving to the south and 40,000 Turkish Cypriots relocating to the north (Figure 8.2). In effect, Cyprus was partitioned into two political entities: the Greek-Cypriot controlled portion recognized by the international community as the Republic of Cyprus; and the Turkish-Cypriot controlled

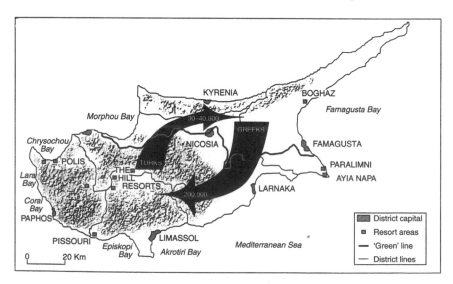

Figure 8.2 Cyprus – redistribution of populations following the 1974 Turkish invasion.

entity, which despite declaring its independence unilaterally in 1983 as the Turkish Republic of Northern Cyprus (TRNC), has been recognized only by Turkey and a handful of Muslim states (e.g., the Maldives and certain ex-Soviet republics).

The war crippled the Cypriot economy. Both the agricultural and manufacturing sectors were severely damaged by the hostilities (Christodoulou 1992). The refugee problem generated a pressing need for housing and employment. Moreover, the island had to contend with the shut-down of Nicosia International Airport, since this lies right on the United Nations-controlled buffer zone, the so-called Green Line. The closing of the airport struck a major blow to the island's tourist industry since it effectively destroyed communications with Europe. Moreover, the tourism sector was almost wiped out because over 80 per cent of the existing accommodation stock and almost all the hotels under construction were lost (Lockhart 1993).

For over two decades since the war, the fortunes of the two separate areas have been widely divergent. While the Republic of Cyprus has successfully used tourism to engineer a dramatic economic revival and become one of the most affluent societies in the Mediterranean rim, the Turkish-Cypriot area has struggled to shake its economic dependence on Turkey and, thus, has been forced into a peripheral status. The next two sections focus on the characteristics of tourism development in each of the island's two entities, paying particular attention to the industry's impacts.

Restructuring the tourism product in the Republic of Cyprus

Between 1977 and 1987, the Republic of Cyprus emerged as the fastest growing tourist destination in the Mediterranean, witnessing an average annual growth rate exceeding 18 per cent (Gillmor 1989; Vassiliou 1995). This growth was truly astonishing, considering that, in 1975, tourist arrivals had fallen to just 47,000, representing 18 per cent of the 1973 total (Ioannides 1992; Lockhart 1993).

In 1988, approximately one million tourists visited the Republic, and by 1995 their number had reached more than 2 million, representing a total of over 20 million tourist bed nights (CTO 1995). By 1998, the number of bed spaces in the southern part of the island had reached almost 87,000 with 90 per cent of the capacity concentrated in the coastal resorts of Larnaka, Limassol, Paphos, and the area surrounding Ayia Napa (Figure 8.3) (CTO 1998).

The tourism sector represented approximately 20 per cent of the Gross National Product (GNP) and receipts amounting to C£870 million (US$1.7 billion) accounted for 40 per cent of total revenues from the export of goods and services (CTO 1998; *The European* 1996). During the same year,

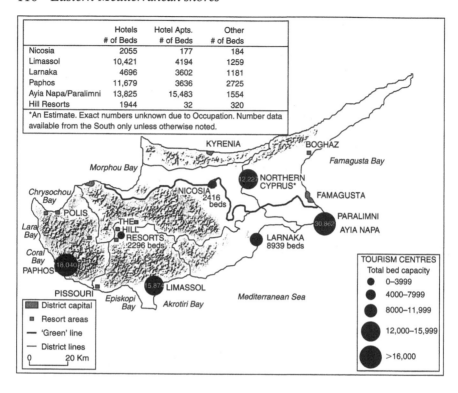

	Hotels # of Beds	Hotel Apts. # of Beds	Other # of Beds
Nicosia	2055	177	184
Limassol	10,421	4194	1259
Larnaka	4696	3602	1181
Paphos	11,679	3636	2725
Ayia Napa/Paralimni	13,825	15,483	1554
Hill Resorts	1944	32	320

*An Estimate. Exact numbers unknown due to Occupation. Number data available from the South only unless otherwise noted.

Figure 8.3 Distribution of hotel/hotel apartment bed capacity 1995.

the hotel industry employed 40,000 persons, representing roughly 14 per cent of the workforce. Indeed, in the period 1980–92 the hotel, restaurant, and trade sector recorded an increase from 33,000 to 64,000 workers (Vassiliou 1995). In total, by the mid-1990s, over 25 per cent of the Republic's labour force was either directly or indirectly dependent on tourism-related employment (CTO 1995; *The European* 1996; *Phileleftheros* 1992). Furthermore, the rapid construction of tourist accommodation in the last two decades has fuelled a boom in the island's construction industry and all other related sectors such as 'cement, bricks, aggregates, furniture, steel structures ... as well as a wide range of services relating to tourism and construction' (Vassiliou 1995: 43). Finally, another indirect effect arising from tourism is that it has served to promote recognition of Cypriot products such as local wines and cheeses and, thus, created new markets for these commodities in northern European countries.

The phenomenal growth witnessed in the tourism industry over the last two years reflects to a major extent the efforts of the Republic's government to revive the economy soon after the 1974 war. Since 1975, consecutive *Emergency Economic Actions Plans* made the development of tourism a top priority. These plans recommended continuing ties with players in major markets, particularly tour operators, in northern Europe (United Kingdom, Scandinavia, West Germany) and re-establishing the island as an international tourist destination (Ioannides 1992). A series of infrastructural programmes, such as the development of a new international airport in Larnaka followed by one in Paphos, and the provision of attractive incentive packages to refugee hoteliers to build new establishments in coastal areas of the south stimulated tourism's rapid growth.

Originally, the most intensive tourism-related development took place in the area to the east of Limassol, as this region was able to capitalize on existing urban infrastructures. By the late 1970s, however, refugee entrepreneurs, inspired by favourable government incentives, had focused their attention on the development of the Ayia Napa region in the island's southeastern corner (Lockhart 1993). Remarkably, by the mid-1980s, this rural backwater had been transformed into a booming urbanized area which controlled the largest share of the island's accommodation capacity.

Since then, tourist facilities have cropped up along major stretches of the southern and western coastline, dramatically transforming the coastal landscape. Most recently, the focus of tourism development has shifted to the Paphos and Polis regions as developers constantly seek out new locations for their hotels and apartment complexes.

Evaluating the impacts of uncontrolled growth

Until now, an interesting feature of the tourist accommodation industry in the Republic of Cyprus is that it has not been characterized by high levels of foreign involvement (Ioannides 1992). Only a handful of transnational hotel companies operate properties (usually through management contracts) around the island, accounting for a very small proportion of total bed capacity (CTO 1996). The overwhelming majority of hotels and other accommodation establishments are owned and operated by local entrepreneurs. This phenomenon contrasts starkly with findings from other island economies (e.g., in the Caribbean) where a large proportion of the accommodation stock is often controlled by foreign companies (Britton, R. 1979; Britton, S. 1982; Debbage 1990; Pearce 1989; Poon 1989).

To a major extent, the hotel ownership profile in Cyprus reflects long-term governmental policies that have sought to limit direct foreign investment while seeking to promote a locally-controlled tourist industry through emphasizing the training of a skilled domestic workforce, including hotel managers. The advantage of local ownership and management, and the fact that the tourist industry relies heavily on local construction companies, the endogenous production of hotel equipment, and the provision of local foods (Wilson 1992) has been that the leakages witnessed in other island destinations (Britton 1978) have not occurred.

Nevertheless despite these positive characteristics, the aggressive and largely unregulated development of tourism in coastal areas has not come without major costs. Government officials, industry analysts, nongovernmental organizations, and residents have become increasingly alarmed about the negative economic, sociocultural, and especially the environmental problems associated with the sector's rapid growth. Furthermore, many tourists are also becoming disillusioned with the Cypriot tourism product as indicated in recent surveys (CTO 1994). There is a danger that the island's largely conventional mass tourism (sun, sea and sand) image may eventually lead to a loss of popularity in northern European markets as increasingly sophisticated tourists seek out less spoilt and out of the ordinary destinations (Krippendorf 1992; Poon 1993). A possible reflection of the south's waning popularity as a tourist destination is that in the last few years the number of arrivals seems to have stabilized around the 2 million mark and arrivals from some of the traditional markets (e.g., Britain) have actually declined (CTO 1995). These events combined with stagnant per capita tourist expenditures indicate that the Republic of Cyprus displays the classic characteristics associated with the consolidation stage of the tourist resort cycle (Butler, R. W. 1980, 1993; Ioannides 1992).

The industry's economic benefits have thus far masked the extreme vulnerability arising from excessive dependence on a single industry (Syrimis 1990). For instance, many businesses in coastal areas depend too heavily on sales to tourists and are, thus, overly susceptible to the industry's fortunes (Best *et al.* 1989). Moreover, the spatial distribution of tourist services and accommodation favouring the coastal areas has accentuated chronic regional economic imbalances. With just 13 per cent of total area of the south, the coastal region accounts for 40 per cent of the population and 45 per cent of the non-farming work force (Constantinides 1991). The towns of Limassol, Larnaca and Paphos are home to 57 per cent of the entire urban population (Table 8.1). In addition to serving as the location of more than 90 per cent of the Republic's tourist activity, the narrow coastline hosts a number of large-scale activities that form the backbone of the economy. It is hardly surprising, therefore, that the northwest region of Laona and many mountainous areas with their narrow, primarily agrarian economic base, and an underdeveloped tourist sector have witnessed large-scale depopulation (Ioannides 1995).

In the early 1990s, labour shortages plagued the rapidly growing tourism industry as well as the overall Cypriot economy. For instance, in 1992 the Cypriot economy and particularly the construction and tourism sectors faced a shortfall of approximately 12,000 persons. The deficit was especially evident in the low-skill, low-wage positions (e.g., chamber maids, bellboys) (Ioannides 1994). Moreover, in the hotel and restaurant sectors there were shortages of waiters, bartenders, cooks, and reception staff. These employment deficiencies led to substantial increases in labour costs to a point where these were outpacing the economy's productivity (Vassiliou 1995). For example, the proportion of payroll costs had risen substan-

Table 8.1 Proportion of population and activities located in Cyprus' coastal zone 1990*

Activity	% of total
Population	40
Urban population	57
Tourist accommodation	93
Power stations	100
Oil refineries	100
Oil storage	100
Chemical industries	100
Cement factories	100
Airports	100
Proposed national parks	100

Source: Ioannides 1991, p. 41.

* Data refers to Greek-Cypriot controlled part of the island only.

tially to 40 per cent of total gross revenue in the higher class hotels, meaning that most now operate with marginal profitability (Andronikou 1993). The labour shortages led the government to pass legislation allowing the import of guest labourers on a temporary visa (*Phileleftheros* 1992). Ironically, in more recent years, the stagnation in tourist arrivals has led to a situation where, for the first time, some hotels had to lay off personnel during the off-peak season (Interview with representative of the CTO 1996).

Declining occupancy levels are yet another increasingly serious problem arising from the rapid development of hotel accommodation, an issue which is particularly troublesome during the off-peak season. In the early 1990s, hotels in Ayia Napa which had summer occupancy levels approximating 100 per cent, witnessed winter occupancies as low as 20 per cent. Average occupancies were approximately 60 per cent (CTO 1993). Oversupply of accommodation has become more problematic in the last two years and by 1996, many developers had begun converting their facilities into office buildings in the hope of recouping at least part of their investment costs (Information supplied by the CTO 1996). Uncharacteristically, hoteliers have joined the call for a ban on additional construction of tourism accommodation to safeguard their already razor-thin profit margins (*The European* 1996).

Inflationary pressures have also resulted from the speculative development activities in coastal areas. This is because 'rich tourists can afford to buy items at high prices, and thus retailers increase their prices and provide more expensive goods and services' (Witt 1991: 41). The excessive demand for coastal land, especially during the 1980s, led to a dramatic land value hike. The price of land in Ayia Napa, for instance, increased 5053 per cent in real terms from C£2000 per plot in 1975 to C£103,072 per plot in 1991 (Information supplied by the Department of Land and

Surveys 1992). The annual rate of increase of land prices averaging 45.6 per cent, far exceeded the annual rate of inflation for the same period (around 15 per cent). By 1995, land prices in certain coastal areas were as high as those in the Cote d'Azur resulting in 'a shortage of building plots for housing purposes' (Vassiliou 1995: 46).

Beyond these negative economic impacts, tourism is also commonly associated with undesirable sociocultural effects (Cohen 1988; Kousis 1989). Nevertheless, until now the Greek Cypriot authorities have generally underplayed the industry's negative social impacts, citing the indigenous population's long-term western outlook and highly educated profile (Andronikou 1979). Policy-makers argue that because Cypriots have already adopted a western lifestyle, they are not adversely affected by their face-to-face encounters with northern Europeans. Regardless of these assertions, however, there have been incidents indicating cause for concern (Andronikou 1987). Increasingly, many locals regard the gradual dismantling of Cypriot culture brought upon by invading, 'immodestly attired' mass tourists displaying loose sexual mores and little interest in anything beyond the island's resources of sun, sea and sand. Often, the heavy presence of tourists in the island's key resorts has led the Cypriot youth to adopt 'different sets of values on morality, style of dressing ... in comparison with prevailing traditional attitudes, and as a result the bonds of closely knit families are in some cases being loosened' (Witt 1991: 44). In one extreme case, residents of Ayia Napa were reportedly so outraged by the complete takeover of their village by tourism activities that they abandoned the area to build a new settlement further inland (Jensen 1989).

Though the Republic of Cyprus accounts for a small percentage of the total number of tourists visiting the Mediterranean region, because of its small population it displays one of the highest ratios of tourists per inhabitant in the whole of Europe (approximately 3.3 tourists per Cypriot citizen) (World Bank 1992). Since the early 1990s, the Ayia Napa and Paphos regions have been particularly plagued by severe visitor overcrowding (Ioannides 1994).

The rate at which once quiet, largely rustic, coastal areas have transformed into sprawling concrete jungles of tourist facilities, plus tourism's environmental threat to much of the native fauna and flora, have alarmed planners and environmental conservation groups. For example, sea turtles have already been driven from the Ayia Napa beaches because of excessive tourism development. Similarly, the nesting grounds of the rare loggerhead and green turtles along the beaches of northwest Cyprus (e.g., Lara Bay) are currently threatened by increasing numbers of day-trippers (Ioannides 1995). In less than two decades (1973–91), a stretch of approximately 48 km of once pristine coastal land has been lost to urban and primarily suburban development. The ribbon-like construction of hotels, tourist villas and apartments, and second homes accounts for the major part of this new coastal development (Tables 8.2 and 8.3).

Table 8.2 Development along the Cypriot coastal zone 1973*

Coastal front	Length (km)	Urban (km)	Suburban (km)	Pristine (km)	Remarks
Paralimni	9	–	–	9	Inland village.
Ayia Napa	16	–	–	16	Inland fishing village.
Ayia Napa-Liopetri	7	–	–	7	No development.
Larnaka	19	7	2	10	High/medium density.
Larnaka-Zygi	30	–	–	30	No development.
Limassol	20	9	3	8	High/medium density.
Pissouri	2	–	–	2	3–4 restaurants.
Paphos	26	–	2	24	One coastal hotel.
Akamas	33	–	–	33	No development.
Latsi	4	–	–	4	No development.
Polis	6	6	–	–	No coastal development.
Polis-Pomos	9	–	2	7	Some tourist development.
Total	181	22	9	150	

Source: Ioannides 1991, p. 42.
*Refers only to the southern part of the island.

Table 8.3 Development along the Cypriot coastal zone 1991*

Coastal front	Length (km)	Urban (km)	Suburban (km)	Pristine (km)	Remarks
Paralimni	9	–	7	2	Medium-density tourist and second-home development.
Ayia Napa	16	–	6	10	Medium- to low-density tourist development.
Ayia Napa-Liopetri	7	–	–	7	Minor development.
Larnaka	19	10	9	–	High to medium urban density, medium-density tourist development.
Larnaka-Zygi	30	–	5	25	Low-density second-home development.
Limassol	20	13	7	–	High urban density, high tourist density.
Pissouri	2	–	1	1	One large hotel, apartment complexes.
Paphos	26	2	10	14	Low- to medium-density tourist development.
Akamas	33	–	–	33	No development.
Latsi	4	–	–	4	A few restaurants.
Polis	6	6	–	–	Almost no tourist development.
Polis-Pomos	9	–	3	6	Village expansion.
Total	181	31	48	102	

Source: Ioannides 1991, p. 43.
*Refers only to the southern part of the island.

A World Bank report (1992) asserts that 'excessive strip development along coastal areas has raised concerns that the carrying capacity of several coastal areas may already have been exceeded' (27) and that 'coastal degradation [has taken] the form of loss of vistas and ocean views, of aesthetic pollution by high-rise hotels and tourist apartments in ribbon-type formation without character and focus' (31). Unfortunately, ineffective development control has created a mishmash of incompatible land uses and buildings in both urban and rural areas (Jensen 1989; Panayiotou 1989; Travis 1980). The clutter of unsightly advertisement billboards adds to this 'architectural pollution' (Pearce 1989).

The provision of adequate infrastructure such as roads, sidewalks, water supply and sewage treatment plants has not matched the rapid pace of tourist accommodation construction (UNDP 1988). While there is an oversupply of accommodation and restaurant facilities, the provision of open space and/or recreational areas (e.g., tennis courts, parks, sports centres) is inadequate. 'In more popular tourist areas, the surfacing of roads, collection of refuse and treatment of sewage lag behind the burgeoning new developments of aggressive real estate operators' (Jensen 1989: 10).

Through building their own private harbours and groynes, hotels have regularly upset the dynamic state of the beaches, in turn leading to wide-scale erosion. Other environmental problems have resulted from the lack of proper waste disposal systems in many areas. Although many of the newer hotels along the coast have their own waste treatment facilities, sewage often seeps into the sea either because of system breakdowns or, more commonly, because during the tourist peak season the high volume of waste exceeds the soil absorption capacity (World Bank 1992). Numerous tourist-related activities have caused overpumping of the wells in the Ayia Napa region, in turn leading to the intrusion of seawater and chemicals from fertilizers in the fresh water supply. As a result, fresh drinking water now has to be trucked in from other parts of the island (Ioannides 1994).

In most parts of the island, water shortages have become acute in recent years as the tourist industry places an enormous burden on already scarce resources. Despite the construction of large dams and a major water conveyor, a series of severe droughts in the 1990s has led to periodic water rationing, which affects thousands of households around the island (Mansfeld and Kliot 1996). In September 1991, water shortages were so severe that the government had to take the extreme measure of recruiting specially converted oil tankers to import water from Crete (Ioannides 1994).

Increasing awareness of the major problems associated with tourism's rapid evolution has led the government and the CTO to enact a series of measures geared towards safeguarding the industry's future position and maximizing its economic returns, while better preserving the human-built and natural environments. These include the 1990 *Town and Country Planning Law* and the CTO's *New Tourist Policy* (1990).

While the former measure's scope extends well beyond tourism, it specifically requires local authorities to take the responsibility for approving plans designating tourism development zones that meet stringent criteria set in the island-wide strategic plan (Constantinides 1991). Unfortunately, however, the effect of the *Town and Country Planning Law* has been minimal at best. Building moratoria instituted in the early 1990s to ease the passage of the new legislation did not impact on previously approved applications for hotels and hotel-apartments (World Bank 1992) nor did they apply to ancillary tourist facilities (Lockhart 1993). As a result, the speculative construction of second homes, retail establishments, and other informal tourist facilities has continued at a high pace in the last few years.

The CTO's *New Tourist Policy* which recommends, among others, the development of a higher quality, diversified tourism product has also had minimal impact partly because of weak enforcement, but more importantly because the reality of oversupply of accommodation has dictated the continued need to cater to mass tourists participating on cheap package tours. It has also become obvious that the ramifications of some of the policy's objectives have not been well thought through. For example, one of the suggestions for diversifying the product and attracting a higher spending clientele calls for the development of golf courses, a strategy which will clearly place further strain on the island's already meagre water resources.

Tourism's environmental problems pose serious challenges to the industry's future success in the south. Officials are becoming increasingly concerned that the number of repeat tourists has decreased in recent years as many seek out more diversified, less environmentally-damaged destinations. It is interesting to note that some of these tourists are now been lured to northern Cyprus which promises a more pristine, out-of-the-ordinary landscape (information supplied by the CTO 1996).

Tourism development in the north: overcoming the obstacles?

Until now, tourism development in northern Cyprus has followed a remarkably different trajectory to the one witnessed in the Republic. In 1994, the Turkish-Cypriot part of the island received approximately 350,000 foreign tourists (about 17.5 per cent the number of arrivals in the south), the overwhelming majority of whom were short-term visitors from the Turkish mainland (Bicak and Altinary 1996; *North Cyprus Home Page* 1997). By contrast, tourists from Germany and Britain, the two leading European markets, accounted for just 24.9 per cent of total arrivals. There are only two hubs of tourist activity in the north, Kyrenia which maintains its status as a resort from the pre-war days and Boghaz, a community a few miles to the north of Famagusta (Figure 8.3).

Tourism's slow growth since the war may seem somewhat surprising, considering this area had much of the pre-war tourism accommodation stock. Nevertheless, only a fraction of the hotels were taken over by Turkish Cypriot entrepreneurs, while most of the facilities have remained abandoned for the last 23 years. This is·particularly the case in Famagusta's tourist area which the Turkish army has declared a militarized zone and, thus, out-of-bounds to civilians. Moreover, during the early years after the war, the north's tourist industry faced problems such as 'lack of local expertise and capital ... [and] ... bottlenecks in transport and marketing' (Lockhart 1994: 371). Despite a substantial growth in capacity during the late 1980s and early 1990s, by 1994 the entire area still had fewer than 8000 beds, compared to roughly 76,000 in the south (Bicak and Altinary 1996; CTO 1993; Lockhart 1994).

Tourism's economic impact has been minor compared to the miracle witnessed in the south. By 1994, net earnings from tourism amounted to $172.9 million representing less than 4 per cent of the region's GDP. Importantly, the GNP per capita amounted to just $3010 whereas in the Republic it had risen to approximately US$12,500, mainly because of tourism's boom. In 1990, approximately 3900 persons were employed in the hotel sector, accounting for just 5.1 per cent of the north's gainful employment (Mansfeld and Kliot 1996). Less than 10 per cent of jobs in northern Cyprus are in tourism and trade while agriculture maintains its position as the region's leading economic activity, accounting for one third of available jobs (*North Cyprus Homepage* 1997).

Despite the region's growing popularity as a tourist destination (Lockhart 1993), a number of factors have stalled the sector's growth. Among them was the fact that unlike its Greek Cypriot neighbours, the Turkish Cypriot administration did not stress tourism development in the immediate aftermath of the 1974 war. Instead, during these early stages, policymakers placed heavy emphasis on creating a robust agricultural sector to serve the predominantly rural population. Secondary priority was placed on developing a small manufacturing sector concentrating on textiles and food processing and packaging (Mansfeld and Kliot 1996; *North Cyprus Homepage* 1997; Scott 1995). The absence among Turkish Cypriots of a significant entrepreneurial class (and thus expertise) on a par with the one that existed in the south (Christodoulou 1992) did not help tourism's growth either (Lockhart 1993).

It was not until the late 1980s that the authorities changed their tune and began focusing more closely on tourism development. Recognizing that tourism could be the economy's principal 'locomotive sector' (Scott 1995: 389), policy-makers finally passed an incentives law (the *Tourism Promotion Bill*) in 1987. While the law has induced a number of Turkish Cypriot expatriates to invest in the tourism sector, by contrast foreign transnational hotel companies and other developers have been reluctant to become involved in what they perceive to be a politically and economically fragile environment. It is important to note that the contribution of expatriate investment in terms of hard currency returns has not been as great as originally envisioned, since a main attraction for investors in the first place was that they could repatriate their earnings to Turkey or other countries (Mansfeld and Kliot 1996). Also, not all development projects have been successful. Most notably, following bankruptcy, the holiday complexes initiated by the British-based Polly Peck company were never finished.

Another obstacle to the sector's growth has been Turkey's clear unwillingness to support tourism development in the region, a result of the prevailing negative public opinion in the economically depressed mainland towards any form of aid for northern Cyprus (Atun, referenced in Mansfeld and Kliot 1996). More importantly, however, it is the lack of political recognition by the international community that has inhibited northern Cyprus' tourism development. Since 1974, the United Nations has only acknowledged the Republic of Cyprus as the sovereign government for the whole island. By contrast, northern Cyprus has suffered the effects of an embargo as all its entry points have been branded illegal by the Greek Cypriot government. Tourists visiting the north are prohibited from crossing the Green Line. Moreover, because there are no foreign embassies in northern Cyprus, many potential travellers are reluctant to visit fearing the lack of refuge in the case of emergencies.

Foreign tourists visiting the Republic of Cyprus are also dissuaded from crossing over the Green Line to the north. 'The maps produced by the Republic's cartographers state that the northern area is "inaccessible" and under Turkish occupation' (Lockhart 1994: 371). Similarly, a message posted on *Kypros-Net* asks tourists not to 'visit occupied northern Cyprus and thus financially, morally, and politically provide support for the illegal and barbaric occupation of our homelands' (1997a). While officially the Republic of Cyprus allows foreign visitors to cross over into the Turkish Cypriot side for day-long visits as long as they do not bring any purchases back, few tourists ever take advantage of this opportunity (information supplied by Public Information Office 1997). The checkpoints along the buffer zone have been closed a number of times in recent years due to a heated war of words between the two sides. Even when the checkpoints are open, pressure groups on the Greek-Cypriot side try to persuade tourists not to visit the north by emphasizing some of the negative aspects (e.g., crime, disease) they may encounter there.

International civil aviation regulations prohibit chartered and scheduled airlines from flying directly into northern Cyprus unless they first touch down at an airport in Turkey. This means that travellers on a package tour from northern Europe must fly to northern Cyprus via Izmir or Istanbul. The required stop adds considerable travel time, making flights to the region fairly expensive compared to travel costs to other

Mediterranean destinations (Lockhart 1994). Another turn-off for potential travellers is that tour operators do not provide travel insurance on packages featuring northern Cyprus (Interview with a representative of the Republic's Public Information Office 1997).

In recent years there has been a proliferation of tour operators and travel agencies featuring package tours to northern Cyprus. A search on the Internet, for instance, indicates about 50 companies, mostly niche or destination specialists, in the United Kingdom, Germany, Israel, and other European countries selling tours to the region. None of the major tour operating firms (e.g., Thomson's, Kuoni), however, have included northern Cyprus in their brochures, since they do not want to run the risk of disrupting their traditionally strong ties with Greece and the Republic of Cyprus (Lockhart 1994).

Barriers to tourism development in the north

In 1994, despite a slight decline in overall arrivals, there was a substantial increase in the number of non-Turkish tourists (22 per cent). In all, a total of 95,115 tourists from northern European markets and Israel were recorded. These tourists spent on average eight to nine nights, providing much needed foreign currency to the Turkish Cypriot economy though their numbers are 'still not at the desired level' (Bicak and Altinary 1996: 929). By contrast, tourists from the mainland still comprised an overwhelming proportion of the total. Many Turkish visitors were actually short-stay 'barter' tourists, visiting northern Cyprus for the sole purpose of purchasing goods which they can sell for a profit in the mainland. These visitors did not contribute significantly to foreign exchange earnings since the region's currency is tied into the Turkish lira. Moreover, rather than staying in hotels, most Turkish visitors stay with friends or in cheaper lodgings such as guest houses (Lockhart 1993).

The average length of stay of non-Turkish tourists does not compare favourably with that of international visitors in the south (approximately 12 days) (CTO 1994). Consequently, hotels have had trouble attracting tourists, and occupancy rates have remained persistently low at about 30 per cent (Lockhart 1994). The tourism industry suffers from extreme seasonality with over 50 per cent of the visitors arriving between July and October. This is considerably worse than the seasonality characterizing the south's tourist industry (Ioannides 1994). In the latter case, authorities have been able to extend the tourist season in part by diversifying the product and attracting a small but growing number of alternative tourists (e.g., hikers, golfers, conventioneers). Conversely, attempts by Turkish Cypriot authorities to diversify the tourism product have thus far been frustrated. For example, efforts to attract international conferences have met with limited success because of inadequate convention facilities and, more importantly, non-recognition by the international community.

On a positive note, it appears that tourism's slow growth has spared the region the most serious social problems witnessed in the south's popular resorts. Similarly, environmental problems have been minimal. As Lockhart (1994) maintains, ribbon development does not appear to have seriously affected any of the resort areas. Since only a small number of large-scale hotels have been constructed, architectural pollution has been kept in check. The unspoilt nature of much of the coastline has prompted authorities to promote the region as an ecotourist destination. Nevertheless, it is unclear whether the region will be able to avoid the environmental woes that have become commonplace in southern resorts. Despite the existence of a planning and zoning framework, weak enforcement of regulations may not, in the long-run, deter the emergence of yet another 'tourist monster' on the island (Mansfeld and Kliot 1996: 199). The Turkish Cypriot authorities should heed the lessons resulting from uncontrolled tourism development in the resorts of Ayia Napa, Limassol, and Paphos; otherwise, the lure of tourism's economic growth potential could, in the short-run, overshadow environmental concerns (Wilkinson 1989).

Lessons and conclusions: strategies for the future

The preceding case study clearly demonstrates the disruptive effect serious crises can have on a particular destination's resort cycle. In the case of Cyprus, war and partition have created two profoundly separate tourist landscapes. On the one hand, since 1974 the Greek Cypriot south has seen its tourist industry boom, becoming the principal driving force behind a miraculous economic revival. On the other hand, Turkish-controlled northern Cyprus has struggled to develop its own tourist industry. The result has been that tourism in each part of Cyprus presently demonstrates two entirely separate stages of the resort cycle.

A mature tourism product, declining growth rates of arrivals matched by waning occupancy rates, and the emergence of serious environmental and sociocultural problems including capacity constraints, indicate that the south has truly entered the consolidation stage of its resort cycle. Indeed, there is a growing worry among policy-makers and industry representatives alike that if drastic measures are not taken soon to differentiate the tourism product and mitigate environmental problems, the destination will sooner or later stagnate and eventually decline. By contrast, northern Cyprus still displays characteristics of earlier stages of the model. While Kyrenia and Boghaz could be said to have reached the late involvement to early development stage, other areas have not yet emerged from their exploration phase.

Tourism's divergent trajectories on each part of the island since 1974 indicate the resort cycle's susceptibility to internal and external contingencies (Butler 1993; Ioannides 1992). On the domestic front, while the government of the Republic of Cyprus aggressively pursued tourism development as an economic growth strategy in the immediate aftermath of the 1974 war, the Turkish Cypriot authorities were slow to appreciate the sector's potential, preferring instead to focus on agricultural and manufacturing development. Moreover, while the Greek Cypriot community already had in place a large entrepreneurial class with the necessary skills to support the development of a robust tourism sector, the Turkish Cypriot community has historically lacked these traits. On the international front, the embargo imposed by the global community on northern Cyprus has severely limited that region's ability to enter the development stage of its resort cycle.

What, then, does the future hold for the island's tourist industry? Each area faces unique challenges. While the south desperately needs to address the environmental repercussions of mass tourism, northern Cyprus must find ways to circumvent numerous obstacles and enhance its image as a destination. The danger for northern Cyprus is that economic growth objectives could surpass environmental and other constraints and, thus, eventually the problems witnessed in the south could be replicated.

In what has become an increasingly competitive market, a serious concern for policy-makers on both sides of the island is the threat posed by more diversified and environmentally pristine tourist destinations in other parts of the world (e.g., islands in the Indian Ocean and the Pacific). Already, Cyprus, like many other Mediterranean destinations, has witnessed declining shares of arrivals as tourists seek out-of-the-way, unspoilt areas further afield. A possible way to face up to this particular challenge is for both sides to set aside their prejudices and market the island as a single destination. After all, despite their political differences, other neighbouring countries (most notably Egypt and Israel) have featured in jointly-promoted tour packages. Such a joint approach makes economic sense especially for long haul tourists (e.g., the United States, Canadian and Japanese markets). Although, given the current political tensions between the two sides, the chances of such a co-operation presently seem remote, recent discussions concerning the Republic's application for admission to the European Union offer cause for some cautious optimism (*Kypros-Net* 1997b). Collaboration between the two sides, however unlikely it may seem currently, would inject new life in the island's tourism product and hopefully rejuvenate it as a tourist destination.

9 Political transformation, economic reform, and tourism in Syria[1]

Matthew Gray

This chapter was originally drafted and submitted in September 1999, before the death of President Hafiz al-Asad. Note that it therefore refers to al-Asad in the present tense, as there was insufficient time to change the text between the death of the late President and publication of this book.

Introduction

In contrast with the majority of the Mediterranean littoral states, Syria's tourism sector is fledgling. In comparison with the states of southern Europe, it does not receive a large number of European or North American tourists. In comparison with other Arab states of the southern and eastern Mediterranean, tourism does not occupy a primary place in the economy or in the economic development agendas of its government. Only Libya and Algeria (the latter as a result of current political instability) receive fewer Western visitors.

Table 9.1 Tourism to Syria 1988–97.

	Number of visitors	Tourism income (US$)
1988	421,000	266,000
1989	411,000	374,000
1990	562,000	320,000
1991	622,000	410,000
1992	684,000	600,000
1993	703,000	758,000
1994	718,000	1,149,000
1995	815,000	1,338,000
1996	830,000	1,206,000
1997	842,000	1,250,000

Source: World Tourism Organization, quoted in 'wazir al-siyyaha al-suriyy dawood: suriyya qadwa fi al-siyaha al-bayaniyya al-arabiyya'. 'Syrian Minister of Tourism Dawood: Syria is a Positive Example of Arab Regional Tourism,' *al-Iqtisad wa al-A'amal*, Special Issue 'GATT and Arab Tourism', Year 20, May 1998, p. 70.

Despite being in its infancy, Syria's tourism sector is expanding rapidly and is beginning to make an important contribution to the economic development of the country. The number of visitors doubled in the decade 1988–97, while the contribution of these visitors to Syria's economy increased five-fold. This growth, while impressive, is from a very low base, as prior to the 1970s 'the tourism sector [in Syria] had hardly existed'.[2] For a number of years, and especially in the 1990s, the Syrian government has attempted to increase the size and value of the tourism sector, meeting with limited success until only recently. Syria offers an interesting case study of the ways in which political imperatives, and changes to the political economy of a state, may make an impact on tourism development, especially as the tourism sector has featured prominently in Syria's economic liberalization and social development programmes.

This chapter investigates the political dynamics of tourism in Syria, from a political economy perspective. It outlines the reasons why the Asad regime has nurtured and encouraged tourism, the ways in which it has done so, and how the components of the Syrian political economy have reacted to, and influenced the outcomes of, the development of tourism. It argues that the regime has employed a 'carrot-and-stick' approach to the economic actors and elites in the tourism sector, linking them with the regime and its goals while providing incentives through a guided economic liberalization of the tourism sector and the economy more broadly. Political factors have partially stifled the economic potential of tourism, however the ability of tourism to ameliorate economic hardship, and the pain of economic change and reform, has nonetheless been important. The growing emphasis of the Asad regime on tourism is indicative of this, and of a broader attempt to deal with both endogenous and exogenous economic change. The place of tourism in Syria, and the accompanying political dynamics associated with it, says much about the political economy of the contemporary Middle East and about regime responses to changing economic and social conditions.

The state and the role of tourism in the economy

Since attaining power in 1970, President Hafiz al-Asad has presided over a gradual but very significant transformation of Syrian society and economy, and even over the political structure of the state. The socialist policies of economic development introduced during the 'radical' Ba'ath period (1966–70) have been altered or in many cases abandoned, in favour of a more liberal economy emphasizing economic pluralism (al-ta'addudiyya). This has included a more balanced relationship between the public and private sectors, manifested in the growing importance of joint sector enterprises, but in recent years has also included elements of economic liberalization and reform.

Simultaneously, and perhaps as a result, the regime's social priorities have also been transformed. The traditional working class bases of support, which brought the Ba'ath party to power in Syria, have gradually become a less salient feature of the regime's base of political legitimacy. In place of a populist orientation, the regime has responded to, and increasingly nurtured, an expansion in the middle and professional classes and a small number of wealthy business elites. By employing a neo-patrimonial political model, which in economic terms incorporates a political symbiosis between the state's leadership and socioeconomic elites, the Syrian political leadership, and especially Asad, has sought a closer relationship with key businesspeople in the tourism sector and elsewhere in the economy.

The logic of the state in encouraging tourism

There are particular reasons why governments throughout the world support the expansion of tourism. First, the potential for tourism to generate foreign currency is a principle ambition of governments, all the more so in states which have artificial or controlled exchange rates, or which are, often as a result, suffering balance of payments problems. Second is the fact that tourism is labour intensive, and creates employment throughout the economy; tourists spend money on hotels, transport, and meals, and also on a wide variety of other goods and services across the economy. Third is the fact that the tourism industry does not, on the whole, require expensive or complex technology or a highly-skilled workforce; foreign language skills and a pool of people with generic management and service industry experience is typically all that is required in the initial stages of tourism development, and most states have such a pool of people available. Further, with the exception of a small number of complex projects such as operating an airline, investment in tourism is not comparatively expensive, and will often return a profit reasonably quickly. Finally, many states, including Syria, already have in place the basic and most important requirements for the development of the tourism sector; a pleasant climate, attractive scenery, historical sites, and friendly people. In other words, governments often feel that their state possesses an untapped economic resource, or even a comparative advantage in tourism, and decide to take advantage of it.

In the case of Syria, these factors are important in explaining the regime's emphasis on tourism. The government's ambitious programme of tourism expansion – which includes the creation of 120,000 new jobs in the industry by the year 2000, and the aim to receive five million tourists by 2010 – is an indication of the perceived economic benefits attached to tourism.[3] But more specifically, there are several other reasons why the Asad regime is encouraging tourism.

The first is that, unlike some other states such as Jordan, Syria's tourist attractions are plentiful, are spread throughout the country and are, for the most part, easily accessible. The sites which are most commonly visited by foreign tourists include the cities of Damascus, Aleppo and Hama, the Roman ruins in the desert at Palmyra, the Crac des Chavilliers in the rural central-west of Syria, the Euphrates river, the Mediterranean coast, and numerous small villages. The nature of Syria's tourist attractions mean that foreign (and especially Western) tourists visiting Syria tend to be middle-aged, stay longer in Syria than in many other states, have high incomes, travel in groups, and have a strong interest in historical and cultural attractions. So, although Syria receives relatively few Western tourists, perhaps as few as 100,000 a year, these tourists are very lucrative and spend their money throughout the country. Since the economic development of rural Syria has been a major goal of successive Ba'athist regimes, including Asad's, it may account in part for the emphasis placed on tourism.[4] Former Tourism Minister al-Shamat stated in 1995 that the regime's tourism goals included 'Encouraging new styles of tourism such as ... winter and desert tourism ... celebrating tourism festivals in all seasons [and] ... encouraging popular and youth tourism...'[5] indicating the importance placed on tourism development in rural and regional areas.

A second factor explaining the targeting of tourism by the Asad regime is its politically safe nature. There are few members of the regime with vested interests in or against tourism, unlike sectors such as agriculture or industry. Further, tourists themselves pose little threat to the stability or popularity of the regime. Tourists rarely have any substantial impact on the politics of the host state. In fact, more often than not tourists are kept away from the people in the host state, except for brief, orchestrated meetings such as in the souq.[6] In cases where some political impact may result from tourism, such as by tourists taking photographs which depict Syria as an underdeveloped state or which contain political overtones, tour guides usually attempt to discourage this behaviour.

In addition, tourism usually contributes to traditional industries, which otherwise may not be viable. Few tourists visit any country without buying a souvenir or momento. In this respect, Syria has become a popular shopping destination, particularly for nationals of Russia and the former USSR, Iran, and Western Europe. As a mode of development, with positive economic and political implications for the regime, Miyoko Kuroda has argued that traditional industries – by implication, nurtured or maintained by tourists – may provide an area in which 'late developers' often have a comparative advantage and in which they typically excel.[7]

Finally, the private sector is relatively eager to enter the tourism sector, and 'there is unanimity that the potential is enormous – especially in tourism....'[8] The private sector's willingness to invest and participate in tourism is the result of several characteristics of tourism generally, and its treatment under Syrian law in particular. There is the potential for Syrian tourism to continue to expand over the coming years and, as mentioned, its size doubled between 1988 and 1997.[9] The private sector is also attracted to tourism by the ease with which the sector can be entered. Start-up times for tourism projects are shorter than for industry, returns are greater than for agriculture, and less specific skills are required to manage tourism projects compared to other sectors. The most important motivation for the private sector, however, has been economic liberalization.

Economic liberalization and tourism: the first Syrian infitah (1973–81)

Under Asad, there has been a gradual economic liberalization of the Syrian political economy. Syria's infitah, or economic opening, policies have been far less extensive or spectacular than those in Egypt or Tunisia, and have had ta'addudiyya rather than a rapid development of the private sector as the stated aim of reform. Nonetheless, a considerable number of reforms have been introduced over the past two decades or more, with many of these having had a considerable impact on the economy. There have been two periods since 1970 where Asad's government has introduced rapid or substantial economic liberalization; the first infitah of the late 1970s, which particularly targeted the tourism sector, and the second infitah from the mid-1980s to the present.

The first infitah emerged as al-Asad sought ways to differentiate his presidency from that of his predecessors. There were a handful of reforms in the early 1970s, but they were limited in scope, and applied mostly to industrial activity. In 1977 and 1978, however, tourism became a primary focus of economic liberalization, with the creation of mixed sector companies in the industry. Most significant were Law Number 56 of 1977, which led to the formation of the Arab Syrian Company for Touristic Establishments (ASCTE) by the Syrian businessman Uthman A'idi, and Law Number 41 of 1978 which created the Syrian Transport and Tourism Marketing Company (TRANSTOUR) under the prominent businessman Sa'ib Nahas. A'idi and Nahas, along with hotel owner Abd al-Rahman al-Attar and a handful of wealthy businessmen in other sectors of the economy, have become known as the 'new rich'[10] business elites of Syria, as a result of the favourable treatment received under the laws of the first infitah.

The common characteristics of these laws, and of many which followed in the 1980s, was the creation of mixed sector companies, with the government handing over to the private sector the managerial responsibilities of the enterprise.[11] The government maintained a minimum 25 per cent interest in the company, but its role was usually limited to the provision of capital such as land, property, and access to utilities. Such joint ventures enjoyed exemption from the otherwise extremely rigorous and controlled currency exchange rules, and exemptions were also granted from income taxes and some duties for up to seven years. A key benefit to the private sector from these laws came from the fact that joint venture companies were largely protected from competition, since they were established through specific laws, and not as the result of a general law liberalizing the economy, or an economic activity, more broadly. They therefore represented liberalization in an extremely limited form, although such enterprises certainly 'represented an element of infitah'[12] in that the role and prominence of the private sector in the economy was increased, and there was an implicit acknowledgment from the government that an increased role and visibility for the private sector was desirable.

The more general economic reforms during the first infitah had little impact on the numbers of wealthy foreign tourists visiting Syria, despite the continual optimism of the regime in claiming that tourism had the potential to become one of the three most important sources of foreign income and employment. Real and perceived instability in the Middle East, as well as some perceived instability within Syria, provided little incentive for tourists to visit Syria during the 1970s.

The limited extent of the first infitah suggests that its motives were predominantly political. The 1970s were, above all, an era of regime consolidation in Syria, with Asad attempting to reinforce his leadership and establish his own brand and mode of economic and social development. The goal in the 1970s was to achieve rapid and popular economic growth, which accounts for the limits on the first infitah and the importance of tourism in it. The Syrian economy remained under the dominance of the public sector, as borne out by Asad's assertion – a continual assertion, until ta'addudiyya replaced it – of the public sector being 'the leading sector'.[13]

The second infitah: 1986 onwards

Syria's second infitah was completely different in its origins to the first. Whereas the reforms of the 1970s found their origins in Asad's attempts to stamp his own philosophy and leadership style on the Syrian political economy and on Syrian society, the second infitah was the regime's response to a serious economic crisis which began in 1986. The Syrian economy performed poorly in the 1980s[14] – a decade which also witnessed profound changes in the international economic and political system, which were to have an equally profound impact on Syria.

The crisis of 1986 was predominantly one of foreign exchange and balance of payments, after many years of declining income from trade and remittances, growing government debt, low foreign investment, and a fall in oil prices. Added to these problems was a rapidly growing population (3.4 per cent per annum), and drought.[15] The result was economic stagnation, with a growth rate of only about one per cent per annum in the mid-1980s – compared to around 10 per cent per annum in the 1970s – and therefore the need for significant changes to the structure of the economy.

As the crisis of 1986 and afterwards began to unfold, the government made a series of macroeconomic reforms designed to counter Syria's economic stagnation. The reforms included exchange rate adjustments, trade liberalization, an expansion of the private sector, pricing reform, and a reduction in government subsidies on commodities and utilities. Although not aimed directly at the tourism sector, the reforms did have some impact there. The role of the private sector in tourism, especially in hotel and restaurant management, increased markedly during the late 1980s and into the 1990s. Exchange rate reforms, including those of 1993 which established a 'neighbouring countries rate' for some transactions of S£42 to the US dollar – as opposed to the previous single rate of S£11.25 to the US dollar – reduced many of the costs incurred by tourists visiting Syria.

The key tourism reform of the second infitah was Resolution 186 of 1986 by the Supreme Council for Tourism, which increased the role of the private sector in developing tourism facilities in Syria. A number of private hotels and tour companies appeared at this time, and the late 1980s were also the period of most rapid growth for the Cham Palace hotel group.[16] Resolution 186 allowed investors who were establishing a tourist facility to by-pass laws on the importation of raw materials, tools, and manufactured equipment – especially for the construction of luxury hotels and facilities – and also to gain exemptions and favourable treatment on taxes and customs duties.[17] Partly reflecting these reforms, there was a moderate increase in the number of Western tourists which Syria received during the late 1980s, until the outbreak of the 1990–91 Gulf crisis.[18]

Perhaps the most significant reform of the second infitah was the promulgation of Law Number 10 of May 1991 for the Encouraging of (Productive) Investment,[19] aimed at increasing direct foreign investment in Syria. The law offered a number of incentives for any investment project in excess of 10 million Syrian pounds which generated employment,

Table 9.2 Visitors to Syria: Arab and others (1997)

Country	Number
Lebanon	791,572
Jordan	477,233
Saudi Arabia	202,622
Egypt	58,619
Kuwait	58,619
Bahrain	33,131
Algeria	29,073
Sudan	27,115
Yemen	17,939
Tunisia	15,506
Total arrivals – Arab countries	1,711,429
Total arrivals – others	620,199
Total number of arrivals	2,331,628
Percentage Arab arrivals	73.4%

Source: World Tourism Organization, quoted in 'wazir al-siyyaha al-suriyy dawood: suriyya qadwa fi al-siyaha al-bayaniyya al-arabiyya'. 'Syrian Minister of Tourism Dawood: Syria is a Positive Example of Arab Regional Tourism,' *al-Iqtisad wa al-A'amal,* Special Issue 'GATT and Arab Tourism', Year 20, May 1998, p. 7.

resulted in exports, and transferred technology or expertise.[20] Law Number 10 included exemptions from the strict Foreign Exchange Law Number 24 of 1986, which placed severe restrictions on foreign currency transactions. Other incentives included customs and duties exemptions, exemptions from company taxes for up to seven years, and the freedom to repatriate profits overseas.

Law 10 was obviously designed to increase investment in Syria, but the exact outcomes and, more importantly, the sources of anticipated investment under the law are open to considerable speculation. The introduction of Law 10 was most likely an attempt to capture funds from expatriate Syrians, and from regional investors, rather than the often quoted Western investment. One observer contended that Arab investment and Syrians abroad 'might well be the government's intention, as it is not keen to see the country inundated by foreign interests due to its political sensitivities and the vested interests of the establishment, and because it implies an erosion of state control and sovereignty'.[21] The potential investment which could be raised from Arab sources, especially the Gulf states, was probably the key ambition of the law. Strengthening this assumption further is the fact that Syria's tourism sector is dominated by Arab visitors, who account for almost three-quarters of tourist numbers, and not by the small but valuable Western tourists which the country receives.[22]

Regardless of the government's intentions in promulgating Law 10, its initial results were mixed. Small and medium-sized businesses, and Syrian expatriates especially, have used the law to their advantage and have invested modest sums in different areas of the Syrian economy, including the tourism sector. Restaurants and medium-priced hotels have been created or expanded as a result of the law, especially by those seeking short-term investments.[23] Equally prominent, however, have been abuses of the law, most notably in car rental companies.[24] Nonetheless, there have been important investment projects initiated under Law Number 10.[25]

In the second infitah, the tourism sector was liberalized during a time of economic constraint and hardship, when the regime was particularly short of rentier income. Although tourism does not technically fall into the category of a rentier industry, many of its characteristics are similar. This would suggest that the emphasis on both liberalization and tourism during times of state financial austerity, especially the direct liberalization in the tourism sector, may be an attempt to breach the state's financial gap between income and (socially-oriented) spending. The part-privatization of tourism, the emphasis on large foreign (read: hard currency) investments, and the reluctance to unify Syria's exchange rates all point to this likelihood.

Political elites, social forces, and the politics of tourism development in Syria

As a result of Syria's highly centralized, authoritarian political system, it is tempting to look at the state as the only, or the overwhelmingly dominant, force in tourism management and development. In fact, while the regime is an important actor in Syria's political economy, other actors and social forces have had a strong impact on the course which the tourism sector has taken in the past two decades. Further, there is only limited validity to a study of the state, if undertaken in separation from the state's social bases of support or from other political groups in society and in the political economy. In the case of Syria, there are two important groups, in particular, which have defined the position and orientation of tourism and tourism policy within the broader political economy: Syria's business elites, and the social forces – classes, especially – which through their relationship with the regime, and with the political elites, have played a role in tourism.

Syria's business elites and tourism

In the Syrian economy generally, and notably in the tourism sector, a handful of wealthy businessmen have come to prominence under the Asad regime. These include Sa'ib Nahas, Uthman A'idi, and Abd al-Rahman Attar. In political terms, the relationship between these men and the al-Asad regime is one of 'symbiosis',[26] although given the patrimonial system of leadership in Syria, Asad maintains a direct political dominance over his businessmen clients.

Sa'ib Nahas,[27] a Shi'a businessman, entered the tourism industry in 1965, when he established 'Nahas Travel and Tourism', a company which represented the East German airline Interflug in Syria. Nahas gained business contacts in East Germany and elsewhere after beginning his career in his father's textile company. Nahas Travel and Tourism subsequently expanded its airline representation activities, and also organized tours abroad for Syrians, and inbound tours for Arab and Western visitors. The most important step for Nahas was during the first infitah, when Law Number 41 of 1978 established TRANSTOUR. TRANSTOUR is a joint-sector firm involved in several tourism areas. One is transportation, especially limousine and car rental under the trading name Europcar. Another is in tourism marketing, and the organization and delivery of tours and organized travel in Syria. By the late 1980s, it was the largest tour company in Syria. Finally, TRANSTOUR undertakes tourism investment projects; designing, financing, and building tourism infrastructure such as the al-Sayedah Zeinab complex, a four-star resort at Amrit, as well as other resorts, hotels and other facilities in tourism areas such as Palmyra. A branch of TRANSTOUR also manages hotels and facilities owned by the government. TRANSTOUR is the largest and best-known tourism firm in Syria, with the possible exception of Uthman A'idi's Cham Palaces and Hotels firm.

The Cham Palace group of hotels was established in 1983, after A'idi's Arab Syrian Company for Touristic Establishments (ASCTE) – the first of the mixed sector companies to be established – came into existence during the first infitah through Law 56 of 1977.[28] Most of Cham's hotels, which now number seventeen and supply some 7530 beds,[29] were built between 1987 and 1990 after the government surrendered its monopoly in the construction industry. Once the government monopoly was removed, many hotels were constructed quickly – in less than 12 months – despite the government having spent years on construction efforts in housing and hotels.[30] One unique feature of the ownership of the Cham Palace hotels is the fact that about ten per cent of stock is owned by company employees – some 6000 shareholders out of a total of more than 19,000 – 'in what can be described as a unique experiment in popular capitalism.'[31] Cham Palaces has increased its value fifty-fold since its establishment, and supplies about 80 per cent of Syria's four- and five-star hotel beds. The company is currently planning to expand into Lebanon and Sudan.[32] A'idi, a Sunni, is probably the wealthiest entrepreneur in the Syrian tourism industry, and is often considered the most prominent.

Less prominent is Abd al-Rahman Attar, a Sunni with interests in two four-star Damascus hotels, and in Orient tours and car rental. He is also active in the Chambers of Commerce.[33] There are also some smaller, but still significant, business actors in the tourism industry, especially in hotel ownership and management, and car rentals. Nahas, A'idi, and Attar, together with other key businessmen and bourgeoisie elites, form what Syrians refer to as 'al-tabaqa al-jadida' or 'the new class'.[34]

These new Syrian business elites have certain characteristics which distinguish them not only from other classes in society, but also from other elites. Their social evolution from members of the merchant and bourgeoisie classes to business elites has usually been sponsored by the president, or occasionally by a political actor very close to the president. Few are members of the old elite families of pre-revolutionary Syria; their predominantly Sunni make-up is indicative of their urban, merchant, and usually religious backgrounds. This is in clear contrast to other elites, such as Ba'ath party officials, the senior ranks of the military, and senior bureaucrats and managers of state-owned enterprises. These elites often owe their positions to their ethnic or religious origins, their social backgrounds, or their managerial or technical ability, the latter often being part of the 'technico' class of which Waterbury[35] has written and which are especially prominent in Syria's state-owned enterprises. Unlike these groups, which gained their status earlier and have traditionally formed the backbone of the regime's elite support, the private sector elites which have emerged in the tourism sector form an extension to the regime's base of elite support, while also being intertwined with other elites through a joining of the public and private sectors.

The symbiotic, though imbalanced, relationship between Asad and the new class offers several insights into the tourism sector, and its position in the political economy of Syria. The most important aspect, which highlights the symbiotic relationship between Asad and the private sector elites, is that they exchange resources and services in which the other possesses a deficiency. The government, or more precisely, the patrimonial leader such as Asad, maintains a monopoly or dominance over political decision-making, and often control over social mobility as well, due to the nature of the political economy of an authoritarian, patrimonial regime such as that of Syria. The president nurtures his relations with key individuals, including the business elites, for several reasons, including the need to maintain economic growth and development, changes in the regime's social bases of support, and the need for solidarity and cohesion beyond that created by blood ties through ethnicity or tribe.

In turn, business elites owe their position, and their future, to Asad and his regime. In the late 1970s and the 1980s, economic development along state-led lines failed to deliver economic growth or improved standards of living. As a result, the private sector, and Syria's business elites in particular, '[have been and] are being invited . . . to provide the dynamism and to stimulate the growth that their economies lack.'[36] The state provides the opportunities for entrepreneurs by allowing entry into previously closed fields of activity, and by legislating for monopoly or oligopoly business conditions in many sectors of the economy. This explains the enshrining into law of TRANSTOUR and ASCTE, rather than the complete or competitive liberalization of the tourism sector during the 1970s and 1980s. Controlled liberalization has meant high prices and profits, which in turn forces the businessmen to continue to rely on the regime for their privileged position.[37]

Besides the threat of losing their position, the political behaviour of these businessmen is kept in check in other ways. One is by the regime informally placing limits on the length to which business elites can influence policy. On occasions when a businessman oversteps his political bounds, he is often arrested or discredited by the official press as a profiteer or as corrupt. A'idi, Nahas, and Attar were all arrested in 1977 after a campaign against corruption. The campaign was intended largely as a warning to these businessmen, who had grown wealthy very quickly, not to translate their wealth into political power or aspirations. As recently as 1996, A'idi was accused of corruption. In the 1990s, however, an encroaching limit on the regime has been the fact that Syria's economy increasingly needs investment and business confidence, and the regime cannot be seen to act capriciously or to be hostile to wealthy actors in the economy.[38] That campaigns against the business elites have had little impact on their positions, and that many have been arrested several times but for very brief periods, highlights the 'carrot-and-stick' approach of the regime in its relationship with key business elites.

The regime also maintains some control over these elites by officially or unofficially allowing them exemptions to laws.[39] For example, Sa'ib Nahas is said to have permission to dismiss staff at will, despite the extremely strict labour laws which protect staff from arbitrary dismissal.[40] Some of these concessions are granted because of the government's involvement in joint-sector firms – hence the lenient labour laws for many of those enterprises – although in the case of Nahas it supposedly extends further than just the joint-sector TRANSTOUR. Favours such as these place businessmen in a position where loyalty and assistance to the regime has been rewarded with gifts that can easily be retracted. One method by which businessmen can protect themselves from attacks such as that of 1977 is by establishing joint-sector firms with the government. These public–private enterprises represent a relationship between the businessmen and the regime, and between the businessmen and the public sector more generally. For the bureaucracy, which often plays a role in organizing many of the opportunities seized by the businessmen, joint-sector enterprises are an opportunity for some individuals to venture into the profitable private sphere, while the regime gains economic and, by extension, political benefit.

The central position of tourism in the Syrian economy, therefore, is both indicative of, and a result of, the type of relationship which exists between the state and the business elites, with a peripheral role played by the bureaucracy as well. The guided liberalization of tourism in Syria has created and nurtured the new class of wealthy business elites, and tourism has featured prominently because of its politically non-threatening nature and because of its financial attractiveness to the private sector. The attractiveness of tourism, to the state and the private sector alike, suggests that tourism as a mode of development in Syria has been at least partly an afterthought in the economic development planning process. Tourism would, quite obviously, not feature as prominently were it not as profitable or easy to enter. Its pivotal role in development, therefore, can be explained as much in political terms as by economic ones.

Other actors and forces and tourism: the bourgeoisie and the urban and rural working classes

Beyond the political and business elites which guide tourism policy and dominate its most senior and visible positions, other groups and actors have played a role in tourism, have been beneficiaries of its expansion, or have featured in the government's policy planning for the sector.

The urban middle classes and working classes have both been important in this respect. As mentioned earlier, some of the chief beneficiaries of tourism are the traditional merchants and craftmakers of the souqs, or markets, of Syria. The crafters, in particular, are often in disappearing industries where the demand created by tourism, and only tourism, has kept their trade profitable; for example woodworkers, calligraphers and some artists. The regime has long sought to maintain the support of these groups, which have traditionally been politically active and are overwhelmingly of the Shi'a, not Sunni, sect of Islam. The methods of maintaining their support are similar to the methods used with larger businesspeople; by keeping their businesses profitable and by linking the regime to that profitability.

The wealthier, but still middle class, urban Syrians also feature as players in tourism. Many Syrians, in both the private and public sectors, are beneficiaries of tourism growth and development. This group, and their role in the tourism sector, has increased considerably in the past few decades: 'Between 1970 and 1991 ... from 140,000 to around 410,000 economically active persons and their families ... from around 9 per cent to 11 per cent [of total population].[41] Much of their economic activity has been in trade, transport, and services, not in manufacturing or industrial activity, mostly because of the profitability of the former. Although members of the private sector middle class are often religious and socially conservative, suggesting they might otherwise oppose uncontrolled tourism, they often have a direct or indirect reliance on tourism and the money tourists spend, meaning that they have not actively, if at all, opposed the tourism sector and its development. Those directly involved in tourism, such as restaurant proprietors, owners of small factories selling products to tourists or the tourism sector, or as owners of rental car companies or large shops catering to tourists, are not surprisingly especially supportive of tourism development.[42] Some political activity is also undertaken on behalf of the private sector middle class by the new business elites, and the views of these two groups are rarely at odds. For example, in 1993 Sa'ib Nahas was interviewed on Syrian radio, which 'in a highly symbolic interview ... allow[ed] him to speak for almost an hour ... about the private sector, its performance and its advantages'.[43] The relationship between the private sector middle classes and the business elites, through institutions such as chambers of commerce and industry and through economic contact between the two, has increasingly converged their political and economic interests. As the private sector middle class has become increasingly involved in tourism, their reliance on the business elites, and the latter's relationship with the regime, has likewise increased. In other words, the regime's relationship with the private sector middle classes has not been through democratization or direct inclusion in the decision-making process, but rather a 'trickle-down' of economic favours and a 'trickle-up' of economic requests and ideas, the majority via the business elites.[44]

The only direct input into economic decisions by the private middle class is through limited regime consultation with chambers of commerce and industry, and with the parliament (majlis al-sha'b). Institutions such as chambers of commerce provide the opportunity for the private sector middle class, among other groups, to voice their views on matters such as tourism policy. Increasingly, merchants and members of the private sector are entering parliament, and while also under tight control there, are consulted by the government on economic policy and are able to relatively freely outline their views and ideas on economic development matters.

Of particular importance to the regime as a traditional base of support, rural Syrian society has also been influential in the tourism sector, serving as a tool by which the government has attempted to develop rural areas and therefore maintain its popular support and legitimacy. Assisting this goal is the fact that Syria's tourism sites are spread throughout the countryside, including many of the most popular sites such as Palmyra, the Crac de Chavelliers, and the Euphrates river. This has allowed for employment creation in rural areas, and a wide distribution of the wealth gained from tourism, which in turn has been followed by political benefits for the regime.

The number of rural Syrians employed in areas related to tourism, such as retailing, traditional crafts, in regional transport, or as guides increased slightly during the 1980s, partly in response to a small expansion in overall tourism and a greater emphasis on regional tourism. Evidence from other Middle East states, with an applicability to Syria, suggests that rural workers are increasingly turning to non-farming sources of income, whether or not they also remain as farmers or agricultural workers in some capacity.[45] From the regime's perspective, it is socially and politically preferable to have rural Syrians work in the tourism sector rather than to struggle financially in traditional farming roles. Some rural workers have been quite successful in transferring their economic activity and their skills to tourism. Furthermore, there has also been an expansion in the size of the rural middle class, which has also played a greater role in the tourism sector. To maximize the economic benefits of tourism for rural Syria, the regime has emphasized rural Syria in its marketing of tourism. Both foreign tourists, and internal tourists from the major cities, have been encouraged to travel to the countryside, especially to major sites but also for desert expeditions and cultural festivals.[46]

Conclusion

The political, economic, and other dynamics which impinge upon the development of tourism in Syria, and upon tourism policy-making, are numerous. International and domestic factors, both within and outside of the state's control, play a role. However, this paper has concentrated on the domestic political dynamics which have influenced the economic liberalization of tourism. In Syria, tourism has played a remarkably strong role in the political process, and in the government's relationship with social elites and with the broader polity. It is this pivotal role of tourism in public policy, and especially in the process of regime maintenance and consolidation, which is so important in Syria compared with other Arab or Mediterranean states.

Under Asad, the regime has gradually replaced some of its traditional reliance on popular legitimacy with a more focused relationship with major private sector elites. The way that this has occurred in the economic liberalization programme, in response to the economic imperatives of the 1980s and 1990s, is a particular case in point. Liberalization has created new economic elites, which the regime has nurtured and tightly controlled lest their economic power translate into greater political power.

The political methods of Asad have not changed dramatically since his ascension to power, even though the structure of political relations has undergone some substantial changes. In the 1990s, Asad has continued to rule Syria, and to conduct elite political relations, by the same methods as in previous decades; through patron–client relations, informal decision-making processes, and in the broader political system through a combination of co-optation and repression. Despite this, however, some aspects of Syrian politics have changed. Although elite political relationships remain patrimonial and informal, new members have entered the patrimonial web, most importantly the wealthy business elites. These individuals, along with a growing private sector middle class and a revitalizing bourgeoisie, have expanded as a result of the impact of economic liberalization on society, and more recently because of the state's need for support in its economic development agenda.

In the immediate future, the Syrian tourism sector offers considerable potential as a path towards economic development. With a rapidly rising population, and economic stagnation in other economic sectors, a focus on new sectors such as tourism is becoming increasingly necessary for both the Syrian economy and, by extension, for the continued legitimacy of the government. How well the government can manage the development and liberalization of tourism, and the ultimate results of an increasingly influential private sector and middle class, will determine the future of tourism in Syria, and of the country's economic and social development more generally.

Notes

1. This is an amended and enhanced version of an earlier paper which appeared as 'The Political Economy of Tourism in Syria: State, Society, and Economic Liberalization', *Arab Studies Quarterly*, Vol. 19, No. 2, (Spring 1997), pp. 57–73.
2. Sylvia Pölling, 'Investment Law No. 10: Which Future for the Private Sector?' in Eberhard Kienle (ed.), *Contemporary Syria: Liberalization between Cold War and Cold Peace*. London: British Academic Press, 1994, p. 16.
3. See 'wazir al-siyyaha al-suriyya Abu al-Shamat: khuttatuna istaqbal 3,5 milayiin sa'ih al-'am 2000', *al-Iqtisad wa al-A'amal*, Special Issue, Year 16, March 1995, pp. 59–62.

4. On Ba'ath agrarian policies and the emphasis on rural development, see Raymond Hinnebusch, *Peasant and Bureaucracy in Ba'thist Syria: The Political Economy of Rural Development.* Boulder: Westview Press, 1989; and Volker Perthes, *The Political Economy of Syria under Asad.* London: I. B. Taurus, 1995, Chapter 3.
5. 'wazir al-siyyaha al-suriyya Abu al-Shamat', pp. 61–2.
6. One study highlighted this in the case of Egypt, and many of the Egyptian characteristics can be equally applied to Syria or other developing states. See Tim Mitchell, 'Worlds Apart: An Egyptian Village and the International Tourism Industry', *Middle East Report,* no. 196, (September–October 1995), pp. 8–11 and 23. See especially the subsection entitled 'Enclave Tourism', pp. 9–10.
7. See Miyoko Kuroda, 'Economic Liberalization and the Suq in Syria', in Tim Niblock and Emma Murphy (eds), *Economic and Political Liberalization in the Middle East.* London: British Academic Press, 1993, pp. 203–13.
8. Alan George, 'No Going Back', *The Middle East,* November 1996, p. 20.
9. See the table in 'wazir al-siyaha al-suriyy, Dawud: suriyaa qadwa fi al-siyaha al-bayaniyya al-arabiyya', *al-Iqtisad wa al-A'amal,* Special Issue, Year 20, May 1998, p. 70. Further, between 1986 and 1993 the growth rate of Syria's tourism sector was 19.6 per cent per annum. See E. Riordan *et al.,* 'The World Economy and Implications for the MENA Region', *Economic Research Forum Working Paper Series,* Working Paper Number 9519. Cairo: Economic Research Forum, 1995, p. 23, Table 9.
10. Raymond Hinnebusch, 'Syria,' in Niblock and Murphy, *Economic and Political Liberalization in the Middle East,* p. 193.
11. The details which follow are taken from Pölling, 'Investment Law No. 10', p. 15; Sylvia Pölling, 'The Role of the Private Sector in the Syrian Economy: "Law No. 10 for the Encouragement of Investment" Against the Background of Ongoing Economic Liberalization and Market Deregulation', Speech to the conference on Economic and Political Change in Syria, School of Oriental and African Studies, London, 27–28 May, 1993; and from author's interviews in Damascus, May 1996.
12. Perthes, *The Political Economy of Syria Under Asad,* p. 53.
13. Moshe Ma'oz, *Asad: The Sphinx of Damascus. A Political Biography.* New York: Grove Weidenfeld, 1988, p. 79.
14. On Syria's poor economic performance in the 1980s, see Volker Perthes, 'The Syrian Economy in the 1980s', *Middle East Journal,* vol. 46, no. 1, (Winter 1992), pp. 37–58.
15. On the crisis of 1986 and the origins of the second infitah, see Nabil Sukkar, 'The Crisis of 1986 and Syria's Plan for Reform', in Kienle, *Contemporary Syria,* pp. 26–43; and Steven Haydemann, 'The Political Logic of Economic Rationality', in Henri Barkey, (ed.), *The Politics of Economic Reform in the Middle East.* New York: St. Martin's Press, 1992, pp. 11–39.
16. Author's interviews in Damascus, June 1996.
17. 'wazir al-siyyaha al-suriyya Abu al-Shamat', pp. 61–2.
18. Christian Schneider-Sickert and Andrew J. Jeffreys for Oxford Business Guides, *The Oxford Business Guide: Syrian Arab Republic, 1995–6.* Surrey: Oxford Business Guide Publications, 1995, p. 35.
19. al-Jamhuriyya al-arabiyya al-suriyya (Wizara al-Iqtisad wa al-Tijara al-Kharajiyya), al-Qanun Raqm 10 li-Tashji' al-Istithmar (Dimashq, 1991), [The Syrian Arab Republic, Ministry of Economy and Foreign Trade, Law Number 10 for the Encouraging of Investment (Damascus, 1991)]. For an analysis of Law Number 10, see Pölling, 'Investment Law No. 10', pp. 19–23.
20. At the 'neighbouring countries rate' of S[£]42 to the US dollar, S[£]10 million equates to a minimum investment of US$238,095.

21. Pölling, 'Investment Law No. 10', p. 21.
22. See, for example, the statistics provided in 'wazir al-siyyaha al-suriyy dawood: suriyya qadwa fi al-siyaha al-bayaniyya al-arabiyya' ('Syrian Minister of Tourism Dawood: Syria is a Positive Example of Arab Regional Tourism', *al-Iqtisad wa al-A'amal,* Special Issue 'GATT and Arab Tourism', Year 20, May 1998, pp. 70–72 (especially the table on p. 72).
23. Author's interviews in Damascus, June 1996.
24. Under such schemes, an individual establishes a firm, buys a number of cars under the provisions of Law Number 10, and then leases them to relatives and friends, thereby evading the otherwise enormous taxes and duties on importing the vehicles. Author's interviews in Damascus, May and June 1996.
25. Pölling, 'Investment Law No. 10', p. 21.
26. The idea of 'symbiosis' is partly based on John Waterbury, *Exposed to Innumerable Delusions: Public Enterprise and State Power in Egypt, India, Mexico, and Turkey.* Cambridge: Cambridge University Press, 1993, especially Chapter 8.
27. The following information on Sa'ib Nahas's career and business interests is drawn from 'Rajul al-Mu'sasat Saib Nahas: al-Siyaha tahtaju ila mal wa rijal . . . wa thiqa' ['Man of Business Enterprises Sa'ib Nahas: Tourism Needs Funds and Men . . . and Confidence'], *al-Iqtisad wa al-A'amal,* Special Issue 'Tourism and Prospective Peace', 18, May 1996, pp. 50–2; a detailed brochure issued by Nahas Enterprises which outlines the history and activities of its firms; and from author's interviews in Damascus, May 1996.
28. Pölling, 'Investment Law No. 10', pp. 16–17.
29. 'al-Sham li-al-finadiq: al-Sharaka al-Funduqiyya al-Ra'ida fi al-a'alim al-Arabi' ['Cham Hotels: The Leading Hotel Company in the Arab World'], *al-Iqtisad wa al-A'amal,* Special Issue 'Tourism Without Borders', Year 16, March 1995, pp. 39–42.
30. Pölling, 'Investment Law No. 10', p. 17.
31. Pölling, 'Investment Law No. 10', p. 17.
32. Author's interviews in Damascus, June 1996.
33. Author's interviews in Damascus, May 1996.
34. Volker Perthes, 'The Bourgeoisie and the Ba'th: A Look at Syria's Upper Class', *Middle East Report,* no. 170, (May–June 1991), p. 31.
35. Waterbury, *Exposed to Innumerable Delusions,* Chapter 8.
36. Ibid., p. 212.
37. See Pölling, 'Investment Law No. 10', and also 'Suriyya: Ihtimam A'ali' ['Syria: Global Concerns'] in 'Finadiq al-Sharq al-Awsat: al-Khassa al-Akbar li-Safar al-A'amal' ['Middle East Hotels: The Greatest Contribution is to Business Travel'], *al-Iqtisad wa al-A'amal,* Special Issue 'Tourism Without Borders', 16, March 1995, p. 30.
38. Author's interviews in Damascus, May and June 1996.
39. Perthes, 'The Bourgeoisie and the Ba'th', pp. 35–6.
40. Author's interviews in Damascus, May 1996. These labour laws mean that, with the exception of some joint-sector enterprises, permanent employees can only be dismissed in very extreme cases such as bankruptcy of the firm or because of theft by the employee.
41. Perthes, *The Political Economy of Syria under Asad,* p. 101.
42. Author's interviews in Damascus and at regional tourism sites in Syria, May and June 1996.
43. Perthes, *The Political Economy of Syria under Asad,* p. 236.
44. Author's interviews in Damascus, May and June 1996.
45. Alan Richards and John Waterbury, *A Political Economy of the Middle East: State, Class, and Economic Development.* Boulder: Westview Press, 1990, p. 178.
46. 'wazir al-siyyaha al-suriyya Abu al-Shamat', p. 61.

10 Tourism in Lebanon

Transformation and prospects

Hyam Mallat

Introduction

Although Lebanon's small area (10,400 km²) forces it to be incorporated with some of the smaller nations, its climate and geography have made it quite favourable to the development of human activity as well as one of the most ancient populated countries in the Middle East.

One of the interesting things that can be noted as to the general layout of the country is the way that harmony has been maintained; the absence of large fluid networks, the mountains have been inhabited and cultivated for several centuries, the roads wind around the hills finding people everywhere that they decide to settle; Lebanon's legendary hydraulic wealth has earned it the nickname of the water tower of the Middle East; the vegetation, which has been cropped down a bit since its rich and grandiose past, remains relatively green and wooded.

The Lebanese climate is characterized by two juxtaposed regions: the Mediterranean-type climate along the coast and the continental-type climate in the Bekaa Valley, behind the first range of mountains.

Days of precipitation are so rare as to impede human activity. In fact, there are 75 to 80 days of rain in Beirut, 78 in Ksara, 55 in Baalbeck and the southern part of the country.

This low number of days of precipitation is, however, at the origin of its hydraulic wealth: the average yearly rainfall is estimated at 850 mm and reached as high as 1050 mm in 1967.

As for the temperature, the temperate maritime air masses coming from the west cool the summer heat and warm the cold winter. However, the mountains stop their progress in the high plateau of the Bekaa.

The coast has only frozen twice in 60 years ($-1°$ in 1907, $0°$ in 1942) and the temperature during the summer peaks at $35°$ only about once a year.

The influence of the sea is equally obvious in the Bekaa region where it does not freeze often: 24 times a year in Ksara, 47 in Rayak, and only 11 in Fakhe.

In general,

It would not be rash to say that Lebanon's climate is one of the most favourable to humans. In fact, pure sunlight in abundance, and the circulation from the west blows in a sea breeze, which calms the rigor of the summer and winter. The human organism that lives in this climate does not have to exhaust itself fighting its surroundings, and is not dispensed for as many efforts at adapting to hot and cold without which it would wilt. The Lebanese have the advantage of being able to change air without leaving their country. If the coast was a little less humid and the interior a little less dry, and if the Khamsin never returned all of man's wishes would be fulfilled.

All of these things have historically made Lebanon a very densely populated political, commercial, and economic hub in the Middle East; however, one cannot forget that this situation is due to the adventurous spirit that have pushed the Lebanese to conquer external markets.

The sea has traditionally been an annoyance and only the political complexities – linked to the long stay of the Ottoman Empire and the creation of colonial empires, then the sovereign states – forced the Lebanese to exploit this area and to consider it the centre of economic activity.

So, men have been able to respond to adversity and leave their mark on the landscape and have made it a centre for habitat and work in spite of the physical elements and historic constraints. This is why, in the beginning of the twentieth century, Lebanon with all its natural resources (mountains, coasts, archaeological ruins, environment) was ready to profit from the new tourism that was moving towards the Middle East.

However, this evolution in the economic and social activity of Lebanon has occurred due to diversification of objectives which leads to disparity in the overall economic and social growth.

Thus it would be futile to distinguish tourism from the other major sectors in Lebanon since it is the economic, social and urban areas that command the Lebanese society. It is the objective and global perspective – its organization in the contemporary world requires a clear understanding of the positive and negative factors and how they influence the whole economic and social structure.

This means that tourism is not inseparable from the societal evolution in Lebanon and in the rest of the Mediterranean world which has experienced deep unrest over the past decades. The resulting repercussions on citizens' lives, and their importance, still remain to be studied, especially in relation to the attitudes and motives of the citizens.

This is why the tourism statistics cannot be fully appreciated without an implicit reference to the vast regional demographic, economic, and financial movements, since group tourism as well as leisure perspectives appear, in reality, as the conquest of the techno-industrial contemporary civilization.

Making the most of an advantageous situation in the Middle East,

Lebanon, with a population of three and a half million inhabitants, offers the following touristic possibilities:

a   A coastline that covers 225 km on the Mediterranean sea where the tourist can take advantage of beach resorts and all that goes along with it.
b   Alpine or mountain tourism occurs in the summer, with the diverse summer centres as well as in the winter with the skiing centres that are scattered all over the mountains (Faraya, The Cedars, Sannin).
c   Business tourism, represented by the financially and economically dynamic Beirut as well as by the diverse investments pertaining to commercial companies' equipment and reconstruction.
d   The presence of about 20 universities and higher level institutes forces a growing number of students to move to Beirut to take advantage of the numerous educational programmes offered by the country.
e   Cultural tourism – this one is without doubt the most revealing because of the lure that it exerts on the tourists and visitors. In fact, Lebanon has several archaeological and historic vestiges going back to antiquity, and each site usually presents several different strata of past civilizations which allow the visitor to visualize the historical evolution that took place over several centuries. Of the many sites which exist, some are considered as part of our global culture by UNESCO and should therefore be stated:

- *Byblos:* invaluable site both historically and archaeologically, there are remains here from the Neolithic period all the way to the end of antiquity: round homes from the stone age and walls, temples, and tombs from the Bronze Age.
- *Saida:* Ancient city of Sidon. This Phoenician city is mentioned in the Bible and by Homer, it is characterized by vestiges from the time of the Temple of Echmoun – god of health. The old village and the Saida port, with their monuments and their buildings from the Middle Ages as well as from the present remain major centres of interest for tourists.
- *Tyre:* This ancient site possesses ruins dating back to the Cananean, Phoenician, Assyro-Babylonian, Roman, and Byzantine periods. This walled space has several religious and civil buildings including: paved roads, aqueducts, monumental passage ways, race tracks, amphitheatre, houses, cemeteries, with enormous tombs, etc. Moreover, the race track of Tyre and its joint buildings have remained beautiful and well conserved.
- *Baalbeck:* The reputation of Baalbeck is well known for the sheer size of its Roman ruins as well as for their architecture. The buildings reflect one of the major periods in Lebanon's history. Since 1955, barring the period from 1975 to 1996, the International Festival in Baalbeck has begun to regain importance and can be

expected to soon become a centre for art and culture in the world.

- *Tripoli:* A city that was founded in ancient times and which is presently characterized by relics of Arabic ottoman monuments and crusaders.
- *Beirut:* Since the downtown area was destroyed during the war, it became the object of archaeological excavations under the aegis of the International scientific community. The research accomplished, as well as the discoveries, will help to place Beirut in a historical context.

Besides these important sites there are also several secondary sites such as Khalde and Jiye rivers, Sarafand, Batroun, Balamand, and so forth.

f   Finally, there is a large group of tourist sites along the coasts, mountains and even in the interior of the country which serve, combined, to attract an increasing number of tourists and visitors. Mentioning just the northern part of the country we have the region of Jounieh, the Harissa mountain, the Zahle region, area of the Cedars of Lebanon mentioned in the Bible, which go back hundreds of years, and the regions bordering the great Cedars.

If it appears as though we have only briefly sketched the importance of tourism in Lebanon it is to be able to show the diversity that exists to attract tourists, visitors, researchers, scholars, journalists, businessmen, investors, etc. Moreover, the preparation of tourism requires, especially in our contemporary societies, an established political policy based on documented facts, a smooth and frictionless relationship between all the interested countries, and a policy that takes an active role in protecting the environment without which tourism in the Mediterranean would cease to be profitable.

Thus, any study will have to take into account the synthetic axes of Lebanese society over the last couple of decades – all of this while predicting points of growth that might in the future be influential as far as the evolution of tourism is concerned. Finally, its capacity and potential will also be analysed.

Synthetic axis of the contemporary Lebanese society

Even though Lebanon's economic evolution over the past decades has been characterized primarily by strong growth until 1975, followed by a standstill from 1975 to 1990 due to the war, this presents certain major characteristics which synthesized as follows:

a   Over the last 30 years Lebanese society has experienced huge demographic growth, accompanied by rapid consumption. This has been slowed down by limited available geographic area due as much to the parcelling of real estate as to the monetary investment in the real estate sector.

b This rapid consumption of geographic space causes a progressive reduction in the amount of agricultural area and a rapid urbanization, concentrating more than 80 per cent of the Lebanese population in the large metropolitan areas along the coast: Beirut, Tripoli, Saida, Tyre, Jounieh, Antelias, etc., or the interior: Zahle, Baalbeck, etc., which also polarizes the suburbs.

c The rapid consumption of space, due as much to urbanization as to the rapid movement to the service sector in Lebanon, forces a response to the consumption usually coming from the outside. This has forced a social transformation in this direction, boosting the standard of living and changes in lifestyle.

Thus, the developments in the tourism sector depend on several interdependent variables and can, in the long run, assure or cripple growth – in the sense that modern tourism is a variable in its manifestation and diversified in the population that practises it.

So, there is an effort being made to control the pace of urbanization to avoid the possibility that too many people move to an area forcing an irreversible consumption of favourable land. This is why a strong tourism industry depends heavily on a strong environmental policy as well as good tourism planning.

Lebanon's economic structure, as it is, would not be able to attain its true potential unless the sociological approach authorized the distinction between the possible choices in order to recognize tourism and provide its protection and its importance beyond and contingency at the moment. It is because of this that we can analyse the present and future variables of tourism and can conclude in favour of a strong policy for tourism, anchored on the objective.

The success of this policy as well as the evolution of tourism depends as much on financial factors, the level of revenue and the purchase power, as on what Lebanese society offers in response to a demand made by a young population. It is therefore necessary to place current tourism, the supply and the various possibilities of the tourism sector to be able to make better predictions for the needs of the future.

So, a study of the present situation in the tourism industry should attempt to respond to certain questions relative to capacity and level of reception because the tourists' parks, in hotels and infrastructures, are linked to the increase of gross domestic product.

The tourism industry in Lebanon has gone through three different phases over the latter part of the twentieth century:

a The first period was from 1950–75 where, due to several local, national and international socioeconomic factors, tourism in Lebanon

Table 10.1 Study of the GDP and tourist receipts 1964–73

Year	GDP (in millions of Lebanese Pounds)	Tourism receipts (in millions of Lebanese Pounds)	% T.R./CIP
1964	3200	203.6	6.36
1965	3623	257.7	7.3
1966	3867	301.9	7.8
1967	3820	240.4	6.8
1968	4273	350.3	8.4
1969	4565	387.7	8.4
1970	4866	430	8.8
1971	5399	580	10.7
1972	6365	610	9.5
1973	8137	880	9.9

1 US$ = 2.50 Lebanese Pounds.

went through a period of rapid growth and the role of tourism in the growth of gross domestic product can be studied as follows:

b   The second period, from 1975–90 is when, because of the war, tourism stopped almost completely. During this period the only visitors from abroad were the reporters and journalists looking for information.

c   The third period started in 1991, with the end of the war allowing Lebanon to progressively regain its place in the world market, at the same time as the Ministry of Tourism, in co-operation with the French government, the World Tourism Organization and the United Nations Development Programme, developed and presented the plan for reconstruction and development of Lebanon's tourism in 1996.

We will, therefore, focus our attention on the elements of this third phase to present the current situation of the tourism industry in Lebanon.

State of tourism in Lebanon

Any idealistic approach to tourism must start in a multidimensional study of the elements and factors of a sector in rapid expansion but with certain problems linked to the structure of the hotel industry, tourist movement and the general economy of the tourist industry.

Role of the hotel industry

The general tendency for the hotel industry is to respond to tourist movements polarized into international, regional, and local clientele.

Table 10.2 Distribution of the number of hotels by Mohafazat and by category

Category Mohafazat	International	4-star	3-star	2-star	1-star	Total number	%
Beirut	1	13	17	5	2	38	16.2
Mount Lebanon	3	17	37	56	28	141	60.0
North Lebanon	–	3	6	15	14	40	17.0
South Lebanon	–	–	–	1	3	4	1.7
Bekaa	1	–	4	2	5	12	5.1
Number	5	33	66	79	52	235	100.0
Total %	2.1	14.1	28.1	33.6	22.1	100.0	

Source: 1995 Lebanese Department of Tourism Hotel Guide.

Table 10.3 Distribution of the number of hotel rooms by Mohafazat and by category

Category Mohafazat	International	4-star	3-star	2-star	1-star	Total number	%
Beirut	153	1433	943	200	43	2772	26.9
Mount Lebanon	198	1044	1760	2004	684	5690	55.2
North Lebanon	–	103	227	488	267	1085	10.5
South Lebanon	–	–	–	263	–	263	2.6
Bekaa	80	30	180	79	130	499	4.8
Number	431	2610	3100	3034	1124	10,309	100.0
Total %	4.2	25.3	30.2	29.4	10.9	100.0	

Source: 1995 Lebanese Department of Tourism Hotel Guide.

The general state of the hotels in Lebanon, as it is currently, reflects the present trend and growth of tourism. This is why, besides the global statistics approach, another qualitative analysis will help clarify the situation of the hotels in Beirut and in the other areas of the country.

Over all, as can be deduced from the tables above in Lebanon in 1995 there were 235 hotels with 10,309 rooms distributed in the following way:

The hotel industry constitutes one of the major components of any tourist policy; beside its material aspect, it is undeniable evidence of the effort being made by the society in order to improve tourism – because tourists require and demand that a certain level of comfort be provided. The actual role of hotels shows the clear predominance of two- and three-star hotels that are scattered over several regions – the only exceptions are the large hotels constructed in Beirut and other main cities in the country.

Moreover, according to the studies done for the reconstruction plan and tourism development in Lebanon, the level of occupation can be estimated at 55 per cent. This is about 2 million nights spent in hotels. The clientele is about 45 per cent European, 35 per cent from Arabic countries, 15 per cent from Lebanon and 7 per cent from other places – about 75 per cent of these are businessmen.

Table 10.4 Distribution of hotels and capacity by category

	Number of hotels	Percentage	Number of rooms	Percentage
4-star hotels	33	14.1	2610	25.3
3-star hotels	66	28.1	3100	30.2
2-star hotels	79	33.6	3034	29.4
1-star hotels	52	22.1	1124	10.9
Total	235	100	10,309	100

Situation in the tourism industry

The statistics regularly updated before 1975 by the tourism services in Lebanon constitutes an excellent indicator of the tourism industry's situation. During the period of 1975–90, statistics were hard to come by because of the war; however, after 1990 statistics were once again published and were considered as chief indicators of a dynamic tourist industry. This is why we have judged it necessary to present the figures prior to 1975 and then show more up to date information from the WTO Commission for the Middle East.

Tourism prior to 1975

Besides the statistics that were regularly published by the National Tourism Council during this period, we must also examine tourism taking into account the structure of its clientele, their expenditure, and the seasonality of the tourist activity. As far as the structure of the clientele, one can argue that Lebanon has developed a tourists industry from a wide variety of people, as much Arab as occidental. It is with this in mind that one should consider the following table (Table 10.5). It shows, besides the obvious fluctuations, that there is a permanent clientele that has some differences, including taste, goals, customs, but who have all been drawn by the Lebanese culture.

The period between 1991 and 1997

With the end of the war in 1990, tourism once again began to return to the country, as shown in the report by the World Tourism Organization entitled 'Tourism Market Trends – Expanded Middle East 1986–1996'.

Commenting on these figures, the aforementioned article stated:

> If we look at the individual destinations of the Expanded Middle East (region), we notice that Lebanon generated the highest receipts per arrival, estimated at US$1702. The next destinations with average receipts per arrival of more than US$1000 were Kuwait (US$1453), Israel (US$1341) and Syria (US$1664).

154   *Eastern Mediterranean shores*

Table 10.5 Evolution of income from strangers by nationality 1951–72

Year	Non-Arabs	Arabs except Syrians	Everybody else except Syrians	Syrians
A. The expansion 1951–55				
1951	51,649	37,763	89,412	37,891
1952	76,817	43,736	120,553	95,885
1953	74,425	49,537	123,762	160,833
1954	83,030	69,203	151,230	544,008
1955	98,790	80,338	179,128	722,336
Annual rate of growth	+14.75%	+21.6%	+17.6%	–
B. The drop				
1956	94,255	96,560	190,815	668,339
1957	76,316	104,182	180,498	369,095
1958	59,140	38,992	98,132	125,913
C. Expansion 1959–66				
1959	92,600	80,858	173,458	179,626
1960	127,088	105,995	233,083	(300,000est.)
1961	153,295	141,011	294,306	233,035
1962	186,992	144,632	331,623	294,204
1963	195,442	194,103	389,545	340,338
1964	260,349	211,827	472,176	451,265
1965	320,571	279,937	600,508	619,862
1966	360,833	331,351	692,184	812,259
Annual rate of growth	+19.6%	+20.3%	+19.9%	+21.4%
D. Change 1967–72				
1967	268,208	247,020	515,228	702,891
1968	283,456	426,545	710,001	790,510
1969	317,379	459,756	777,135	810,050
1970	288,097	534,250	822,347	863,852
1971	396,601	619,171	1,015,772	1,241,625
1972	470,973	577,186	1,048,159	1,232,903
Annual rate of growth	+11.8%	+15.2%	+13.5%	+8.2%
1973	409,294	552,000	961,294	1,900,000

Table 10.6 Tourist arrivals

Year	Tourist arrivals	Percentage rise
1992	178,000	
1993	266,000	49.44
1994	335,000	25.94
1995	410,000	22.39
1996	420,000	2.44

Table 10.7 Tourists receipts (US$)

Year	Tourists receipts	Percentage rise
1993	600	42.86
1994	672	12.00
1995	710	5.65
1996	715	0.70

They go on to say:

> International tourists arrival to Lebanon amounted to 420,000 in 1996 and international tourism receipts to US $715 million. These amounts represent an average annual growth rate of 16.5 per cent for arrivals and 6 per cent for receipts during the period 1993–1996. The latest data available indicates a sustainable growth of both arrivals and receipts since 1992. Lebanon enjoys the highest receipts per tourists arrival of all EME destinations.... It is expected that in 1997 the number of international tourists arrival will reach half a million.[1]

We can better understand the need to study a plan for the reconstruction and the development of tourism in Lebanon, by realizing the diverse objectives linked to the reconstruction and implementation of a general plan for the development of the tourism industry.

Future outlook

Modern tourism has become the pre-eminent international industry. To better predict future possibilities in Lebanon, we should consider its relation to the Mediterranean region. In most of these countries tourism is accepted because of its triple effect on the economy: source of a strong currency, area for new jobs, and town planning and environmental policy.

This is why the future of tourism in Lebanon depends on its ability to attract clientele, taking advantage of the countries endowment in touristic infrastructure. In this way the plan for reconstruction and development in 1996 predicts the following things:

1 Adopt a tourist development policy based on two axes, the first a product-market axis and the other a spatial arrangement axis. In the first case, it would act according to the recommendations of the study, and place Lebanese tourism in the areas where it has an advantage such as business tourism, summer and winter tourism, regional cultural tourism, and interior tourism. In the second case, we see that

Lebanon is a country with an old civilization, and where the archaeo-logical ruins and the sites present evidence dating back to antiquity, some Lebanese cities have their tradition well anchored in history, such as Sidon, Tyre, Byblos, Tripoli, Baalbeck and many others. Regional schemes should be made to outline these areas to preserve the natural environment as well as the culture and history of the country in order to be able to protect the coast which covers more than 200 kms.

2   The established plan predicts orienting the development in a way to maximize its contribution to a growing economy while still preserving and reinforcing the specifics of a liberal Lebanese economy.

3   Get those in public office to attach a great deal of importance to their efforts to promote the tourism sector and in encouraging investment and participation from the private operators.

4   Development of qualified personnel for all the areas of tourism through professional learning centres, of which several already exist and have proven their worth.

5   Develop a systematic policy for editing, promoting and marketing tourist products and assuring a qualitative improvement of these products.

Conclusion

To be able to situate in a concrete manner the position of tourism in ref-erence to the society and the Lebanese economy in general, it is necessary to describe the actions over medium and long-term periods.

Actions over a short- and medium-term

The urgent choices, options, and possibilities of tourism:

a   The destiny of Beirut will develop in relation to the tendency for luxu-rious and business tourism to concentrate. The logical option would be to promote the evolution of this tendency especially when the growth is presently noticeable.

b   Control and arrangement of the coastline. As dependent as it is on a generally recognizant policy, it also requires adequate decisions – in much the same way as the hotels in the mountains or even along the coast.

c   The conservation of natural resources and beaches requires an appro-priate policy of arrangement responding to a demand that will increase over the next few years.

d   Reinforce the human expansion – of which several infrastructures have already been placed. These have to gain a certain importance not only by education but also by letting workers in the tourism sector know that tourism in general depends on the ability for the local society to welcome them.

e   Start a series of actions to renovate the hotels outside of Beirut, based on studies which would take into account the social cost and the economic cost of the projects. In this way the present situation of the hotels demands that one take into consideration several factors, the establishment of hotels and the periods of exploitation.

f   Inaugurate certain actions of which certain ones have already been studied because it is important to note that in a country like Lebanon, where a liberal economy rules the spirits as much as the investments, only the success of certain centres can prove the role of the state and the administration on a socioeconomic level. It is in vain to think that everything should be organized because it is more important to succeed in a few strategic operations (profitable) which will have beneficial effects on other sectors.

g   Reinforce the Ministry of Tourism's administrative structure to respond quickly and effectively to the requirements of tourism. When the professional formation and control of the tourism industry accelerate at the same time, it makes for a more flexible and efficient arrangement.

The first action must be to make an inventory of the information, the publicity and cultural changes that are possible for which they would develop an adequate tourism promotion.

Long-term actions

1   To promote an environmental code – because tourism, due to the weakness of the capital requires studied and equitable decisions.
2   Exhaustive, frank, and scientific studies, presented in publications setting the possibilities in this domain.
3   Think up certain, original and ingenious formulas to stimulate an influx of tourists.
4   Execute studies on how the urban area can be changed to promote, not only local tourism, but also international tourism.

Lebanese tourism, after having suffered certain negative consequences associated with the war 1975–90, regained its place in the national economy. Here, Lebanon has become a country which is quite dependent on tourism.

In fact, the interaction of the tourism policy with the various sectors, principally the environment, requires certain well-determined plans to be executed for the enterprise to succeed.

This is why, besides the statistics presented, which are still limited – since the base is still under reconstruction – the strength of Lebanese tourism over the last few years shows Lebanon's place in the tourism market of the Middle East and the importance of co-operation amongst all the countries in the Mediterranean to be able to arrange their human resources and the materials in the most efficient way.

Note

1. WTO. 1997. *WTO Commission in the Middle East Tourism Market Trends 1986–1996*, p. 34.

11 Acquired tourism deficiency syndrome

Planning and developing tourism in Israel

Yoel Mansfeld

Introduction

The impact of tourism development on the social, economic and physical environments of a given community has been at the centre of academic tourism research during the last three decades. A wide range of case studies have looked into the various positive and negative impacts, in an attempt to identify and explain the causal relations between development and impact characteristics. In doing so, these studies have tried to demonstrate how positive results should be enhanced, and how negative results could be avoided. Although many such case studies were policy- and planning-oriented, many countries and tourism regions opted for ignoring the lessons of uncontrolled rapid tourism development. In most cases, such an 'ostrich attitude' led to a combination of negative economic, social and environmental damages, which reflected on the future of the local, regional and/or national sustainability of a tourism system and its host community. Thus, mistakes have been unintentionally and/or unavoidably repeated while trying to exploit tourism resources. This repeated sequence is termed here 'acquired tourism deficiency syndrome' or ATDS. This chapter suggests that Israel, among other Mediterranean Basin countries, has been a victim of such a syndrome.

Since the mid-1960s, many of the developed as well as developing, countries around the world embarked on tourism development processes, seeking quick solutions to various acute social and economic problems. Tourism has been regarded ever since as a potential remedy for problems such as foreign debts, negative balance of payments, severe unemployment rates, stagnated exporting sectors, undiversified economy, extreme social imbalances between people residing in core versus peripheral regions, etc. In this respect, Mediterranean countries have followed this pattern. Since the beginning of the 1960s, tourism development along the Mediterranean basin has been a continuing process (Jenner and Smith 1993; Pearce 1989). This process has been very much in line with Butler's (1980), Doxey's (1975) and Miossec's (1976) evolutionary models of tourism development. While some countries, such as France and Spain, embarked on this process as early as the 1960s, others, such as Turkey and Israel, are still busy saturating empty strips of their coastline with tourism infra- and superstructures. Facing social, economic and political needs, decision-makers and planners still find themselves reluctant to learn from other countries' past experience. Thus, Portugal did not change its tourism development strategy following the saturation and other negative impacts that had emerged from the uncontrolled and over-development of the Spanish coastline. The same goes for Turkey, which ignored the results of rapid tourism development in Greece and, in the mid-1980s, went into a massive development process of its pristine coast (Aktas 1995). By turning a blind eye and adopting such a development approach, many Mediterranean countries are now facing a complex array of problems and negative impacts (Hermans 1981; Pearce 1989). Moreover, while tourism has been developed far beyond social and environmental carrying capacity thresholds, the real beneficiaries of tourism have not always been the local communities – localities that have given up so much of their culture, self identity and their leisure and recreational resource (Butler 1980; King 1994).

Israel, unfortunately, found itself in a similar trap. Being a world centre for the three monotheistic religions, the Israeli tourism product is somewhat different from that of most Mediterranean counties (Kliger and Shmueli 1997). Thus, a large proportion of its inbound tourist flow is based on two distinctive market segments: pilgrimage and cultural tourism. In the past five decades, Israel has been facing a constant need to diversify its economy and achieve economic growth, while overcoming the problems of negative balance of payments. It also needed to face a major undertaking – social and economic absorption of massive immigration waves. Tourism development could have been a major contribution towards achieving this end. However, despite its strong tourism potential, this country is still far from an optimal exploitation of its tourism resources. Some of the reasons for this lack of effective exploitation lie in factors beyond the local tourism industry's control. The on-going security situation in the Middle East, for example, occasionally hampers international tourist flows (Mansfeld 1994). The low-cost tourism products in Egypt and Jordan also impinge upon Israel's ability to compete in the eastern Mediterranean region. However, the long term growth trend of inbound tourist flows to Israel proves that despite these constraining circumstances Israel still has a strong attraction potential (see Figure 11.1). Therefore, the question is why has this country failed to achieve further growth of its tourism yields? The hypothesis put forward in this chapter is that this failure has been a result of Israel's shortsighted tourism development policy, which has, unfortunately, initiated other countries' mistakes.

Thus, using the Israeli tourism industry as a case, the aim of this chapter is twofold:

Acquired tourism deficiency syndrome 161

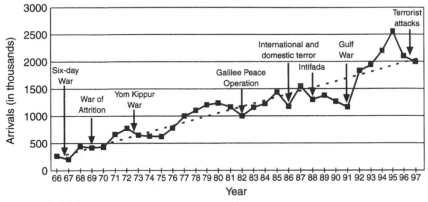

Source: Israel Bureau of Statistics, Selected Years

Figure 11.1 Tourist arrivals in Israel: long-term trend and major security crises 1966–97.

1 To characterize the evolving Israeli tourist landscape and the factors that have shaped it.
2 To unveil the reasons why Israel has become a victim of ATDS despite being capable of learning from other Mediterranean countries' bad experience.

In order to illustrate the various dimensions of Israel's ATDS, this chapter will use the case of tourism development in Eilat – Israel's southernmost winter-sun resort.

Tourism development in Israel – a historical overview

The incubation period

Tourism started its initial stage immediately after the establishment of the state in 1948. At that time, it was hardly regarded as an economic activity but rather as means to expose the country to potentially new immigrants from the Western and more affluent world (Blizovsky 1973; Nathan 1973). The period between 1948 and 1964 was dedicated to the development and organization of tourism administration that could facilitate tourism development. As early as 1949, just one year after the establishment of the state, Israel started to deal with procedures to accommodate Christian pilgrims visiting the Holy Land. This involved the expansion of roads, hotels and hostels, and the improvement and rehabilitation of religious sites. The first governmental tourism body was established in 1952 as a Tourism Centre within the Ministry of Commerce. It had three main responsibilities: promotion, investment co-ordination and distribution of tourism information through its offices in Paris and London.

During the 1950s, the Israeli government established two governmental corporations to co-ordinate and promote further development. One was the Government Tourism Corporation (in 1955), which has been responsible for initiating and co-ordinating the development of tourism infrastructure ever since. The other is the Tourist Industry Development Corporation (in 1957) which has been responsible for making loans and other financial assistance to tourist enterprises interested in tourism investment in Israel (Nathan 1973).

In 1964 the Israeli government decided to upgrade its tourism administration and established an independent tourism ministry. This decision opened a new era in Israel's effort to exploit its tourism resources, enhanced the regulation of the Israeli tourism industry and formed a basis for control and planned tourism development process. In retrospect, this has not been always the case and the regulation, as will be elaborated later on in this chapter, has been confined mainly to issues of training, standards of hotel operation and the travel agencies sector in Israel (Nathan 1973).

In search of a 'takeoff' point

The territories annexed by Israel following the Six Day War in 1967 opened new opportunities for the Israeli tourism industry. Tourism attractions of Jerusalem, Hebron, Beit Lehem and Jericho became part and parcel of Israel's enlarging tourist product. The Israeli government realized that in order to take fast and comprehensive advantage of its new attractions, it must convince transnational corporations and local entrepreneurs to invest their money in Israeli tourism (Shaari 1973). To assure a positive business climate, as in many other developing countries, the Israeli government committed itself during the 'takeoff period' (1967–72) to the following:

a to cover the infrastructural work (land and utilities) needed to facilitate further tourism development by the private sector in those regions prioritized by the government (see Figure 11.2), and
b to introduce a comprehensive incentive programme to reduce the risk taken by the private sector. This programme included grants and/or long-term loans, the possibility to purchase the state-owned land without tender and with long-term loans, and state and local tax exemptions (Belizovski 1973).

This comprehensive incentive package stimulated the development of new hotels and facilitated the accommodation of the growing number of tourists arriving in Israel during the 'takeoff period'. For the first time, a regional approach to tourism development was introduced and incentives were granted based on prioritizing regions representing a higher level of attractiveness for international tourists (Shaari 1973). This was part of a development policy based on the following principles:

- intensified development and expansion of existing resort and recreation centres;
- establishment of new tourism regions;
- encouragement of year-round travel flows to Israel;
- development of new tourism products such as desert tourism and health tourism.

It should be clarified that, although all parties involved in tourism development at that time sought this development policy, it was not based on a tourism development master plan on national and/or regional levels. Lack of such a plan opened the gates for potential and actual uncontrolled tourism development.

Towards controlling the tourism development process: planning versus real world dynamics

In 1972 the government decided to prepare a national plan for tourism development in Israel until the year 1985 (Ephratt 1993). For the first time the Ministry of Tourism decided to invest in order to effectively exploit the state's tourism resources. Effectiveness at that time was perceived by the Ministry as increasing the number of tourist arrivals regardless of regions' and/or attractions' carrying capacity thresholds (Shaari 1973). The plan had three main sections:

- forecasting of future demand for both domestic and international tourism;
- an economic plan based heavily on investment guidelines and not on marketing plans; and
- a physical plan defining location priorities and type of infrastructure to be developed.

The preparation of this plan involved various government ministries, and the Ministry of Tourism co-ordinated the teamwork. The plan was submitted to the government for approval only in 1976. While preparing a master plan, which was supposed to provide guidelines for tourism development in the entire country, most of the actual development until 1976 took place in Tel Aviv and Jerusalem (see Figure 11.2). Thus, by the mid 1970s, most tourism services, accommodation facilities and developed tourism attractions in Israel were located in the most populated part of the country. The evolving tourism array at that time failed to integrate high tourism potential with the actual development of tourism growth centres (Litersdorff and Goldenberg 1976). For example, apart from Tel

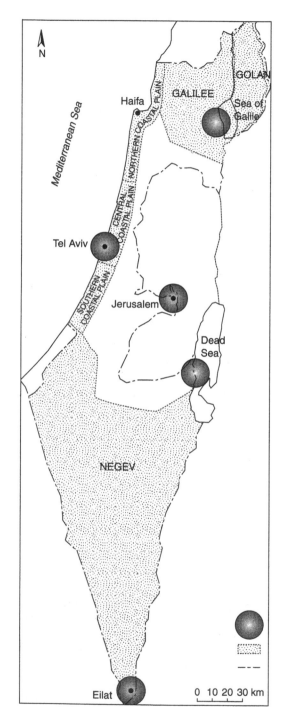

Figure 11.2 Israel tourism growth centres and main tourist regions.

Aviv's coastal strip, potentially attractive destinations such as the Dead Sea, the Galilee region in the north, the Negev desert and the Judea desert were not developed into fully-fledged tourist regions. The years 1975–86 were characterized by an average growth of 8.8 per cent in tourist arrivals. During this period Israeli tourism proved its ability to adjust itself to fluctuating trends of tourist arrivals (see Figure 1). These were mainly a result of unsteady economic conditions on the global level, as well as turbulent security situations in Israel and the entire Middle East. These trends left their imprint on the structure of the inbound tourist flow to Israel. While the market share of North American tourism to Israel diminished, European inbound flows grew steadily. This trend was accompanied by a decline of Jewish visiting friends and relatives (VFR) visits, the mild growth of Christian pilgrimage and the sharp increase of 'winter sun' tourism to Eilat (Mazor 1987).

During this period, the Ministry of Tourism managed to extend Israeli tourist space by developing other regions besides the traditional urban tourism centres of Jerusalem and Tel Aviv. This was part of an effort to implement the first tourism master plan which advocated the dispersal of further tourism development into less populated peripheral regions. Thus, while Tel Aviv and Jerusalem still attracted 56 per cent of the bed nights, the tourist centres along the Jordan rift, namely Eilat, the Dead Sea, Arad and Tiberias, became substantial peripheral sub-centres of tourism activity. Together, they attracted 32 per cent of the total nights generated by international tourism in 1985 (Mazor 1987). This shift towards peripheral destinations has marked the end of highly concentrated tourism activity which put substantial infrastructural pressure on the two metropolitan centres and did not contribute to the economic and social development of Israel's periphery.

Winds of peace twisted with violence

The uprising of the Palestinians in 1987–92 and the Gulf War in 1991 hampered Israel's further growth in tourist arrivals and tourism infrastructure (see Figure 11.1). However, once the Middle East embarked on a peace process in 1991, the demand for Israel as a tourist destination grew sharply. This caused further development of the Dead Sea area and of Eilat together with the introduction of rural tourism into Kibbutzim and Moshavim to facilitate diversification of their economic base. As many of these settlements are located in peripheral regions of Israel, this process contributed again to population dispersal and to reducing the urbanization process. In 1994, Israel hosted two million tourists and forecasts estimate that by the end of year 2000 three million visitors will arrive. The long-term growth of the tourism industry has been interrupted occasionally during this period by various security situations. The consequent decline in tourist arrivals at such times has helped to sweep the carrying capacity questions under the rug. In other words, there was already, at the beginning of the 1990s, a need to introduce concepts of sustainable development and carrying capacities into the planning process. Evidence of environmental degradation and negative social impacts in the wake of uncontrolled tourism development (e.g., Eilat, Tel Aviv and Tiberias) already existed. However, due to the occasional periods of decline, the ministry and the private sectors were always preoccupied with solving the marketing problems, assuming that as a result of the fluctuating nature of this industry such threshold capacities would not be reached. The truth is that until recently, environmental and social impacts of tourism development as planning and implementation issues have been ignored (Goldenberg 1996). Only the latest tourism development master plan, submitted for government approval in 1996, paved the way to integrated development policies based on the environmental, economic and social considerations.

The evolution of tourism planning in Israel

Tourism planning in Israel started at the beginning of the 1970s on a national scale only. It involved a preparation of a national tourism master plan, which was submitted in 1976. In 1987 a new and updated national plan was introduced. Still, no detailed regional or local-level plans accompanied this manifestation of the country's policy towards tourism development. Regional and local tourism development plans were only introduced in the late 1980s. These plans were not a result of planning based on vertical integrated approach. Instead, they focused on specific local and/or regional needs, disregarding the state's overall policy towards tourism development. When the most recent national tourism master plan was in preparation (1994–96), the already existing regional and local level plans served as imposed constraints. Below is a brief summary illustrating the principles of each national master plan, its advantages and drawbacks.

The 1976 master plan

One of the main drawbacks of this plan was its operational target set for 1985. In such a dynamic industry, and given the turbulent situation in the Middle East, a plan based on ten years operational framework is simply not workable. No matter how accurate the planner's forecasting tools are, in such geopolitical settings one simply cannot come close to the real figures. Therefore, any planning based on such long-term predictions becomes totally irrelevant.

The aim of this plan was defined as:

a   an effort to provide development guidelines that will enable the government to increase its revenues in hard currency; and b   to meet the growing recreational needs of Israel's domestic market (Litersdorff 1976).

The planners had assumed that Israel would enjoy a growing demand for its tourism products based on its unique product mix. The planners' orientation was merely economic. Thus, new tourism growth centres recommended for future development in this plan were selected according to their estimated economic potential. Social, cultural and/or environmental considerations were not incorporated into the planners' strategy. The end result from a physical perspective was a plan advocating two major growth centres – Tel Aviv and Jerusalem. Various sub-centres were also suggested along the Mediterranean, along the Dead Sea, the Rift Valley centres (Tiberias, Arad and Eilat), the medieval town of Akko and finally Zefat in the Upper Galilee (see Figure 11.2). At the end of the day, this master plan did not materialize. The Israeli tourism map remained highly centralized. Tourism infrastructure for the domestic market was hardly developed. Alternatively, the government encouraged the erection of mainly four and five-star hotels confined to the two major metropolitan centres, namely Tel Aviv and Jerusalem. Consequently, profits from tourism and new jobs generated by the further development of this industry were not evenly spread. Moreover, residents in these two metropolitan areas began to realize the environmental and social cost involved in such a spatially-imbalanced development process.

The 1987 second master plan

In the mid-1980s, it was already evident that the 1976 master plan was obsolete. It could be assumed that the lesson would be learned and the target year for the new national tourism master plan would be adjusted to no longer that five years, but this expectation was only partially fulfilled. The development policy in this plan was based on a ten-year supply and demand forecast, yet the operational framework for development was based on a five-year period only. This was based on the assumption that, in a Middle East geopolitical context, it is literally impossible to forecast the demand for tourism services in Israel.

The major problem with this master plan was not its time framework but rather its rejection by the Ministry of Tourism. All along the preparation process, which took three years, there had been a lack of communication between the ministry and the planners. Thus, the end result was a much more applicable master plan but one which is not politically acceptable. The government refused to approve the plan for various reasons. First was lack of a governmental budget to implement this plan. Second, along with maintaining the growth of the major tourism centres in Israel, the plan proposed the decentralization of tourism development and the development of small-scale tourism attractions all over the country. Thus, the plan recommended the development of tourism in Haifa – the capital of the North, in the peripheries of the Golan Heights and in the upper Galilee, in the Western Galilee coastal zone, Judea desert, the Dead Sea, and the coastal zone in-between Haifa and Tel Aviv (Mazor 1987) (see Figure 11.2). In other words, the plan encouraged the advantages involved in further development based on dis-economies of scale rather than the development of the existing attractions in the metropolitan growth centres. The Ministry of Tourism could not accept as a legitimate economic policy the idea of spreading small amounts of incentive grants over too many small-scale tourism development projects. Consequently, the ministry had to supply regional and local governments, potential developers and entrepreneurs with a workable development policy on regional and local levels. The development rationale behind these small-scale plans was not derived from a national plan but from an existing need to extend the capacity of the Israeli tourism industry. Thus, in 1996, when the third and most recent national master plan for the development of tourism was ready for government approval, it was already constrained by development processes that stemmed from those local and regional policies.

The 1996 master plan

The most recent master plan, submitted to the government at the beginning of 1996, has revolutionized the way tourism planning in Israel was treated by various government ministries. For the first time, all government agencies involved in planning on a national level joined together to take an active part in the planning process. Moreover, the multi-disciplinary team of planners sought an integrated planning process. Economic considerations no longer formed the planning rationale and, hence, the emerging planning policy. Alternatively, principles of sustainable development based on control of the physical, social and economic impacts of future tourism development were introduced. The main aim of the plan, as formalized by the planners, was to 'provide guidelines for future development of the tourism and recreation system in Israel. Such a formula should be sensitive to this system's interrelations with other planned systems (such as the economic, social and physical systems) in order to assure further constructive development of this industry' (Goldenberg 1996).

The specific targets of this plan were as follows:

- guaranteeing land reserve for future tourism development;
- the development of highly potential tourism regions;
- functional definition of tourist sites;
- determination of location, spatial arrangement and capacities of all types of accommodation facilities according to anticipated demand and measures carrying capacity;

- definition of major tourism routes;
- definition of the inter-relations between this master plan and other national master plans (Goldenberg 1995).

Based on the history of tourism planning on a national level in Israel, one would reaffirm that the recent tourism master plan is based on revolution-ized planning concepts. But is it a real and substantial change? In the past the acquired deficiency syndrome of the Israeli tourism planning was characterized by faulty forecasting frameworks and overemphasis of the economic component in those plans. The results of such a policy is very clearly illustrated by the example of tourism development that has taken place in Eilat.

Acquired tourism deficiency syndrome – the case of Eilat

General background

Eilat is a small development town located in Israel's southernmost corner, bordering with Egypt and Jordan (see Figure 11.2). Its development after the establishment of the state of Israel was part of a government decision to disperse the Israeli population from the crowded Mediterranean coastal plain to the periphery. Eilat is not just a peripheral development town. On the one hand, its geographical location represents extreme arid settings and, consequently, difficult living conditions. On the other hand, Eilat is a meeting place for the beauty of the north tip of the tropical Red Sea and the surrounding red granite mountains. The environmental qual-ities of such a location and its warm winter weather have made it a highly attractive destination for both domestic and overseas tourists. Yet, since the establishment of the town in 1949 and until the beginning of the 1960s, tourism was not regarded there as a legitimate economic activity. At the beginning of the 1960s, Eilat introduced the tourism sector but only to a very limited extent. During this decade, ten per cent of the workforce was employed in this sector (Karmon 1963).

During the 1970s the State of Israel had good reasons to turn Eilat from a deserted small development town located 'at the end of the world', into a bustling resort town specializing in winter-sun tourism (Azariahu 1993). The government's interest to establish a highly developed tourist sector in this town stemmed from both national and local economic needs. On the national level, Israel realized that tourism resources in Eilat could attract enough European tourists to yield substantial revenues in hard foreign currency and contribute to the diversification of its exporting industries. At that time, all-inclusive tours offering Europeans winter-sun holidays were in high demand. However, the supply of such resorts located within a reasonable 'product range' from Western Europe has been limited. Realizing the lack of intensive competition, and Eilat's competitive environmental qualities, the government was convinced to prioritize tourism development in this town (Achituv 1973).

On the local level, the Israeli government had to deal in the 1970s with a major economic problem, namely, the closure of the Timmna copper mines north of Eilat. This closure made 400 skilled workers redundant and had a destructive influence on the occupational confidence of Eilat's residents. The emigration of skilled workers and their families from Eilat, the high unemployment rates and the emerging negative business climate called for an immediate government action to save this fragile local economy. Tourism seemed at that time to be the appropriate economic remedy. The availability of such unique natural tourism resources encouraged the government to make an explicit commitment to take an active part in this venture. It erected a governmental company, *Eilat Shore Development Corporation,* which has developed the entire infrastructure. This company has also been in charge of selling state plots allocated for the development of hotels, man-made attractions and other tourist services, to both Israeli and overseas investors. In order to create an appropriate business climate and to attract foreign investors, the government assembled together a comprehensive incentive package under the law for the encouragement of capital investments. These incentives included long-term low-interest loans, grants, reduced local rates and panial government financing of marketing campaigns (Blizovski 1973).

The development policy adopted by the government in pursuit of this end has only partially yielded the anticipated results. In fact, since the late 1970s, Eilat is facing a rapid tourism development process. Unfortunately, this process is not accompanied by any strategic economic plan. Ever since, the development policy adopted by the Ministry of Tourism has been based on building as many hotel rooms as possible, under the assumption that Eilat could use thousands of hotel rooms. Further more, the bigger this resort town grew, the more tourists would come and generate increasing income for the benefit of the state's economy and the local residents' standard of living.

What then were the real economic consequences of this policy? In the mid-1970s, the government managed to bring tourism in Eilat to a 'take off' point by attracting substantial foreign investments. The breakthrough came mainly from a British hotelier by the name of David Lewis. Profits made by his Spanish hotel chain, lberotel, were invested in a local hotel chain named Isrotel. Through this company David Lewis erected a few four-star hotels concentrated in a tourist 'ghetto' east of Eilat's residential area. These hotels surround a small marina and are located on or close to the Red Sea waterfront. David Lewis brought to Eilat the concept of a self-contained hotel resort. In such hotels, guests are not expected to go beyond the hotel's main door in order to shop or use other tourist services; the idea is to offer them most tourist and travel services within the hotel compound. Thus, Isrotel's supply of thousands of hotel rooms generated inbound tourist flow to Eilat. Both domestic and international markets created new jobs and foreign currency earnings. The question, though, is to what extent did this rapid development lead to economic success? If we examine the recent trend, generally speaking, one can claim that since 1991 Eilat has managed to 'attract' a growing number of bed/nights. This is clearly a positive trend in terms of the town's economic performance (see Figure 11.3).

However, when the figures on bed/nights are compared, based on nights generated by domestic versus international tourists, a different picture emerges. Figure 11.3 shows that Eilat still caters primarily to the domestic market, which does not contribute to Israel's foreign currency earnings. The Israeli government together with Eilat's Hotel Association has been highly active in promotional and marketing activities in the European-generating markets. However, despite this on-going yearly investment in promotion of around $5 million dollars, Eilat's international bed/night segment has never exceeded 40 per cent (Mansfeld 1996).

Moreover, since the mid-1980s, Eilat has faced the rapid development of its accommodation capacity. Figure 11.4 shows the rate of development of hotel rooms in Eilat since 1985.

According to the Ministry of Tourism and Eilat's local municipality, by the year 2000, Eilat will reach a capacity of 14,000 hotel rooms. This is a reliable estimate based on the number of hotels that are already under construction and others already approved by the local planning authorities. As Table 11.1 shows, Eilat represents rapid growth in room capacity not only in terms of local standards. Thus, the highest percentage of the hotel rooms to be completed in Israel by the year 2000 are under construction in Eilat (see Table 11.1).

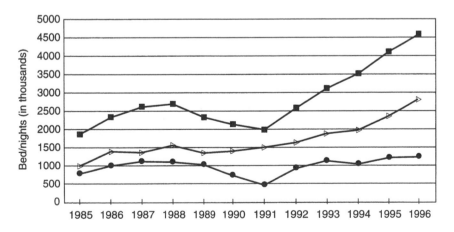

Figure 11.3 Eilat's bed/nights: international versus domestic tourists.

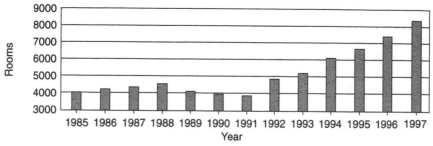

Source: Israel Bureau of Statistics, Selected Years

Figure 11.4 Eilat's accommodation capacity.

Table 11.1 New hotel rooms to be completed by the year 2000

Location	Number of rooms	% of total
Eilat	5700	28
Jerusalem	4000	20
Dead Sea	3700	18
Nazareth-Tiberias	2000	10
Tel Aviv	1100	6
Other locations	3600	18
Total	20,100	100

Source: IBS, *Tourism and Hospitality Services Quarterly*, selected years.

This uncontrolled capacity growth has led local interest groups in recent years, to openly declare their concern. Eilat's mayor also stated his intentions to halt the further development of new hotels (Mansfeld and Czmanski 1995). Yet, in spite of all these declarations, nothing could stop or control this growth. In this sense, Eilat has been repeating the same 'ritual' that took place in many other Mediterranean countries (e.g., Spain, Portugal, Turkey, Cyprus and Greece), that is, first carry out an uncontrolled development process, and then try to deal with the negative consequences that emerge in its wake.

The impacts of rapid tourism development in Eilat

The rapid growth rate of Eilat's accommodation capacity has led to various economic, social and environmental impacts that are discussed below. First, tourism growth generated a number of new jobs. For many years now there has been literally no unemployment in Eilat. This is a unique situation if one compares Eilat with other Israeli peripheral developments. To what extent has this full employment helped to build a strong and stable local community? In the case of Eilat this 'blessing' has negatively affected the local residents. Today, around 70 per cent of Eilat's workforce is either directly or indirectly employed in tourism (Mansfeld 1996). As tourism dominates the local economy, should a crisis (economic, geo-political, etc.) occur, the majority of the locals would have to leave town, as no other employment solutions are available. If one analyses the social profile of the workforce engaged in tourism occupations, the picture is not too encouraging. Most of those working in Eilat's hotels and attractions are low-skilled or unskilled workers. Thus, Eilat lacks a well-stratified local community. Around 38 per cent of this workforce consists of single-parent families, i.e., divorced mothers who moved to Eilat hoping to improve their economic situation. These families place a heavy burden on the local welfare services. Most of those employed in tourism earn minimum wages; they leave upon realizing that tourism is a very low-paying sector. This occupational mobility does not usually take place within Eilat's tourism industry and, therefore, involves migration. On average, people stay in Eilat for two to three years. In such circumstances it is impossible to develop a strong community commitment.

Moreover, the local tourism industry suffers from a very poor occupational image. Consequently, locals (mainly the young generation) who were born and raised in Eilat prefer to move to the Tel Aviv metropolitan area, hoping to find better occupational opportunities. Who then fills this gap? The Israeli government took two steps to ensure a constant flow of necessary workforce, even with increased room capacity. The first was to provide army veterans (aged around 20–22) with financial incentives to move to Eilat and work there for one year. The second was to allow hotels to employ foreign workers (mainly from African and Asian countries). These two measures have helped mitigate the excessive demand for a workforce although at a certain cost. The foreign workers generated various negative impacts. First, many Eilati parents perceived them as a potential threat to the local young generation and felt intimidated by their presence in residential areas. Second, these workers settled for extremely low wages even by Israeli tourism standards. Thus, locals having higher wage expectations felt discriminated against and reluctant to accept jobs in this sector. Another problem with the foreign workers is that the hoteliers covered all their food and accommodation costs. Such benefits allowed them to send all their salaries in hard currency to their families back home. Thus, Eilat's economy did not benefit from their purchasing power while they were contributing to a 'leakage effect' (Gruber 1992).

The locals' attitude towards foreign workers in Eilat is just one aspect of the threat tourism poses to their social carrying capacity. Social antagonism towards tourism in Eilat has been developing in similar fashion to Butler's (1980) 'life cycle' and Doxey's (1975) models. In other words, as the tourism development process in Eilat entered a rapid and uncontrolled stage, locals changed their attitude towards tourism from euphoria through realism to antagonism (Mansfeld and Ginosar 1994). Locals saw

Eilat growing from a small resort town into a saturated agglomeration of hotels concentrated in two separate tourist ghettos on the beach. On the one hand, they realized that the limited wealth generated by tourism is not evenly spread within the socioeconomic mix of the local community. On the other hand, the fact that most available jobs in Eilat are generated directly or indirectly by tourism made them aware that their future economic situation is not that promising. As early as the beginning of the 1990s, social stress related to tourism in Eilat became empirically evident. Using a 'Value Stretch' model Mansfeld (1992) characterized and measured the level of this irritation. Based on these findings he anticipated that it is just a matter of time before the latent social antagonism in Eilat would become an extrinsic phenomenon. Sure enough, since then, three different spontaneous lobbies have indicated locals' dismay about the consequences of over development. One lobby aimed its activity against the introduction of casino gambling into Eilat. Locals developed a coherent public opinion against gambling, knowing that this activity would bring various types of social perversion to the town. This lobby claimed openly that such a small-scale resort had not reached an economic threshold to justify the introduction of gambling into the town. Therefore, together with its anti-gambling campaign, this lobby called for maintaining Eilat as a small and manageable resort town (Mansfeld and Czmanski 1995).

Another lobby became active against Eilat's deteriorating environmental qualities. Eilat's uncontrolled tourism growth did not take into account the fragility of two ecosystems – that of the Red Sea and that of the desert mountains surrounding the town. These two ecosystems have already suffered major degradation as a result of tourist activity beyond their environmental carrying capacity. For example, tourists staying in Eilat consume on average three times more water than locals. Because the development process was never thoroughly planned, nobody took this fact into account when Eilat's sewage system was constructed. Thus, as early as 1988, the local sewage-recycling centre could not deal with the growing amount of sewage. Ever since, a growing overflow of untreated sewage is channelled directly to the Red Sea, causing immense damage to underwater fauna and flora. Thanks to pressure put on the local authority and on the Israel Water Company by the local environmental lobby, and after a long struggle, this problem was solved. However, the damage was done and the natural qualities of the Red Sea will never be as attractive as they once were. Yet again, an acquired tourism deficiency syndrome took place. The Israeli government and the local authorities preferred, as in many other Mediterranean countries, to turn a blind eye, to avoid planning and control and to deal with the negative consequences *post factum.*

A third lobby addressed the problem of uneven commercial activity in Eilat. Backed by the local Chamber of Commerce, this lobby tried to assure local merchants access to commercial businesses located in the heart of the tourist areas. As indicated earlier, the hoteliers have predominantly controlled the tourist areas of Eilat. These people made every effort to convince tourists to remain in this area while consuming tourist goods and services. Many merchants living in Eilat invested their savings in opening small businesses based on the growing tourism industry. However, both the municipality and the proprietors of the tourist industry blocked such access. Thus, the economic boom was confined to areas in town where the locals have no say or influence. Consequently locals have never been the real economic beneficiaries of tourism in Eilat (Gruber 1992).

Locals' frustration with the tourism industry also stemmed from the fact that they realized that they had to 'pay the price' by giving up their recreational resources, but they themselves gained nearly nothing in terms of improved quality of life. Indeed, Eilatis have sacrificed their local recreational assets in the name of economic development. Their beaches have been turned into an international meeting place of sun worshippers, snorkelers and scuba divers. The pristine coastal strip where the desert meets the Red Sea was turned into a concentration of concrete high rises. Authentic landscape qualities have become standard international architecture that characterizes many over-developed sea resorts.

One of the most striking phenomenon in Eilat is the huge difference between the appearance of the tourist areas and that of the residential neighbourhoods. Locals realized that while their neighbourhoods were badly kept, most of the local taxes were spent in maintaining the tourist areas. This anomaly should be evaluated in light of the fact that, for most years, local hotels paid hardly any rates, as part of the incentive policy aimed at attracting investors. This fact added to the anger of the already frustrated local population (Mansfeld and Ginosar 1994).

This situation together with all the other phenomena characterizing Eilat's tourism development process, stimulated locals' deep antagonism towards tourism. But has such an antagonism developed also amongst tourists visiting the town? In order to answer this question one has to differentiate between the domestic and the international markets. For the domestic market, it has shown a growing commitment to the Eilat tourist product for many years. Israeli tourists preferred Eilat over other domestic destinations because it always transmitted an international tourist flavour (Azariahu 1993). However, in recent years the gap between the cost of staying in Eilat and that of staying in alternative 'sun and sand' destinations in the eastern Mediterranean basin has widened. Eilat becomes more and more expensive to Israelis. Therefore, to date, a growing number of them opt for the Greek Islands, Cyprus and/or Turkey instead of an expensive holiday in Eilat. Thus, more foreign currency 'leaks out' by Israelis travelling overseas while Eilat sees diminishing occupancy rates generated by domestic guests.

Will the international market balance this forecasted decline in domestic tourism to Eilat? Recently, the future has not looked too promising. There are various reasons for this pessimistic view. First, interviews with European travellers visiting Eilat prove that their level of satisfaction from the Eilat tourist product is deteriorating. Tourists claim that Eilat looks more like a building site than a tourist destination. They criticize the deteriorating environmental situation and claim that Eilat is a saturated resort. Many tourists point at the landscape transformation that has taken place in Eilat and blame the developers for their insensitivity in this respect. Complaints were also made on the quality of service and the lack of attractions to keep them busy should rain and wind replace sunny weather during winter months (Fleischer and Mansfeld 1995). Furthermore, complaints are made not only by the tourists but also by tour operators in the generating markets that sell Eilat in Europe. For example, the German tour operators have taken Eilat out of their brochures because of the poor environmental situation in and around the town. German tourists have raised their requirements in terms of environmental standards. If a given destination does not maintain high and uncompromising standards in this respect, it is taken out of the alternative German destination map. And finally, one has to bear in mind that, while Eilat is maintaining an uncontrolled tourism development process, the neighbouring countries also try to seize their portion of the market. Egypt is developing its Red Sea coast and offers environment-friendly hotels together with cheaper packages. It is only a matter of time before Jordan upgrades its tourism infrastructure in Akaba, and will then also compete with Eilat. Since 1994, and as part of the peace process in the Middle East, Israel and Jordan have promoted the idea of regional co-operation and the establishment of one economic zone in the Red Sea tip. This idea, which incorporated a new regional tourism zone, gave rise to much hopeful expectation (Government of Israel 1994). However, these plans have not yet materialized. Apparently, the turbulent situation in the Middle East does not provide the necessary grounds for the fulfilment of such policies. In fact, what currently remains from all those peace documents and euphoric plans for regional co-operation is the possibility of border crossing and visiting the Red Sea tip on a regional basis. Tour operators in Europe already sell Eilat as part of a regional package incorporating Akaba and the Sinai coast. Thus, a regional rather than local holiday will cause Eilat to lose some of its profits from international tourism. Tourists will stay for shorter periods and spread their spending money over three countries. What future can one, therefore, see for a resort that turns a blind eye to what is happening in and around it, and rapidly expands its accommodation capacity, yet is going to lose clients because of lack of appeal and growing regional competition?

Summary

The thesis proposed in this chapter suggested that tourism planning and development in Israel has been a victim of 'acquired tourism deficiency syndrome' or ATDS. This syndrome is a result of a policy that ignores other countries' past experience in tourism planning and development. Furthermore, it was claimed that many countries' tourism development policies were based on acquired false propositions regarding the impact of uncontrolled and rapid tourism development. Using a review of tourism planning in Israel and tourism development in its southernmost resort of Eilat, the chapter exposed how Israel fell into this trap. When analysing the Israeli case and looking for lessons that can be implemented in future rational development policies at a Mediterranean level, the following recommendations and observations must be addressed:

- Adopting a tourism development policy based on the idea that what is good for one country is good enough for others, might yield disastrous economic, social and environmental consequences.
- Mediterranean countries should establish formal communication channels so that information on the consequences of tourism development in their resorts will be thoroughly studied. Platforms for exchange of information on planning strategies, development policies and the management of tourist attraction in the Mediterranean are imperative if this region is willing to learn from its past mistakes, rather than maintain an acquired tourism deficiency syndrome.
- Mediterranean countries should carefully re-evaluate the need to structure their tourism development policies on foreign investments. The Israeli case proved that opening a window of opportunity for transnational and/or foreign investors does not bring real economic and social benefits in the long run. Moreover, such foreign corporations occasionally prefer to repatriate their profits, to stay away from any community involvement and are hardly sensitive to the fragility of the socio-cultural and environmental texture of resort areas.
- The role of tourism planning has to be repositioned in countries which have potentially competitive and attractive tourism resources and at the same time suffer from acute economic problems. In such circumstances, planning has always been ignored or never properly implemented. Planning in such a situation should not be regarded as a delaying mechanism but rather as a quality assurance device.
- Mediterranean countries should no longer embark on national tourism planning. This is, in most cases, a long and costly process that can hardly be implemented. Alternatively, the planning process should concentrate on the regional and local level.
- In Mediterranean countries that face occasional social, political and/or security situations, tourism should never become a leading exporting industry. The possibility of immediate decline in inbound tourist flows as a result of such situations can cause huge damage to locals and proprietors who are economically reliant on tourism.
- Social, cultural and environmental carrying-capacity thresholds must

be measured in every resort or region that undergoes tourism planning and development. The case of Eilat proves that one cannot simply assume that all such resorts represent the same level of sensitivity and resistance towards uncontrolled development.

- National economic interests leading to local and/or regional tourism development must be redefined in light of local interests and carrying-capacity thresholds.
- And finally, when measuring the cost and benefit of tourism development in Mediterranean countries, it is paramount to make sure that a certain portion of profits made are reinvested in upgrading the quality of life and standard of living of the host community. Furthermore, it is essential .to allocate a certain amount of these profits towards the constant need to monitor the impacts of tourism and to maintain tourism activity well below the carrying capacity thresholds.

If most of the above recommendations are implemented, there is a good chance that this region in general and Israel in particular will move away from the acquired tourism deficiency syndrome.

Part III

Southern Mediterranean shores

Tourism development and Islamic fundamentalism

12 Tourism in Egypt

History, policies, and the state

Turgut Var and Khalid Zakaria El Adli Imam

Introduction

The Arab Republic of Egypt occupies the northeastern corner of the African continent and the Sinai Peninsula. It has an area of about 390,540 square miles (1,011,500 km²) including the 23,440 square miles (60,710 km²) of the Sinai. Ninety-nine per cent of the Egyptian population lives on only 4.0 per cent of the land. Most of them are in the Nile River valley and the large, fertile delta of the river. Egypt is bordered in the south by the Republic of Sudan and in the west by Libya. Towards the north lies the Mediterranean Sea and on the eastern coast are the Red Sea, and the Gulf of Aqaba, between Sinai and Saudi Arabia. The Gulf of Suez and the Suez Canal separate African Egypt from Sinai. In Northern Sinai, Egypt's border with Israel was fixed in 1979 by a peace treaty, though the disputed territory did not return to Egypt until 1982 (*Compton's Interactive Encyclopaedia* 1995). The estimated population of Egypt was over 60 million in 1995 and ranks 18th in the world (Van der Heiden 1996).

Egypt is one of the oldest civilizations in the world and is mentioned in Biblical and Koranic texts. Being endowed with historical monuments, archaeological sites and landmarks covering several millennia of civilization, as well as extensive beautiful beaches, it has considerable potential for developing its tourism sector. Besides its diversified topographical and cultural attractions, Egypt's tourist endowments are further enhanced by mild weather for at least eight months of the year (Wahab 1997).

Brief economic history

During the presidency of Gamal Abdel Nasser, the economy of Egypt was radically socialized. Beginning in 1961, foreign trade, banking, insurance, and most wholesale and industrial establishments were nationalized. Those sectors which remained in private hands were placed under heavy regulatory restraints. Industry was expanded and production increased according to a five-year plan. Inadequate foreign investment, a sluggish bureaucracy and the disastrous 1967 Arab–Israeli War subverted subsequent development programmes until a process of economic reform was inaugurated by Abdel Nasser's successor, Anwar Sadat, in the aftermath of the October War of 1973.

By reversing many of Abdel Nasser's policies and opening Egypt to foreign investment, Sadat began a gradual revival of the Egyptian economy which was significantly enhanced by remittances from Egyptians working in the surrounding oil producing countries. The very slow but sure relaxation of import, currency and trade restrictions stimulated Egypt's foreign exchange economy.

Tourism, which had fallen off drastically during Abdel Nasser's time, due to Egypt's anti-western stance and poor tourist infrastructure, was restarted with the privatization of many nationalized tourist facilities.

Sadat's dramatic peace initiative and treaty with Israel transformed the western view of the Arab leader and his country and further enhanced the country internationally, the gesture was motivated by more practical considerations.

Despite the many advances the country has witnessed under President Hosni Mubarak, Egypt continues to suffer from the vagaries of regional instability and its exploding population. Government leaders openly admit that population growth is undermining all efforts toward developing the country's economy. This situation is further aggravated by consumerism.

Servicing a foreign debt over twice the size of the national budget is another negative factor. Under pressure from the IMF and World Bank, Egypt finally began to lift price controls, reduce subsidies and begin to relax restrictions on trade and investment. Tourism represents one of the most lucrative sectors of Egypt's economy but is highly vulnerable to internal violence and regional politics. The government remains hopeful that the oil and gas discoveries in the western desert will produce significant revenues (ArabNet 1998).

Although Egypt is nominally a multiparty democracy with a 454-member People's Assembly and 210-member Advisory Council, the true power of government is held by the President who serves for six-year terms and exercises wide-ranging powers. The People's Assembly approves the budget, levies taxes, approves government programmes and can censure cabinet members.

The Republic is divided into 26 governates or *muhafizat*. Cairo is the country's capital city and the seat of government. In 1971 under President Gamal Abdul Nasser a constitution established Egypt as an Arab socialist country. From 1961 the government of Gamal Abdel Nasser banned all existing political parties except for the Arab Socialist Union (ASU) and for 16 years Egypt was ruled as a one-party state. The multiparty system was reintroduced by Anwar Sadat in 1977.

The Egyptian judicial system is an amalgam of Islamic, French and English law with a hierarchy of courts descending from the Supreme Constitutional Court down to primary and summary tribunals in each of the country's 26 muhafizat (ArabNet 1998).

Hosni Mubarak had been Sadat's vice-president since 1974 and, like Sadat, seemed singularly unimpressive prior to assuming the presidency. At first he continued Sadat's policies but with less flamboyance and more domestic sensitivity. He allowed the publication of Islamic newspapers and downplayed the Israeli connection. At the same time, he accelerated the process of privatization and developed Egypt's tourist infrastructure, which enhanced its lucrative tourist industry. More impressively, he managed to resume diplomatic and trade relations with moderate Arab countries while maintaining the treaty with Israel. By the end of the 1980s, Egypt was once again playing a leading role in Arab politics. Egypt's vital role in support of Saudi Arabia and Kuwait in the Gulf War combined with the death of socialist–communist influence in the Arab world returned the country to the centre of Middle Eastern politics.

However, Egypt's domestic situation is far from stable. The country's economic reforms and infrastructure development cannot keep pace with the population explosion and inflation. Extremist Muslim groups launched a campaign of terrorism against foreigners, which paralysed the government and damaged tourism between 1992 and the beginning of 1994. Security forces broke the main terrorist groups in Cairo and Upper Egypt and the summer of 1994 experienced a spectacular revival of tourism, particularly from Saudi Arabia and the Gulf States.

Although most terrorist cadres have been imprisoned and many have been sentenced to death, the threat to Egypt's stability remains, as Islamic fundamentalism becomes more deeply rooted in Arab societies

Egypt's early attempts to industrialize during the nineteenth century were thwarted by the colonial powers who aimed to monopolize African and Asian markets, while exploiting their natural resources. It wasn't until the early twentieth century that a limited manufacturing sector was developed to cater to domestic demands. The Second World War stimulated industrial growth and the beginnings of a major textile industry. The socialist government of Abdel Nasser emphasized industrial development and established an industrial base, which continues to expand.

Cairo, Alexandria, Helwan and the new industrial cities outside Cairo are modern Egypt's main industrial areas, producing iron and steel, textiles, refined petroleum, plastics, building materials, electronic products, paper, automobiles and chemicals. Apart from textiles, most industrial products are made for local consumption.

Egypt is strongly committed to the use of tax incentives to stimulate foreign investment. The preferred form of tax incentive is the tax holiday. It is clear that Egypt will not abandon its use of tax incentives in the foreseeable future because of its firm belief that they are an extremely important policy instrument in creating an appealing investment climate for investors. Egypt's panoply of tax incentives can be conveniently grouped into three categories: tax holidays granted under the Investment Code by the General Authority for Investment (GAFI) and other government bodies; general investment incentives available to all firms under the income tax law; and selective incentives directed to specific firms under the income tax law. The following paragraphs briefly describe the nature of this tax incentives (Shawki 1998).

Tax holidays granted by GAFI and others

GAFI administers Law 230 of 1989, which allows it broad discretion in granting tax holidays to almost any type of economic activity. The duration of the tax holiday depends on whether the benefiting enterprise operates inside or outside of a free zone. Presently, there are eight free zones in operation. Outside the free zones, approved projects can obtain tax holidays of five to 15 years on corporate income tax and tax on moveable capital revenue. The standard tax holiday is five years, but a ten-year tax holiday is granted for land reclamation projects and projects in new industrial cities and remote areas. Under Law 59 of 1979, investors who establish their operations in one of the new satellite cities receive a standard ten-year tax holiday. This tax holiday is administered by the Ministry of Development, New Communities, and Public Utilities. At present, there are over seven new satellite cities and several more are planned.

In addition, a 15-year tax holiday, which can be extended for an additional five years, is granted to low and middle income rental housing built in a new city.

General investment incentives under the Income Tax Law

General investment incentives for corporations include a 25 per cent investment allowance on new machinery and equipment, a paid-up equity capital deduction, a lower statutory rate for industrial and exporting firms, a rollover of capital gains on sales of real assets, and an interest income exemption on publicly-subscribed bond issues and bank deposit interest.

Selective investment incentives under the Income Tax Law

Under the basic Income Tax Law 159 of 1981, certain investment projects are either completely exempt from tax or are eligible for a five-year tax holiday. Investors engaged in land reclamation could claim a ten-year tax holiday and a tax holiday can be granted to projects utilizing new technology, or to those which increase exports or reduce imports. With a firm and strong political and economic base, Egypt offers investors not only attractive tax incentives, but also the challenges and high yields of an emerging market economy. In addition, Egypt has concluded double taxation treaties with certain countries, which a potential investor should bear in mind when assessing the viability of the Egyptian market (Shawki 1998).

Changing the mode of the Egyptian economy from totally government-oriented to market-driven has required utilizing all sorts of transformation vehicles. As a step in that direction, the Egyptian government has embarked upon a gradual and stable privatization programme with the aim of achieving its goals with the minimum negative impact on the social structure of the nation (El Hayawan 1998).

The continuous drain of resources and the outstanding budget deficit that existed before the serious economic reform steps were introduced was largely due to the unmistakably low performance of the public sector companies. In fact, return on investment in 1994–95, according to El-Hayawan was a mere 1.2 per cent. It was even worse in 1992–94 period, ranging between negative and less than one per cent. (El Hayawan 1998)

So, one might ask where the privatization programme is heading. First, from the time the programme was launched in early 1994 to the present, shares of 27 State Owned Enterprises (SOEs) were publicly offered on the Egyptian stock exchange with a total value of 2.6 billion Egyptian Lira (LE). Four companies were sold to strategic investors, totalling 664 million LE. During 1997, the programme aimed to transfer 40 companies from Law 203 to law 159, meaning that the public offering would exceed 51 per cent of its capital. In addition, 40 per cent of another 12 companies would be offered to the public on the stock exchange by the end of 1998, another 27 companies would be offered through the stock exchange to the public; and, 64 other companies to strategic investors. So, 143 companies were expected to be either wholly or partially privatized during 1997 and 1998. Aside from that, all state owned banks are planning to offer their shares in all joint-venture banks for sale to private investors. It is hoped that the final outcome of the privatization programme will be more active participation of the private sector and the complete activation of the Egyptian stock exchange (El Hayawan 1998).

The role of the private sector reassessed

The government of Egypt has recently taken several steps that demonstrate the government's policy of encouraging investment by the private sector. These steps include acceleration of the privatization programme, a relaxation of restrictions in banking activity and the capital market, and passing and amending investment-positive legislation. While there are impediments to privatization, such as difficulties of labour redundancy, variables of market demand, and the capability of the Egyptian capital market to handle the volume of transactions, there is both interest and capital sufficient for the successful sale of shares in healthy public sector enterprises to the private sector. Privatization in Egypt is taking two forms:

divestment of public sector holdings in production and manufacturing companies; and, encouragement of private sector investment in sectors historically controlled and operated by the public sector, such as electricity, roads, airports, ports, and oil and gas transmission. With respect to the capital market, recently the cabinet decided to repeal the 2 per cent capital gains tax on the sales of shares and to exempt mutual funds from a 40 per cent income tax, which has resulted in an increase in activity. In addition, a new clearing and settlement system is due to come into operation shortly to ensure clearing and settlement of 'transaction plus three days' instead of the current physical delivery system, which takes up to ten days. Concerning investment in the banking sector, the government has approved a statutory amendment to permit majority foreign equity ownership in locally registered banks (McKinney 1998).

Other legislation, such as the Investment Law (Law 230 of 1989), has created a legal climate conducive both to private sector project promotion and project financing. This has been achieved by prohibiting compulsory pricing or limitation on profits on project products, releasing housing projects from compulsory rent control, exempting projects from industrial or commercial tax for specified periods and from tax on dividends arising from such projects, providing for the right to import project equipment and materials, and allowing repatriation of project profits within the credit balance in the foreign currency account of the project. In addition, the Investment Law was amended on March 1 1996 to ease the incorporation process and allow investors to denominate the company's capital in foreign currency. Additional incentives and protections are mandated by statute for projects in new communities, remote areas, tourist ventures, and free zones. International agreements also provide for added private sector encouragement and protection (McKinney 1998).

Egypt is a member of the World Trade Organization, enjoys Most-Favoured-Nation status with the United States, has a preferential trade accord with the European Union, and is a signatory on bilateral investment agreements and double tax treaties with a large number of nations. Egypt and the US have an investment guarantee agreement through the Overseas Private Investment Corporation and a Bilateral Investment Treaty (1992) that ensures nondiscrimination for investors of the two countries and sets legal standards with respect to expropriation, compensation, and international arbitration of investment disputes. The 'Partnership for Economic Growth and Development', a bilateral Egyptian–USA programme begun in March 1995, is also designed to stimulate private sector investment through 'quick start' projects in the area of technology transfer, development of Upper Egypt, and eco-tourism in Sinai, the Red Sea, and Upper Egypt (McKinney 1998).

At the 1994 MENA summit in Casablanca, the future of the Middle East and North African region was discussed (Leheta 1998) and several conclusions were reached with respect to political developments in the area and how they are linked to infrastructure development. The conclusions revealed that the obstacles to integration, which is viewed as a natural result of the comprehensive peace in the region that will lead to the area's stability, development and prosperity, are many. As such, a number of land, air and sea transport projects have been proposed in order to further the integration of the region. A transport network spanning the region is not only in the interest of the residents of the area themselves, but is also in the interest of world trade. Such a transport network requires more than roads, railways and ports; it requires national and international collection and distribution centres having all the required facilities. Connections in the MENA region exist to Europe through Morocco, to Africa from Algeria to Kenya, and from Libya to Chad. However, there are missing links between Yemen and Oman, Egypt and Central Africa, and Africa to Asia along the coastal road. Suggested land transport projects include a highway across the desert from North to South (from Alexandria to Zaire), and a bridge over the Rashid Nile Branch followed by a bridge over the Suez Canal to Sinai, which is currently under construction. These two projects would continue the coastal road from Africa to Asia.

Two other suggested land transport projects include three underwater tunnels south of Port Said (the Egyptian government has already announced its intention to build these tunnels, which constitute an excellent investment and/or finance opportunity for the private sector) and a causeway from Egypt to Saudi Arabia to complete the East–West land road to the Gulf area, which the Egyptian government has also announced its intention to build. With respect to air transport, the numerous airports distributed over the entire area are not prepared for the increase in both cargo and passenger traffic that is expected in the near future. Suggestions for improving air transport are a new airport at the intersection of Taba, Aqaba and Eilat, and a new airport in central Sinai. The suggested airports lie at the crossroads of land, sea and air routes. To handle the expected traffic in many Red Sea ports, a great need exists to renovate and expand; for example, container terminals, storage space and handling facilities. Other suggested projects are a medium-sized port in Gaza, small, specialized ports all along the Gulf of Suez and improvements to Beirut, Tripoli and Safaga ports. In spite of the ideal geographical position of the MENA region as a central location that can perform distribution services for the rest of the world, there is only one distribution centre in the entire area at Jebel Ali in the United Arab Emirates. All of the previously mentioned projects are tools needed to support a major distribution centre in the Suez Canal area, which is viewed as the most important potential distribution centre for the area and for world trade, since all international goods passing from North to South and from East to West must pass through the Suez Canal region. The Sinai area is the connection between North and South, East and West. If this region is properly

Table 12.1 Number of tourists between 1982–June 1996

Nationality	1982	1983	1984	1985	1986	1987	1988	1989	1990	1991	1992	1993	1994	1995	1996
Arabs	618,331	598,680	596,145	564,105	554,181	657,006	659,749	952,209	1,140,231	1,082,340	1,102,942	922,389	931,730	822,899	895,402
Americans	193,065	219,505	227,254	212,049	94,763	148,999	164,141	200,479	179,144	119,863	224,479	187,476	182,378	228,896	269,057
Europeans	524,030	566,848	629,326	629,648	565,881	863,617	1,011,678	1,188,783	1,123,161	889,950	1,664,906	1,205,740	1,243,629	1,811,000	2,321,032
Asians	73,219	83,124	94,332	99,191	83,780	112,099	119,167	143,421	141,010	99,760	187,304	157,854	180,952	219,464	288,320
Others	14,606	29,775	13,403	13,433	12,645	13,232	14,758	18,506	16,571	22,364	27,309	34,303	43,299	51,202	122,131
Total	1,423,251	1,497,932	1,560,460	1,518,426	1,311,250	1,794,953	1,969,493	2,503,398	2,600,117	2,214,277	3,206,940	2,507,762	2,581,988	3,133,461	3,895,942

Source: Central Agency for General Mobilisation and Statistics.

developed, there will no longer be a 'choke point' for international trade. Instead, it will become a take-off point (Leheta 1998).

Recent trends in Egyptian tourism statistics

Between 1982 and June 1996, the number of tourist arrivals went up from 1,423,251 to 3,895,942 which represents an increase of 274 per cent. However, Arabs visiting Egypt increased only marginally, about 33.1 per cent. On the other hand, European visitors increased, during the same period, about 443 per cent. European tourists represent (in 1996) approximately 60 per cent of the total arrivals. Table 12.1 gives the full detail of tourist's arrivals to Egypt. Parallel to the numbers of arrivals, the foreign exchange receipts from tourism between 1982 and 1996 increased approximately 400 per cent to over 3 billion US dollars (Table 12.2) putting the tourism industry in second place after workers' remittances and revenues, but ahead of Suez Canal revenues and oil exports. (Wahab 1997). In addition, as stated by Wahab, Egypt has considerably more domestic control over the generation of foreign exchange through tourism, which has shown a greater responsiveness to favourable domestic policies than other sources of foreign exchange. As an illustration, the 37 per cent devaluation in May 1987 coincided with a 28 per cent increase in tourist arrivals, a 34 per cent increase in nights spent in Egypt and a 133 per cent increase in recorded tourist receipts in 1987–8 over the previous year.

Table 12.3 represents the development of the hotel capacity. Between 1982 and 1996 the number of hotels (including vacation villages and floating hotels) increased from 263 to 761. Currently Egypt has about 132,764-bed capacity. Table 12.4 shows the breakdown of hotel rooms in terms of quality.

Table 12.2 Revenue – Egypt

Year	In million Egyptian Lira (LE)	In million US dollars
1982/83	252.1	304.1
1983/84	238.7	288.4
1984/85	339.7	409.6
1985/86	424.3	315.3
1986/87	518.1	379.6
1987/88	1922.2	885.9
1988/89	2303.8	900.6
1989/90	2761.8	1071.8
1990/91	4937.0	1646.2
1991/92	8401.3	2529.0
1992/93	7918.3	2375.0
1993/94	6001.6	1779.3
1994/95	7802.5	2298.9
1995/1996	10,215.9	3009.1

Source: Central Bank of Egypt.

Table 12.3 Hotel capacity (according to type of establishment)

Type of establishment	1982			June 1996		
	Units	Rooms	Beds	Units	Rooms	Beds
Hotel	214	17,105	33,276	461	44,844	89,133
Tourist villages	2	250	500	85	10,920	21,883
Floating Hotels	47	1509	3223	215	10,947	21,748
Total	263	18,864	36,999	761	66,711	132,764

Source: Ministry of Tourism.
- Increase in number of hotels of various categories.
- Number of tourist villages increased by 42.5 times.
- Number of floating hotels increased by about four-and-a-half times.

Table 12.4 Distribution of hotel capacity (according to category)

Category	1982		June 1996	
	Units	Rooms	Units	Rooms
Five stars	42	7500	124	19,939
Four stars	45	4388	94	10,356
Three stars	63	3071	175	14,316
Two stars	55	1999	143	6043
One star	36	1046	83	2966
Under renovation	22	860	142	13,091
Total	363	18,864	761	66,711

Source: Ministry of Tourism.
- Number of hotels from one to three stars increased by about three times to cater for the increase in middle-class tourism.
- Five and four-star hotels multiplied by about two-and-a-half.

Due to changes in various foreign investment related legislation, over 21.6 billion Egyptian pounds of investment was undertaken between 1986 and June 1996 (Table 12.5). These investments, as expected, will create over 347,700 jobs. About 20 per cent of the workforce in tourism are women. (Wahab 1997). These new job opportunities contribute to reducing the unemployment rate in Egypt.

Although tourism still contributes less than 3 per cent of GDP, it has been since 1987 the fastest growing sector in the economy, in response to favourable economic reforms at both macro and sectoral levels (Wahab 1997).

The role of the state

The Ministry of Tourism is the main authority dealing with tourism in Egypt. The Ministry is organized into four major sections: (a) Planning and Development; (b) Regulation of Tourist Services; (c) Administration;

Table 12.5 Development projects and tourist investments from 1986 to 30/6/1996

Area	Number of projects	Area (thousand m²)	Capacity rooms	Cost (million LE)	Job opportunities
Gulf of Aqaba	146	33,296	47,433	5624	96,355
Red Sea	102	63,086	78,801	8731	176,570
Ras Sedr	23	3423	6877	929	10,315
El-Ain El-Sokhana	115	26,043	40,080	5477	56,244
Arich	1	1000	300	115	450
North Coast	9	4712	5176	737.8	7764
Total	396	131,560	178,667	21,613.8	347,698

Source: The Egyptian Tourism Development Authority (TDA).

and (d) Financial and Legal Affairs. Like most other ministries, the Ministry of Tourism suffers from overstaffing and inadequate technical capability. In order to be competitive in the region and to protect Egypt's unique natural and cultural resources, the Ministry adopted a strategy to support private sector efforts in tourism. The first step was the creation of the Tourism Development Authority in 1991, which draws on private sector expertise to assist the Ministry in guiding and promoting touristic investments. Coupled with recent changes in investment legislation and privatization, Egypt has taken major steps for private-sector oriented and environmentally-sound tourism development (Wahab 1997).

The Ministry also oversees several public sector organizations: (a) Egyptian General Authority for Promotion of Tourism; (b) Public Authority for Conference Centres; and (c) Tourism Development Authority. Under Law 203, the public sector Tourism Authority, which was under the supervision of the Ministry of Tourism in the past, became a holding company for the sector and consists of five affiliated companies including Egyptian General Organization for Tourism and Hotels, Misr Travel Company, Egyptian Hotels Company, Misr Hotels, and Grand Hotels of Egypt (Wahab 1997).

Liberalization and deregulation in tourism

The Egyptian Cabinet and Ministry of Tourism have taken important steps both at the macro and sectoral levels towards liberalization and deregulation of the tourism sector. These include the liberalization of foreign exchange rates; the easing of restrictions on chartered flights; and inviting private airlines to introduce international scheduled flights on routes not served by Egypt Air, the national flag carrier. (Wahab 1997) Realizing that tourism is one of the main cornerstones of the national economy, a comprehensive plan was prepared for the fiscal year 1994/1995 to enhance the efforts in three main tourism fields: (a) development; (b) promotion; and (c) public awareness (El Beltagui 1995).

The Ministry of Tourism was the first to apply the policy of privatization for achieving balance in the national budget. Therefore, a national tourism development strategy based on marketing concepts was drawn. This strategy determined the priority zones, laid down regulations for investors, and gave consideration of protection of natural and cultural resources. The basic elements of this policy are as follows:

1   Enhance the encouragement of projects in areas lacking tourism and accommodation facilities with special emphasis on Sinai and the Red Sea regions.
2   Increase job opportunities for a growing population.
3   Increase the lodging capacity, including hotels, holiday villages, etc.
4   Upgrade the efficiency of the whole tourism sector by developing human resources through education and training.
5   Upgrade various tourist services to meet international standards through supervision of the Ministry of Tourism in accordance with tourist related legislation.

It is interesting to mention that through these measures in 1993, 100 per cent of total investments came from the private sector, with the role of the state reduced to nil (El Beltagui 1995).

In the case of promotion, an ambitious plan was carried out highlighting Egypt's diversified and unique tourist product for the purpose of increasing tourist demand in order to achieve the target arrivals and tourist revenues. One important aspect of this promotional plan was to consolidate Egypt's image on the international tourism map as a venue for convention tourism.

Finally, the Ministry of Tourism enhanced public awareness of the importance of tourism via the mass media. Ten TV spots were produced in 1993–94 and frequently run on major Egyptian TV channels. Also, an arrangement was made between the Ministry of Tourism and Ministry of Education to introduce tourism subjects in the curricula of elementary and secondary schools. (El Beltagui 1995).

As a result of these measures and a clear tourism strategy, including privatization and deregulation, Egypt was able to achieve her 1992 level of tourist arrival and tourist revenues. According to the World Tourism Organization's statistics in 1995 Egypt had 2,872,000 international tourists. The tourist revenues compared to 1994 went up from US$1.4 billion to US$2.8 billion in 1995. (WTO 1995) Considering the international outbound tourist expenditures of US$1.3 billion, the Egyptian tourism balance showed over US$1.5 billion. The trend seems to be growing by the addition of new facilities. As of June 1996, Egypt had 761 hotels and other accommodation facilities with 132,764 rooms with approximately over 250,000-bed capacity (Tables 12.3, 12.4 and 12.5). The occupancy rate for rooms in 1995 was 67 per cent (Wahab 1997). Parallel to the accommodation facilities, as shown in Table 12.6, restaurants, cafe-

Table 12.6 Distribution of public establishments according to category

Type of establishments	1982						June 1996					
	5 star	4 star	3 star	2 star	1 star	Total	5 star	4 star	3 star	2 star	1 star	Total
Restaurants	32	41	90	10	2	175	111	166	255	21	5	558
Cafeterias	25	38	83	4	0	150	6	90	185	40	38	359
Entertainment places	12	29	51	2	1	95	3	9	8	2	0	22
Total	69	108	224	16	3	420	120	265	448	63	43	939

Source: Ministry of Tourism.
- Tourist establishments (Five star category) represent 12.8 per cent.
- Three star establishments represent 47.7 per cent.
- Four star establishments represent 28.2 per cent.
- Two star establishments represent 6.7 per cent.

terias, and entertainment places have also shown a remarkable development between 1982 and 1996. The number of travel agencies climbed from 331 in 1982 to 804 in 1996 (Table 12.7).

One of the objectives of the Egyptian tourism strategy was to create employment through tourism. As reflected in Table 12.8, direct employment in tourism went up from 82,133 to 263,600, more than a three-fold increase, in 1996. Obviously, considering the multiplier impact of tourism revenue on employment, and personal income, the actual impacts on the Egyptian economy are larger.

The momentum for development of tourism is continuing and new projects have been added every day. As stated by Wahab, the private sector is

Table 12.7 Number of travel agencies

Year	Number of travel agencies
1982	331
1983	397
1984	442
1985	501
1986	547
1987	589
1988	637
1989	695
1990	709
1991	721
1992	736
1993	767
1994	780
1995	790
1996	804

Source: Ministry of Tourism.

Table 12.8 Direct employment in the tourist sector

Type of establishment	1982	1996
Accommodation establishments	22,636	116,705
Tourist establishments	51,800	84,600
Travel agencies	3200	18,310
Shops and bazaars	3562	40,000
Tour guides	935	3985
Total	82,133	263,600

Source: Ministry of Tourism and the Tourist Chambers.

Until 1995
• Tourist employment increased 3.2 times.
• Employment in regular accommodation establishments represents 44.3% of the total employment.
• Employment in tourist establishments represents 32.1%.
• Employment in Travel Agencies represents 6.9%.
• Employment in bazaars and tourist shops represent 15.2%.
• Tour Guides represent 1.5%.
[These figures represent direct employment in the tourist sector. However, indirect employment is estimated at about $263,300 \times 3 = 790,800$ workers, which brings the total employment in the tourist sector (direct or indirect) to 1,054,400.]

being encouraged through various incentives to introduce area infrastructure in new tourist regions by large investment companies like Egyptian Resorts Development Corporation (Wahab 1997).

In spite of various measures there are certain constraints for development of tourism in Egypt. In order to remedy the situation, the Council of Ministers recently issued several decisions to ease or to remove some of the constraints. These included:

1 Giving the Ministry of Tourism the right to issue a certificate of exemption from custom duties for equipment and machinery (including spare parts) necessary for tourism investment projects.
2 Facilitating procedures for licensing of tourism investment projects and their renewals.
3 Liberalization and deregulation of air charter rules, especially those charters that are coming from points of departures that are not served by Egypt Air.
4 Facilitating the formation of private sector companies to develop tourist infrastructure in various tourist areas and regions.
5 Simplification of the procedure to issue construction permits for hotels and other tourist establishments within city boundaries.
6 Reduction of passenger docking fees in Egyptian ports by 75 per cent and facilitating the creation of marinas on the Red Sea and the Mediterranean to develop yacht tourism.
7 Preparing an integrated promotional plan for health tourism in areas like Helwan, Aswan, Siwa, etc.

8 Developing an integrated project for tourism health insurance and an air ambulance system in all Egyptian tourist regions and cities.
9 Providing and implementing a comprehensive national programme for raising popular awareness of tourism and co-ordinating efforts between respective government departments and the media.
10 Preparing certain legislative actions that would update provisions of existing laws concerning tourist establishments, tourist chambers and federations (Wahab 1997).

Tourism policy and sustainability

Although Egypt has not yet formulated a binding national policy, according to Wahab, deliberations between the government and World Bank resulted in agreement to a number of goals, namely:

1 Changing the role of the public sector from that of owner/operator to planner/regulator and promoter/facilitator.
2 Deregulating the industry to allow the private sector to operate freely in a competitive environment.
3 Protecting and conserving the unique cultural and natural resources in the tourism areas.
4 Promoting a larger role for the private sector in the design, financing, implementation, ownership and operation of tourism facilities.

Traditionally, in countries like Egypt, the public sector has assumed the role of provider of infrastructure, bearing all the risks of future developments. However, the last goal brings the private sector into decision making, financing, implementing, and finally, the operating of the facilities with all the risks involved (Wahab 1997). The goal of sustainability cannot be achieved without a serious attempt to measure the probable impacts of proposed projects. Recently, an environmental impact assessment study was made compulsory for any kind of tourism development project. The necessary approval cannot be obtained without conducting such a study.

Egypt, besides privatization, deregulation, promotion, and awareness of the importance of tourism by the public, had a serious problem related to terrorism. Although every country is prone to terrorism, Egypt has had more than her share, especially during the past five years. It seems that Islamic fundamentalism has deliberately targeted one of the most important economic sectors in Egypt by attacking tourists. In 1996, an attack by Fundamentalists left 18 Greek tourists dead. A later attack in Luxor killed more than 58 persons and led the government to take even stricter measures to protect Egypt's tourism industry which was expected to bring 4.4 million tourists in 1998 (Morello 1998). Before Luxor, almost every major shopping mall and large hotels had visible security arrangements, including metal detectors and police presence. After the Luxor incident, even the tour buses were required to take necessary precautions like having a police presence in the bus and sometimes a police escort on the way to attractions. With political stability, Egypt is expected, as stated by *Condé Nast*, (Allman 1998) to overcome this hurdle and become a major tourist destination in the near future. It is the prosperity of the middle class that forms a good defence against terrorist attacks. As reported by the Egyptian Ministry of Tourism, on February 26 1998, the State Department cancelled the travel warning for Egypt for the improvement of security in the Luxor area. It is also reported by the Mayor of Luxor the occupancy rate was 34 per cent, the same as in February 1997.

Conclusion

In conclusion, Egypt, after many years of state control and ownership of tourist facilities, began to privatize and develop a sustainable tourism strategy that would protect her unique cultural and natural resources. It seems that the policy of privatization and deregulation is working and, in spite of terrorist activities, the sector is growing rapidly and making a major contribution to the national economy.

13 A dynamic tourism development model in Tunisia

Policies and prospects

Robert A. Poirier

Growth of tourism in Tunisia

Tunisia's approach to tourism is a classic illustration of the package tour concept, as approximately 80 per cent of arrivals come in groups. This approach necessitated the establishment of an elaborate infrastructure constructed in a relatively short period of time. To accommodate the burgeoning demand for facilities along Tunisia's 800 mile coastline, the country launched a major hotel construction programme during the 1970s and 1980s. The expansion was dramatic, making Tunisia one of the fastest growing tourist economies in the world (Table 13.1). With just over a 34,000 bed capacity in 1970, today the industry has grown to a capacity of more than 160,000 (Office National du Tourisme Tunisien 1995) and the Ministry of Tourism anticipates that the total capacity will exceed 200,000 by the end of the century (*Tunisia Digest* 1993). Although the Tunis vicinity has a very large number of hotels, the majority of the bed capacity (about 70 per cent) is found in the prime resort areas of Nabeul-Hammamet, the 'Sahel' (a section of the eastern littoral) and the Djerba-Gabes regions. On the whole, the newer resort establishments are larger and have been purpose-built for the European package tour industry.

The industry has experienced a phenomenally rapid growth, with intense capital investment in the past two decades. In 1977 private

Table 13.1 Tunisia's lodging establishments and bed capacity (1970–95)

Year	Establishments	Bed capacity
1970	212	34,297
1975	273	62,937
1980	319	71,529
1985	420	93,275
1990	508	116,534
1991	532	123,188
1992	563	135,561
1995	912	161,498

Source: ONTT (1995).

investments in Tunisian hotels amounted to 10.5 million Tunisian Dinar (TD) but that relatively modest amount was soon overshadowed by a more than forty-fold increase to 449 million TD by 1995 (Table 13.2). The growth during the course of the Seventh Development Plan (1987–91) showed the continued focus on seaside resort development which has marked, until recently, hotel development policy since the 1960s.

Tunisia's proximity to Europe and the relatively larger discretionary incomes available to Europeans makes tourism in Tunisia Eurocentric despite the fact that arrival data suggest a large Maghrebi presence on Tunisian soil. A consistent 90 per cent of foreign entries in Tunisia are from North African littoral (Maghreb) countries (mostly Libya) or Europe. Since the 1970s, Europeans overshadowed all other non-African entries (North American entries are marginal) but that profile changed in the 1980s as the European–Maghrebi gap began to narrow somewhat, excepting the Gulf War year (1991) which showed a large influx of Maghrebi in relation to Europeans (Table 13.3).

More important than the number of entries, however, is the fact that Europeans stay significantly longer and spend more money per capita/per diem than their Maghrebi counterparts. Whereas Maghrebis average about 2.2 nights, the Europeans are staying at a rate of about 10.5 nights (Office National du Tourisme Tunisien 1995). Initially French tourists, due to Tunisia's historic and economic ties with France, were the largest single national group utilizing resorts. Strategies by the Ministry of Tourism to stress the winter 'fun-in-the-sun' appeal of Tunisia to the northern climates has resulted, however, in a large influx of German tourists in recent years. In the 1990s, Germany has led the world's tourists flocking to Tunisia, accounting for 20 per cent of the total foreign entries in 1995, almost double the number of French entries (Office National du Tourisme Tunisien 1995).

North American traffic to Tunisia, which accounts for less than 1 per cent, has shown some encouraging growth, but is especially sensitive to

Table 13.2 Investments in hotels (1977–95)

Year	Investments (in TD million)	Year	Investments (in TD million)
1977	10.5	1987	63.0
1980	31.5	1988	78.8
1981	33.7	1989	109.1
1982	43.0	1990	118.8
1983	74.5	1991	125.0
1984	113.2	1992	202.8
1985	99.0	1994	397.8
1986	85.0	1995	449.1

Source: ONTT (1995).

Table 13.3 European and Maghrebi entries (1977–81; 1988–95)

Year	European	(% of total)	Maghrebi	(% of total)
1977	849,100	(83)	104,300	(10)
1978	887,100	(77)	163,000	(14)
1979	1,068,000	(78)	211,800	(45)
1980	1,110,500	(69)	431,300	(26)
1981	1,146,100	(53)	915,800	(42)
1988	1,681,882	(48)	1,696,581	(48)
1989	1,670,500	(52)	1,443,215	(45)
1990	1,705,400	(54)	1,378,426	(43)
1991	1,086,500	(34)	2,058,721	(63)
1992	1,849,380	(52)	1,598,971	(45)
1993	2,158,851	(59)	1,372,970	(37)
1994	2,415,690	(62)	1,323,408	(34)
1995	2,357,242	(57)	1,640,410	(39)

Source: ONTT (1995).

political issues. The Gulf War (1991), for example, demonstrated the ephemeral appeal of travel to an Arab country for most American vacationers. This conflict resulted in a 37 per cent decline of North American arrivals between 1990 and 1991 (Office National du Tourisme Tunisien 1992). It would take very little to shake the confidence level of a target population already burdened by prejudices and somewhat sceptical about personal security in the Arab world.

For example, tourism in Egypt was adversely impacted by Islamic fundamentalist attacks against tourists in that country in 1993–94, although traffic did rebound later. More recently (September 1997), a terrorist attack allegedly perpetrated by Islamic militants, which killed ten foreigners in their bus outside the Egyptian Museum in Cairo, could result in changes of travel plans for many.

Likewise, the continuing crisis in Algeria could easily precipitate a decline in Tunisian tourism revenues as travellers fear a spill-over of anti-Western sentiments in Tunisian society. The stunning electoral 'victory' of the Front Islamic du Salut (FIS) and subsequent events in that strife-torn land continue to have repercussions across the Maghreb (Tahi 1992). Since 1992 thousands of lives have been lost in the Algerian conflict, including many foreigners, and anti-Western sentiments have risen. Europeans are not immune to the concerns of American tourists; for example, the victims of the recent bus attack in Cairo were Europeans. Italy, a major supplier of tourists in North Africa, has already substantially reduced economic activity with Algeria due to the Algerian crisis (LaFranchi 1993). The Gulf War had an impact on European travel to Tunisia as well. Prior to the war, in 1990, European entries relative to Maghrebi amounted to 54 per cent and 43 per cent of total entries respectively. In 1991, however, the figure dramatically changed to 34 per cent and 63 per cent respectively (Table 13.3).

Tourism and the economy

Assessing tourism's contribution to development is problematic and it is 'generally acknowledged that the evaluation of the tourism industry's performance is a task which has defeated most Third World governments' (Lea 1988). There are a couple of possible explanations for this. First, the political science literature is not without controversy in its conceptual understanding of development (Chilcote 1984). Second, tourism is a complex of activities ranging from transportation, lodging, food, crafts, cultural events, etc. Hence, operationalizing the term poses problems, and sorting out the plethora of variables for valid measures between tourism and development is troublesome. Thus, the effect of tourism on the Tunisian economy is mixed and measuring its 'multiplier' consequences is difficult (Ryan 1991). Given this problem, it is equally difficult to compare 'pre-tourism' and 'post-tourism' development changes solely as a function of tourism.

Tourism, however, cannot be considered outside the framework of the global economy and western structural adjustment programmes which play a major role in driving tourism policy. Prior to the heavy emphasis on tourism, Tunisia's first president, Habib Bourguiba 'envisioned the creation of a modern capitalist economy' (Anderson 1986: 238). Despite Bourguiba's intentions and efforts to stimulate the economy, growth, hence social development, remained sluggish throughout the 1960s. Tunisia remained dependent on foreign aid as deficits grew largely because outmoded state enterprises had 'become a fetter rather than a factor of economic development', and 'the principal cause of the growth of the country's external debt' (Grissa 1991: 120). Under the leadership of a free enterprise activist Prime Minister, Heidi Nouira, tourism became, in the 1970s, a major focus and other investment incentives turned the economy around as Tunisia experienced a positive balance of payments (Anderson 1986). Population increases, however, weakened the growth spurt because there was no alleviation in the rise of unemployment. Job creation priorities, declining oil production, rising debt servicing and structural adjustment requirements drove policy-makers to expand export promotion (Ferchiou 1991). The serious political commitment for job creation developed in the early 1970s paralleled the country's commitment to promote tourism as its leading export commodity. Formalizing this commitment, the country adopted investment laws favourable to labour-intensive projects (Ferchiou 1991). Leading this activity was international subcontracting such as resort construction projects. By 1986 an invigorated private sector in Tunisia and an export driven economy, mostly from tourism, created conditions for growth for the first time since independence (Grissa 1991).

Advocates assert that the tourism sector promotes foreign exchange earnings, reduces unemployment and provides more security than traditional exports. Trade statistics for Tunisia clearly show tourism's important, if not primary, position in the country's economic profile. In 1970, for example, this sector contributed 31.6 million TD in foreign receipts, which covered 48.7 per cent of the trade deficit. In 1994 and 1995, foreign earnings from tourism were at an all-time high of 1.2 and 1.3 billion TD respectively (Office National du Tourisme Tunisien 1996). Since 1970, tourism receipts have, on average, accounted for close to 60 per cent of the country's trade deficit coverage (Table 13.4).

As discussed previously, the Gulf War contributed to a sharp decline in tourism receipts (about 632 million TD), but recovered by 49 per cent the following year to 945 million TD, just short of the 1086 million TD in the previous record year of 1988 (Office National du Tourisme Tunisien 1992). The performance of the sector also continues to look strong relative to other aspects of Tunisia's export economy. With the notable exception of 1991, tourism has consistently held about a 20 per cent average share of the country's total foreign trade (Table 13.5).

In addition to its value as a source of foreign exchange for Tunisia,

Table 13.4 Coverage of trade deficits by receipts from tourism (1970–92)

Year	Imports (TD m)	Exports (TD m)	Deficit (TD m)	Receipts (TD m)	Coverage (%)
1970	160.4	95.8	664.6	31.6	48.7
1975	572.8	345.6	227.2	115.2	50.7
1980	1428.4	904.8	523.6	259.7	49.6
1985	2131.4	1435.1	696.3	415.0	59.6
1986	2303.7	1403.7	900.0	385.8	42.9
1987	2509.1	1770.7	738.4	568.9	77.0
1988	3167.0	2055.5	1111.5	1086.1	97.7
1989	4150.7	2782.0	1368.7	880.7	64.3
1990	4852.0	3086.0	1766.0	827.8	46.9
1991	4789.0	3429.9	1359.1	632.0	46.5
1992	5673.9	3566.7	2107.2	945.0	44.8

Source: ONTT (1992).

Table 13.5 Tourism and foreign trade (1983–92)

Product	1983	1984	1985	1986	1987	1988	1989	1990	1991	1992
Tourism	389.2	357.7	415.0	385.8	568.9	1068.1	880.7	827.8	632.0	945.0
Petroleum	547.6	591.7	568.7	339.4	418.3	330.7	555.5	531.6	488.8	538.4
Olive oil	24.8	57.3	42.8	53.4	65.6	70.7	81.5	106.9	266.8	138.5
Phosphate	26.6	27.0	25.4	25.8	25.9	26.9	32.5	17.4	14.0	27.8
Phosphorus	91.4	91.9	91.2	74.7	69.8	142.0	159.3	156.0	123.1	144.4
Textiles	219.7	223.4	283.4	388.4	496.2	614.3	816.0	1091.3	1212.4	1401.0
Dates/citrus	18.0	21.5	35.1	37.7	46.0	47.0	47.4	54.6	59.3	51.6
Other	617.7	749.5	823.4	809.8	1034.3	1092.4	1494.1	1914.4	1972.6	2069.7
Total	1935.0	2120.0	2285.0	2115.0	2725.0	3410.1	4067.0	4700.0	4769.0	5316.4
Tourism (%)	20.11	16.87	18.16	18.24	20.88	31.85	21.65	17.61	13.25	17.78

Source: ONTT (1995).

tourism has been providing jobs in a society where unemployment remains a serious problem. Currently, the country has a 14 per cent unemployment rate with more than 400,000 unemployed, while an additional 40,000 to 60,000 people are projected to enter the job market to the end of the century (Institut National de la Statistique 1990). Tunisia's past commitment to socialism created jobs which were sometimes wasteful and redundant (Grissa 1991). Privatization, therefore, has led to even more intense political pressure to generate employment to 'prove' the value of the country's new economic philosophy. Between 1962 and 1992 tourism accounted for an uninterrupted rise in direct employment and a decline in the annual average deficit between labour supply and demand (Table 13.6). Since tourism has achieved positive results in generating direct employment, it is understandable that policy-makers find it appealing. The direct employment trends for a recent five-year period substantiate the criticism that job growth is concentrated in the high density tourist enclaves such as the Nabeul-Hammamet, Sousse-Kairouan and Djerba-Gabes regions (Table 13.7).

A 1988 study directed by Office National du Tourisme Tunisien (ONTT) estimated that the number of jobs directly attributed to tourism ranged between 0.88 and 1.2 per hotel bed, a figure not unlike that found in other LDCs (Green 1979; Smaoui 1979). ONTT reports, however, that for 1992 the ratio was well below that range as the addition of 6654 beds to the infrastructure created 2661 direct jobs which translates to a rate of only 0.39. It is, of course, difficult to obtain data on the indirect spin-off of

Table 13.6 Direct and indirect employment from tourism

Year	Direct	Indirect
1987	40,182	100,000
1990	46,614	116,000
1995	63,000	250,000

Source: ONTT (1995).

Table 13.7 Growth of direct employment by tourism region (1988–92)

Region	1988	1989	1990	1991	1992
Tunis–Zaghouan	5948	5877	6131	6281	6246
Nabeul–Hammamet	11,721	12,116	12,882	13,414	14,633
Sousse–Kairouan	9023	9677	9818	10,317	11,160
Monastir–Sfax	6169	6509	6682	6801	7647
Djerba–Gabes	6512	6991	7842	8788	10,197
Gafsa–Tozeur	1416	1612	2034	2389	2728
Bizerte–Tabarka	1150	1126	1225	1285	1613
Total	41,939	43,908	46,614	49,275	54,224

Source: ONTT (1992).

the new positions or on the quality of these jobs but, typically, resort industry employment in LDCs offers rather low-paying, marginally challenging labour as domestics. Females accounted for 16 per cent of all tourism personnel, confirming Enloe's observation that 'they are presumed to be naturally capable of cleaning, washing, cooking, serving. Since tourism companies need precisely those jobs done, they can keep their labour costs low if they can define [them] as women's work' (Enloe 1990). When such a label is given, then the work done by women outside the household is usually recognized as supplementary labour, justifying lower wages for women workers (Mies 1986). There has been very little scholarly research on the specific employment role of women in tourism (Kinnaird and Hall 1994), although much effort has been directed in the broader area of women's roles in economic development (Afshar 1991; Charlton 1984; Sebstad 1989; World Bank 1989).

The situation in Tunisia is similar to that which prevails in most African countries, in that the labour requirements of tourism are especially 'suited' to the prevailing economic conditions characterized by high unemployment, low levels of education and skills. When compared with most other industries, tourism can employ people with relatively little specialization. Thus, it is possible to absorb a large proportion of the workforce from traditional sectors of the economy with a minimum of training.

Tourism creates relatively few managerial and professional posts, and these are often filled from other sectors and/or by specially-recruited expatriates. To counter this problem, several hotel/restaurant management schools and institutes have been set up which have become, in themselves, a growth industry. In 1988, for example, over 300 diplomas were awarded, but that figure more than doubled by 1992 to 667 in various categories of the profession. Today more than 2000 hotel restaurant diplomas are awarded from the several speciality academies in the country (Office National du Tourisme Tunisien 1995).

The expansion of this industry, and others in Tunisia, led to a migration of those from agriculture and fishing who believed that work in either a factory or a hotel was less back-breaking and offered the advantage of a fixed income. Considering that almost all Tunisia's beaches are choice locations for packaged tourism, efforts to maximize that blessing have contributed to a disproportionate growth along the Mediterranean littoral at the expense of the central and southern regions of the country, and often to the detriment of agricultural production (Larson 1991). Attempts to increase the productivity of the domestic agriculture sector to provide food supplies for European palates can impact on women's status and income by increasing financial gains from farming. More often than not, however, patterns of land ownership in Tunisia, as elsewhere in the developing world, disadvantage females and the trend to large-scale mechanization (Larson 1991) further marginalizes rural women producers.

Sociocultural issues

It is difficult to gauge the sociocultural impact of tourism with precision because, unlike most Islamic/Arab countries, Tunisia is divided between a European and an Arab orientation. Former President Bourguiba, and the anti-Islamic reforms of the Neo-Destour socialist movement begun in the early 1930s, was very effective in Westernizing both elites and masses in Tunisian society to the degree that Tunisia is the most 'liberal' of the Muslim societies in the Arab world. Women, for example, have long been emancipated by reforms in status laws achieving rights well ahead of most of their Arab counterparts (Tunisian External Communication Agency 1993). Although traditional dress is still seen among men and women, western clothing is the predominant choice. Furthermore, it is difficult to single out tourism as a factor in sociocultural change when Tunisian society has been influenced for decades by Western television programmes. The predominance of French and Italian language programming has already undermined Arabic language and values, reinforcing the pervasive presence of European culture; thus, arguably, softening Tunisians to be more 'amenable' to tourists.

Notwithstanding the above caveats, however, is the fact that Tunisia is still an Islamic society with values very different from the hordes of European tourists who populate its beaches in search of sun, sand and sex. The cultural impact is real and cannot be taken lightly. Travel plays an important role in Islamic culture because of the required pilgrimage to Mecca and is discussed at length in the Qu'ran. Furthermore, vestiges of ancient traditions among desert nomadic peoples concerning safe travel and hospitality still mark Arab culture. The Islamic view of travel is, according to Din, 'to help instil the realization of the smallness of man and the greatness of God' as well as to promote brotherhood within the Islamic Ummah (community) (Din 1989).

Packaged tourism, however, has secular needs characterized by hedonism, permissiveness and very little cross-cultural understanding and communication. Muslim countries either discourage tourism, e.g. Saudi Arabia, or seek to accommodate it, as is the case in Tunisia, by isolating it from the mainstream (Din 1989). The construction of enclaves, however, does succeed in making the intercultural contact transitory but does not entirely limit the impact. In the many beach enclaves, European values and activities reign supreme. Contact with Tunisians are rarely spontaneous and, when they do occur, are likely to be contrived. Sights which are common in the West, such as scantily-clad visitors on the beach or around the hotel pool, and open affection between men and women, offend many Tunisians. Additionally, tourists knowingly or unknowingly violating rules of propriety in and around mosques and Islamic religious activities provide fuel to Islamic fundamentalists who criticize the excessive Westernization of Tunisian society.

The backlash can be seen in such behaviour as women returning to the veil as a symbolic protest over increasing Western influences and in organized political opposition along Islamicist lines which the government has suppressed, tainting Tunisia's human rights record (Poirier and Wright 1993). More specific to tourism, workers at the seaside resorts live in the restricted world of their elders but work daily in the world of exotic nightlife and semi-nude beaches. According to Waltz, 'no one feels these contradictions more than do young women,' who are 'raised with the notion that their bodies are the symbol of their families' honour' so that the return to Islamic dress 'is perhaps the only course of guaranteed safety' (Walz 1986: 665).

From the perspective of Islamicists in Tunisia, tourism is a cause of adverse cultural impact. Although the Islamic movement is very complex as it ranges along a broad spectrum of thought, it holds in common a concern for the deterioration of Islamic values (Magnusson 1991). In many countries, such as Thailand (Cohen 1982) and The Gambia (Harrell-Bond 1978), where tourism is emphasized, much has been written about the rise of prostitution (Crush and Wellings 1983; Enloe 1990; Graburn 1983).

Although not on the magnitude of Thailand or the Philippines, Tunisia has not escaped this problem. Data on prostitution, however, are nearly impossible to acquire in Arab countries. Nevertheless, Tunisia's prostitution, like The Gambia, is predominantly male prostitution and quite obvious at any of the tourist enclaves. Prostitution anywhere today carries a high risk of AIDS, especially in Africa, as the World Health Organization estimates that half of the world's victims are African, while 80 per cent of female victims of the virus are African (*West Africa* 1991). The full implications of this terrible human tragedy for tourism, and vice versa, are unclear in general and for Tunisia in particular because governments generally are not forthright on this issue.

Environmental concerns

Tourism's impact is not, of course, limited to the human and cultural environment. The idea that 'tourism kills tourism' by a heavy physical impact at a popular site is well known in the literature (Preglau 1983). The 1989 Hague Declaration on Tourism, a manifesto for 'ecotourism', called for a tourism promotion policy that is consistent with 'sustainable development'. Although Tunisia has not yet reached the mythical 'carrying capacity', the growth of the industry is clearly causing concern in some circles. Despite the 'success' of the Rio Earth Summit in 1992 in placing the environment on the global agenda and considering the link between environment and development (Reed 1992; Rogers 1993), few LDCs have well organized environmental movements. Nevertheless, there is a growing body of literature on environmental issues in the LDCs

(Goodman and Redclift 1991; Merchant 1992; Panayotou 1993) but little of it deals specifically with tourism in these regions (Briassoulis and van der Straaten 1992). Tunisia, however, is a country that is mindful of environmental issues and is, in fact, a leader among LDCs in establishing progressive policies to facilitate a peaceful coexistence between tourism and the environment.

As a starting point, it is important to consider that the establishment (in 1991) of Tunisia's Ministry of Environment and Land Management, and the emergence of stricter legislation, highlights the country's desire not to walk the same development path as the West. Establishing environmental policy at the ministerial level shows governmental commitment to improving the quality of the environment and the recognition of the significance that the environment, particularly coastal land and water, has to the tourism industry.

The Ministry of Environment and Land Management incorporates several environmental agencies that have been around for two decades. The decision to place these agencies under the rubric of a ministerial level department gives them greater political and legal authority than ever before. Although the National Office of Sanitation (ONAS), founded in 1974, is the 'principal actor in the battle against water pollution and the protection of water reserves' (Ministere de L'Environnement et de L'Amenagement du Territoire 1993: 12), the first institutional response to the overall environment and its problems, however, was in 1978 with the establishment of the National Commission on the Environment (CNE) which inspired environmental agendas in both the agricultural and economics ministries. CNE was a force in the development of the National Agency for the Protection of the Environment (ANPE) founded in 1988. ANPE was the first autonomous agency with *general* and *intersectoral* authority in environmental matters. It has the double mission of analysing the state of the environment through funding research activities and acting as the law enforcement arm of the state against violators of environmental regulations. In 1992, for example, the agency issued 447 citations of which nearly 55 per cent were litigated in court (Ministere de L'Environnement et de L'Amenagement du Territoire 1993: 13).

A sure sign of the country's political commitment is that, at various international and regional fora, Tunisian President Ben Ali has called on developed countries to accept the formula of debt for nature swaps as a way to help developing nations solve debt problems *and* finance eco-protection projects. Additionally, Tunisia's Eighth Development Plan does earmark roughly 1.4 billion TD for environmental projects, including 598 million TD in direct investments. The money will be spent through a series of ambitious programmes recently launched by the ministry that considers everything from coastline to desert. To reduce soil erosion and desertification that cause annual losses of nearly 50,000 acres of arable land, the government has launched a massive reforestation project

(Project Green) (Ministere de L'Environnement et de L'Amenagement du Territoire 1993: 25). The campaign against desertification involves 20 projects, most of which are completed. In the National Strategy for Reforestation (1990–2000) plan, Tunisia will develop some 2000 km/annum of forest barriers against the encroaching desert and the reforestation of 32,000 hectares/annum (Ministere de L'Environnement et de L'Amenagement du Territoire 1993: 24).

Future commitments of money are also going towards conserving biodiversity and promoting ecotourism as an alternative to sun and sand. Tunisia's desire to focus some tourism development away from the littoral to more adventurous tourism in the Saharan region in the southwest is an example of this commitment. In 1992, areas like Tozeur, Kebili and Gafsa experienced a 60 per cent increase in visitors from the previous year (*Tunisia Digest* 1993: 4). Hotel construction in these areas alone increased by 132 per cent from 1987 to 1995, far outstripping any other zone (Office National du Tourisme Tunisien 1995).

Project Green, which is administered jointly by the environmental, agricultural and health ministries, is intended to conserve biological resources and ecosystems as well as diversify tourism and promote ecotourism. Financed partly by Swedish and German money, Project Green involves national park development and management, and the establishment of environmental museums (Ministere de L'Environnement et de L'Amenagement du Territoire 1993: 25).

Tourism's major impact is that it always involves the requirement of an extensive infrastructure. Unfortunately, the building of roads and the expansion of hotels, major aspects of that infrastructure, can significantly alter the physical environment due, in the latter case, to the problems of sewage treatment and refuse control. In the absence of organized environmental opposition, and perhaps because of that absence, it is interesting to note that the government was seemingly more concerned with the environment's impact on the tourism industry than the reverse. In that regard, the Seventh Development Plan (1987–91) recognized that the *existing* physical environment did 'not contribute to the improvement of Tunisia's image', and had a detrimental effect on the tourism industry. This assessment forced the government to draw attention to the problems associated with public hygiene and called on municipalities to be sensitive to 'the problem of garbage removal, beach maintenance ... and insect control' (Office National du Tourisme Tunisien 1987).

The greatest concern lies in cleaning up Tunisia's overpopulated coastline. The coast is home not only to tourists but to the majority of the country's industries and people. Tunisia, like most LDCs, has experienced rapid urbanization. In 1952, a decade before significant tourism development began, the urban population was 32 per cent. Currently, it is over 60 per cent, with 75 per cent of that population residing along the littoral (Ministere de L'Environnement et de L'Amenagement du Territoire

1993: 66). With well over 80 per cent of the tourism infrastructure concentrated along the coast, it is easy to see what impact tourism's employment attraction has had on Tunisian demographics (Poirier 1995: 159). Demographic shifts of this magnitude inevitably bring environmental problems, particularly in effluents and solid wastes, where none existed before. To reduce the heavy quantities of toxic and other wastes, the government is establishing pollution reduction measures such as wastewater treatment plants and fines for polluters. Other projects include controlling coastal erosion and petroleum spills.

For example, Project 'Blue', mandated by the Environment Ministry is, to the tourism industry, the most important of the government's environmental programmes as it involves measures to protect the Mediterranean littoral. Broad and comprehensive, the project is intended to achieve and maintain bathing water conditions optimal to public health along the coast. Financed partly by state funds and technical assistance from Germany and Monaco, Project 'Blue' establishes a central water quality assessment laboratory in Tunis, five regional laboratories located at principal tourist sites, and 32 water purification stations. The coastal protection plan also involves constructing six effluent control systems that will remove contaminants from the major public beach areas of Tunis and Nabeul-Hammamet (in the planning stages), Sousse and Jerba (on line since 1993), and Monastir and Mahdia (currently under construction) (Ministere de L'Environnement et de L'Amenagement du Territoire 1993: 63).

Coastal water quality, although important in its own right, has little touristic value per se, without paying attention to the quality of the beaches. The government's commitment in this area, under the auspices of Project 'Blue' is also evident, with a US$2 million allocation to maintain clean standards for beaches. This involves appropriate equipment to maintain beach sanitation and the construction of more toilet and shower facilities on beaches (Ministere de L'Environnement et de L'Amenagement du Territoire 1993: 60).

Public awareness

The connective link between issue and political movement involves public awareness. George Lombardo, programme manager for the World Environment Center, which recently completed an environmental audit of Tunisia, argues that there seems to be a rising consciousness among Tunisian industries and people towards the environment (Lombardo 1995). Beyond 'consciousness', he believes that the government is making real moves to clean up the environment. 'Consciousness', of course, inevitably comes into conflict with vested, 'pragmatic' interests. Under economic liberalization policies promoted by President Ben Ali, steps taken to placate the International Monetary Fund and Western banks

(Moore 1991: 67–97), the entrepreneurial spirit has taken hold in Tunisia. Private industries, not unexpectedly, complain of being squeezed in one direction to make heavy investments in pollution prevention and clean-up technologies, and in another to compete under the country's genuine and extensive privatization initiatives (Poirier 1995: 162).

This raises the issue of whether the country's investment in cleaning up the environment should take precedence over investments to eliminate pollution from the start. The Tunisian government does provide legal incentives for new projects to reduce pollution. But what about existing projects? Is it possible for the government to press for more regulation without alienating large sectors of Tunisia's growing industries? Adding additional financial burdens on industry might slow down growth, placing more people on the street in a country where unemployment is already at 14 per cent (Poirier 1995: 164).

The 'consciousness' issue is important, nevertheless, but public environmental awareness will continue to lag behind serious efforts to change attitudes through what are excellent educational materials and programmes provided by the government. At the personal behavioural level, Tunisians have not yet fully acquired habits of recycling; people still routinely throw away waste materials without concern for the aesthetic or environmental consequences.

The government, however, must be applauded for its recognition of the importance of public education to improve the environment. It is necessary to emphasize that the Ministry of Environment and Land Management is only six years old. In that short period of time Tunisia has become a role model with public education programmes that should be adopted in countries like the United States. The Tunisian government has the fortitude for meeting these problems directly, recognizing that tourism and the environment work together for the overall health of the economy.

Government policy dictates that it is the responsibility of government to 'sensitize the public about the problems of environment and development' and that it is 'each citizen's personal obligation to protect the environment' (Ministere de L'Environnement et de L'Amenagement du Territoire 1993: 336). To accomplish this objective the Ministry of Environment and Land Management has launched an intense and comprehensive public education campaign on environmental issues affecting all segments of society but concentrating primarily in the educational system at each level (Ministere de L'Environnement et de L'Amenagement du Territoire 1993: 99). In addition to the use of the state controlled mass media to heighten awareness, the government has adopted a mascot which is a caricature of a fennec fox, indigenous to North Africa, named 'Labib', whose image is emblazoned on billboards, bumper stickers, T-shirts and various beach paraphernalia sold to tourists. The mascot has already captured the imagination of school children and may well become as popular as the Smoky the Bear logo of the US Forest Service.

To further the goal of environmental awareness, the government has launched a major campaign of curricular reform at the level of the primary and secondary schools throughout the country. Environmental modules are now required in the curricula of language, natural sciences, geography and civic education courses. Schools, themselves, have taken a leadership role and have formed over 1000 'environmental clubs' similar to the American 4-H Club model (Ministere de L'Environnement et de L'Amenagement du Territoire 1993: 100).

Conclusion

Given the pessimistic future of the economy in Africa, it is likely that policy-makers will continue to emphasize tourism as has been the case in Tunisia. Unfortunately, there is no consensus among scholars or policy-makers as to the long-term viability of tourism as a sound investment strategy for economic development. A sanguine assessment of Tunisia's success with environmental protection may be premature this early in the process. To date, the country's effort in the environment have largely won praise by outside experts and is certainly one of the most effective in the Middle East (1995). Tunisia's growing tourism industry and the country's dependence on that industry necessitate sound environmental policies. The cost/benefit ratio between perceived or real economic gain and environmental consequences is difficult to gauge. However, Tunisia's approach to environmental policy and balanced tourism appears to be heading in the right direction with future prospects looking positive. Furthermore, Tunisia is seen as a leader in responsible tourism and environmental management as more and more developing countries recognize the need for balance in these policy areas.

Tunisia is likely to continue on this path in the foreseeable future. Larger global issues, like the Gulf War, or regional problems, such as in Algeria and Egypt, indicate Tunisia's vulnerability. Political instability, both regional and Tunisia specifically, could seriously impact future economic growth if policy-makers place too much emphasis in this one sector.

14 The political economy of tourism in Algeria

Yahia H. Zoubir

Introduction

Unlike its neighbours, Morocco and Tunisia, Algeria never exploited tourism as a major economic sector. Despite great potential, due to remarkable natural sites, Algerian authorities shunned tourism for social and religious considerations. Because the country enjoyed important hydrocarbon resources, the political leadership devoted little attention to what could have been an extremely lucrative industry. However, since the mid-1980s, the Algerian regime sought to compensate for the dwindling foreign earnings (due to the drop in oil prices) by progressively liberalizing and privatizing the tourist industry. This chapter examines the Algerian tourist industry since the country's independence. The objective in this chapter is to analyse the reasons that prevented successive Algerian governments from promoting this industry. As this chapter will show, economic necessities were the main reason for the authorities changing their attitude towards tourism. Furthermore, the encouragement of tourism may have come too late; the rise of religious extremism and the violence, which has ravaged Algeria, have hampered even further the development of this industry.

Algeria's tourism potential

Algeria is no doubt one of the Mediterranean countries that offers the greatest potential for the development of tourism. The second largest country in Africa, about three-and-a-half times the size of Texas, Algeria enjoys an ideal location in the southern Mediterranean; it possesses 1200 km (about 800 miles) of beautiful coastline. The variety in topography and climates allows for tourist visits all year round. The sand beaches of Algeria are legendary and so are its Hoggar Mountains in the deep Sahara desert. Furthermore, beautiful oases such as Bou Saada, Ghardaia (in the historic M'Zab region), Biskra, Tougourt, and El Oued are scattered in the Sahara. The beautiful canyons in the Aurès Mountains resemble those of Arizona in the United States. The innumerable vestiges that underscore Algeria's rich history, especially Roman, Arab, and Ottoman, can be found throughout Algeria's vast territory. Impressive Roman ruins dominate the cities of Tipaza and Cherchell in the west and Djamila (near Sétif) and Tebessa in the east. The two natural museums, the Tassili and the Hoggar, contain prehistoric rupestrian paintings; UNESCO has classified these natural museums as 'belonging to universal patrimony'.[1]

Tourism in the colonial era

The French colonial authorities did not develop a tourist industry in Algeria; such a business activity would certainly not have been lucrative in the 1950s because of the bloody war of national independence (1954–62) that Algerians fought against the French. Moreover, the French made no effort to industrialize the country. In fact, Algeria fits the typical model of a colonial economy, whose main sector was that of extractive industries. Raw materials and agricultural produce constituted the primary commodities, which the French colonists shipped to the metropolis. Yet, although the country did not witness any major industrialization drive, the French authorities built a relatively satisfactory infrastructure; the main objective of such infrastructure, of course, was to serve the colonial system that the French colonialists had put in place. The French-built hotels, businesses, administrative buildings, theatres, airports, harbours, museums and, more importantly, one of the most developed railroad system in Africa; with a length of 4300 km (Bennoune 1988: 90).

Tourism in independent Algeria

In spite of its extremely attractive potential, Algerian authorities shunned tourism, at least in its large-scale version, from any development programmes. This policy lasted from the country's independence in 1962 until the mid-1980s. The successive governments were aware that tourism could constitute an important source of revenue, but a number of factors precluded them from developing that potentially lucrative sector. First, Algerians fought one of the bloodiest anti-colonial wars against France, a Western power. The war against the French negatively affected the perception that Algerians held about Westerners, the likely visitors to Algeria. The brutal character of the colonization (1830–1962) and the fierce nature of the war (1954–62) produced a fervent type of nationalism rarely equalled in the Third World. Second, French attempts to destroy Algeria's religious and cultural identity produced adverse effects in that the protection of one's identity became paramount. A sense of Puritanism prevailed throughout the society. Therefore, from both a nationalistic and a religious perspective, tourism represented an unattractive form of enterprise. In fact, many Algerians feared that tourism would represent a neo-colonial form of domination. Another factor which prevented the expansion of tourism immediately after independence was the structural problem Algerians inherited from the colonial era. Indeed, at independence, the tourist industry was virtually non-existent; only 5400 beds (in poor condition) were available throughout the country.[2]

Clearly, whatever their true motivations, successive governments in Algeria saw tourism not only as a threat to Algerian society's cultural and nationalist identity, but also lacked the means and the experience to develop tourism in the aftermath of the country's independence. The determining factors, though, for not developing that industry were primarily moral and cultural. Indeed, many Algerians criticized their neighbours, Morocco and Tunisia, for having allowed Western cultural 'imperialism' to pervert their cultural identity through tourism. Nevertheless, in reality, Algerians were far more fortunate than their neighbours, for Algeria's hydrocarbon wealth (oil and natural gas) spared the country from relying on sectors such as tourism for national development. Furthermore, having opted for a socialist path of development, albeit within the context of Islamic values, Algerian authorities did not allow the development of a heavy presence of Western multinational corporations. Indeed, many MNCs had expressed a genuine interest in building hotels and resorts along Algeria's coastline and in the desert. Algerian leaders pursued a nationalistic and populist approach, which rejected any type of economic activities that would have helped foreigners establish control over the country's natural resources. These and other factors largely explain why Algerians opted for a type of national development which discouraged tourism. Instead, the authorities, especially under Houari Boumediene's rule (1965–78), believed that import-substitution industrialization (ISI) would create the conditions for a self-reliant, prosperous society. The hydrocarbon sector provided the main source of financing for industrializing industries and the creation of jobs. Following the creation of state-owned enterprises (SOEs) and the nationalization of natural resources and other key sectors of the economy, the state became the instrument of economic development.[3] The state was omnipotent; except for the single-ruling party, the Front de Libération Nationale (FLN), and its mass organizations, the regime prohibited the legal existence of autonomous associations and parties. In other words, Algeria built a strong state around a strong military; the authorities impeded the emergence of an independent civil society (Zoubir 1999).

Under Boumediene's rule, the authorities were not oblivious to the potential for tourism. They were fully aware of the benefits that the country could reap should they decide to allow the promotion of tourism. The National Charter of 1976, which contained a strong socialist direction, recognized that 'there are ... many economic factors operating in favour of a rapid and extensive development of Algeria's tourist potential'.[4] The regime viewed tourism in utilitarian and populist terms. One can break down the vision of the framers of the Charter regarding tourism into two categories: national/mass tourism and international tourism. They identified various functions within each of those categories.

National tourism

The model that they proposed was clearly an imitation of what the Soviet Union and the East European countries had instituted. Most of the Algerian leadership believed that national tourism would represent a major source of employment. But they also believed that it was incumbent upon the state to create the proper conditions of leisure and vacation for the employees (and their families) of the public sector, i.e., the majority of the workforce. The authorities entrusted the SOEs with the construction of the necessary facilities for their employees and for the citizens, as well as for foreign tourists. Because SOEs were responsible for the building of tourist facilities, Algeria's rulers were confident that 'workers and their families [would] thus find it possible to benefit from leisure activities adapted to their conditions and placed within reach of their incomes'.[5] The regime hoped that the private sector, which was excluded from other major sectors of the economy, would partake in the development of the tourist industry.

The other rationale for creating a tourist infrastructure stemmed from the need to offer hotel facilities for employees who, because of the new economic expansion, had to move around the country and thus needed places to stay. This objective, however, created a major problem in that tourism in Algeria 'has become hotel management [hôtellerie]!'[6] This is an accurate description of the situation; the dreadful housing shortage in Algeria has become legend. In the 1990s, the resorts along the coast have become housing facilities for the Algerian *nomenklatura*. Only a few housing facilities are reserved for rentals. And even those require strong connections within the bureaucracy.

There existed yet another rationale for national tourism in that 'Algerians, especially those who want to discover and know their country, will be able to make use of the appropriate tourist facilities everywhere'.[7] The true objective, however, as admitted by the framers of the Charter themselves, was 'the currency savings resulting from the fact that the facilities created at home will lead many Algerians to spend their holidays in their own country'.[8] Given that the government had restricted travel abroad and limited the amount of foreign currency Algerians were allowed to take with them, mass tourism became a mere device to prevent the export of foreign currency.

International tourism

Two factors motivated the expansion of international tourism. The first factor was predicated upon the fact that international tourism 'could play an increasingly significant role in foreign exchange earnings, because of the foreign currency which visitors from abroad bring to the country'.[9] The second factor was that international tourism 'opens the door on the outside world and affords a means of communication with other nations'.[10] Whilst the framers might have been sincere about such rationale, the nationalist and cultural factors mentioned earlier compelled them to qualify their appreciation for international tourism. Indeed, they argued that 'the task assigned to the tourist sector in Algeria of obtaining overseas currency ... has to be contained within certain limits in order to preserve Algerian society from inconveniences that can arise from the invasion of developing countries by large numbers of foreign tourists'.[11] Although Algerians are known for their hospitality, they are also quite conservative. Unlike the political elite in Tunisia and in Morocco, the Algerian ruling elite was more prudent in promoting international tourism for cultural, nationalist, and, one might add, patriarchal reasons. Nonetheless,

[T]he Revolutionary Power has launched a major program for the construction of hotels, holiday villages, tourist complexes, and thermal resorts throughout the country, thus increasing the value of the attractions which the country already possesses in this area. Many similar projects will be undertaken in the future. In this way, Algeria will have a modern infrastructure spread throughout the whole country, adapting itself to the seasonal specialities of its main regions.[12]

With the help of the famous French architect, Fernand Pouillon, the state built giant tourist sites, especially near large cities along the coast: Zeralda, Sidi Fredj, Moretti, Club des Pins, Tipaza (all near Algiers); Les Andalouses (in Oran, western Algeria); and a site in Annaba (eastern Algeria). Impressive hotels were also built in Saharan cities such as Biskra, Ghardaia and Bou Saada. The new sites were of good quality, especially in their first few years of existence. The government made major efforts to attract foreign tourists; the Office Nationale Algérien du Tourisme (ONAT), Touring Club, and the Agence du Tourisme Algérienne (ATA) were the two state-owned enterprises in charge of promoting tourism and organizing tours. But because of the artificial exchange rate fixed by the authorities, which provided the Algerian *dinar* with more value than it was really worth, international tourists preferred to go to Tunisia and Morocco, where they enjoyed more competitive deals. Furthermore, Tunisians and Moroccans had more experience in the tourist industry, which represented one of their main sources of revenue.

Algeria's tourism in the era of *infitah* (liberalization)

Houari Boumediene's death in 1978 marked a turning point in Algeria's strategy of development. His successor, Chadli Bendjedid, more pragmatic and better predisposed toward the private sector, geared the country toward a slow policy of *infitah* (opening), i.e., of economic liberalization. Progressively, a new legislation (e.g., Investment Law of August 1982) allowed the national private sector to play a much greater role than under the previous regime.

It is interesting at this stage to ask the question, 'What was the logic of encouraging tourism?' A recent study on Syria (Gray 1997: 57–73) offers an excellent answer to this question:

- importance of tourism in foreign currency earnings, especially in countries with artificial exchange rates or those suffering from balance of payments problems;
- tourism is labour-intensive and creates employment;
- tourism does not require expensive or sophisticated technology or a highly-skilled labour force;
- the state that decides to encourage tourism usually has a nice climate, historical sites, beautiful landscape and amicable people.

As the author puts it, 'governments often feel that their state possesses an untapped economic resource, and decide to take advantage of it' (Gray 1997: 58). These propositions apply to the Algerian case, as well.

The emphasis on tourism coincided with the economic liberalization that began in 1979. In fact, as of 1980, the state encouraged private savings, through financial and fiscal incentives, to be channelled toward investments in tourism. The restructuring of SOEs also applied to tourist enterprises. The authorities created 17 Entreprises de Gestion Touristiques (EGT) [tourist management enterprises]. The EGTs enjoyed, at least in theory, greater autonomy. The authorities sought to create more efficient tourist enterprises whose primary consideration would be profit. The Entreprise Nationale des Études Touristiques [ENETs] complemented the EGTs, ONAT, and Touring Club. The role of the ENETs consisted of studying potential sites and supervising the construction of the new sites. In the decade 1977–87, the state's schools for hotel management trained 4554 students, including 100 African and Middle-Eastern students.[13]

The efforts the Algerian state made to attract tourists produced few tangible results. In 1985, for instance, the total number of tourists reached 407,353. However, out of this total, 168,262 came from Tunisia. Tunisians came to Algeria because most commodities in Algeria were still subsidized by the state during that period. Many products, such as basic food products and pharmaceuticals, were much cheaper in Algeria than they were in Tunisia, or in Morocco for that matter. The second largest group of tourists came from France with 124,510. But this number is misleading, as most of those were Algerians residing in France. They usually came in great numbers for the summer vacation. If we discount 'tourists' from Tunisia and France, the total number of tourists for that year would only be 114,581, a figure much closer to reality. In fact, in the following years, 1986 and 1987, the number of visitors was much lower.[14] In other words, Algeria attracted fewer than 100,000 tourists per year. In 1988, the total number of beds (public and private sectors) available was just over 48,000. Generally speaking, the hotels were run-down and were no match for the quality of the hotels in Tunisia and Morocco, countries which historically have attracted many more tourists than Algeria.

The drop in oil prices and the severe economic situation which shook Algeria in 1986 forced the state to look closer to the potentiality of the tourist industry. In view of the constant deficit that the tourist industry experienced, the authorities decided to promote that sector. The state now attached greater importance to tourism; indeed, the government set a few guidelines for tourism in the second Five-Year Plan (1985–89) that highlighted the growing significance of that industry. Not only did the state planners suggest that the sector be expanded, but the government also made it plain that private entrepreneurs should become more involved in the financing and operation of projects in tourism. The government sought to bring the number of beds available to 100,000 by the year 2000. At first sight, this number does not seem high; but such a figure was extremely ambitious knowing the severe problem that Algeria faced in the construction industry. Algeria has confronted, up to this date, one of the worst housing shortages in the world. That shortage, in fact, has been one of the leading causes of social discontent and riots in the country. The shortage is so important and the population growth so high that whatever the goodwill of the authorities to build new units, the deficit could hardly be alleviated.

In the late 1980s, tourism became an attractive option for the regime; the government was hopeful that tourism would bring the much-needed foreign currency to compensate for the downturn in oil revenue. The authorities, which made tourism an important part of the liberalization scheme, launched a big promotion campaign in Europe and offered competitive prices in the hope, rather over-ambitiously, of doubling the number of tourists by 1989 to between 600,000 and 800,000.[15]

In 1987, the FLN-dominated parliament, the *Assembée Populaire Nationale* (APN), approved the fiscal budget which granted tax-relief measures for tourism, with corporate tax to be reduced to 30 per cent on profits rein-vested in Algeria and 20 per cent on profits for joint-venture companies.[16] The government was genuinely intent on developing the tourist industry, the decentralization of the national company being a case in point. However, foreign investors remained sceptical because the government did not allow for more than 49 per cent foreign holding. Furthermore, the Algerian government's unwillingness to allocate greater investment in that sector, thus showing little commitment to it, dissuaded foreign investors from embarking on major tourist projects in the country.

Despite the scepticism of foreign investors in Algerian tourism, a few joint ventures did develop, especially with Canada (Groupe GPL International), France (PANSEA, part of ACCOR; SOFITEL), Italy (ALFRA), China and Kuwait (Kuwait Algerian Investment Company, KAIC). The Korean Daewoo Corporation exhibited increasing interest in the tourist industry, especially in the construction and management of hotels such as the Hilton, in a joint venture with the local Safex. At the domestic level, a number of private tourist agencies made their appearance in the tourism market.

The Algerian regime's desire to promote the tourist industry was genuine. The primary motive was naturally the need for foreign currency. In the 1980s, the government emphasized international tourism. In 1987, the authorities planned to spend US$4 billion over the next twelve years to build new hotels.[17] The objective was the creation of an additional 120,000 beds. But again, the government's failure to make a major financial commitment to the tourism sector deterred foreign investors from taking any risk, especially with a government whose co-operation was unsure.

Political economy of tourism in the 1990s

Whatever the true intentions of the Algerian government in the 1980s, the political evolution of the domestic situation represented a real blow to the tourist industry. In 1988, riots in Algiers, as well as in other major cities, shook the country. Though tragic, the riots forced the regime to allow a degree of democratization in the country. The process resulted in a multi-party system. The Islamist movement, which grew considerably in the 1980s, emerged as the most powerful social and political force opposed to the incumbent regime. The main Islamic party to emerge in the political landscape was the Islamic Salvation Front (FIS). Though socioeconomic considerations rather than theological motives were the driving force of the FIS, the party espoused an ideology that contained a high dose of religious and nationalistic values (Zoubir 1995: 109–39). The members of the FIS interpreted those values, especially the religious ones, in their most conservative dimension. Municipal and departmental elections took place in June 1990. The FIS won an impressive 54.2 per cent of the vote against 28.13 per cent for the old single-ruling party, the Front de Libération Nationale (Zoubir 1995: 124).

The FIS electoral victory had to have a serious impact on tourism. The FIS now controlled important tourist centres, such as Tipaza and Jijel, a city in perhaps one of the most beautiful cornices in the world. Opponents of the FIS might have exaggerated the intentions of the FIS regarding

the threat the party posed to the secular aspects of Algeria. But it is undeniable that some of the first decisions the new elects made consisted of
banning the wearing of shorts for both men and women in their districts.
In other districts, the FIS representatives demanded that foreign correspondence be in Arabic only. And in yet other communes, the FIS council
members banned the production and consumption of alcoholic beverages. Although the FLN in the past was guilty of the same temptation, the
FIS decisions came at a time when Algeria had shifted toward international tourism. More importantly, the FIS had raised a question which
had a more direct impact on tourism in Algeria: the banning of mixed
beaches. The FIS published an article in its newspaper, *El-Mounqid* (The
Saviour), which stated very clearly that there would be separate beaches
for men and women:

> As to mixed beaches, Islamic education does not allow their existence.
> It is against modesty and morality for a woman to expose herself
> naked [in a two-piece bathing suit] in front of the whole world. . . .
> The FIS must take the initiative and direct [Algerians] to demand that
> there be separate beaches for men and for women. . . . (Al-Ahnaf *et al.*
> 1991: 255–6).

Undoubtedly, should the FIS have come to office and applied Islamic Law
(*Shari'a*) as its members had promised to, the tourist industry would have
been severely restricted. One can only infer from some concrete acts to
bear such an opinion. For instance, the Islamists had forced all restaurants
to close during the holy month of Ramadhan (fasting), not even to serve
foreigners as used to be the case until the late 1980s. In fact, during
Ramadhan in 1990, Islamists launched punitive attacks against the only
restaurant that was opened in Algiers, thus forcing it to close down (Khelladi 1992: 115). This is not the place to compare the moralizing policies of
the FLN with those of the FIS; suffice it to say here that the FIS sought to
carry out much further some of the policies that conservative groups
within the FLN had intended on pursuing. But, despite its many privileges
and influence, the FLN never represented the real power in Algeria. The
military and the high bureaucracy had a greater say in the direction of the
country. The military, the backbone of the Algerian State, always stood as
a modernizing force. One of the reasons why the military, or at least one
of its main factions, opposed the Islamists centred on the issue of modernity. The military, especially the officer corps, held a much more secular
vision of society than did the members of the FIS, even those trained in
Western universities.

The legislative elections held in December 1991 resulted in a large
victory for the FIS. The second round of those elections would undeniably
have seen the overwhelming triumph of the FIS, which would have dominated the legislature. Civilian and military authorities, distressed by such a

prospect, decided to cancel the elections altogether. In March 1992, the government banned the FIS. The High State Council (HCE), the collegial ruling body headed by Mohammed Boudiaf, decided to fight radical Islamism. Boudiaf was assassinated in June 1992 and was succeeded by Ali Kafi, who promised to continue his policies. Both had, in fact, tried to reach out to the Islamists. The mandate of the HCE ended in early 1994 and Liamine Zeroual, a retired general, became the State's President. Zeroual, too, sought national reconciliation and dialogue with all opposition parties. But while he and his collaborators were more accommodating of the Islamists and sought to reach out to the FIS, the so-called 'eradicators' (as opposed to the 'conciliators') had no tolerance for the FIS. However, the authorities in general have tolerated the existence of moderate Islamist organizations which 'reject violence and abide by the laws of the republic'. In other words, the regime would tolerate those parties as long as they renounce the invocation of the *Shari'a* in politics or to have a religious identity. Two Islamist parties, the Movement for a Peaceful Society (formerly Movement for Islamic Society) and Ennahda (formerly Islamic Ennahda), today share 104 out of 380 seats in the National Popular Assembly (APN). Yet it is not clear what their attitude toward tourism might be. What is certain is that these parties would tolerate tourist centres only if those centres are reserved solely for foreigners and would not 'pollute' Algeria's Islamic values. The Algerian legislature is dominated by conservative forces, but it remains to be seen whether the conservative deputies will oppose the passing of laws which favour the growth of tourism.

In the 1990s, Algeria has been marked by a degree of violence not seen since the War of National Liberation. The armed conflict between government forces and armed Islamist groups has resulted in tens of thousands of deaths, the majority of those killed being innocent civilians. Massacres of defenceless villagers and car bombs in urban centres have created a state of insecurity in many parts of the country. This, of course, meant an incredible decrease in the number of tourist visits. The World Tourism Organization has reported that in 1997 the Maghreb (Algeria, Morocco, and Tunisia) received 7.2 million tourists, with Tunisia having the lion's share (55 per cent), followed by Morocco (37 per cent) and Algeria (eight per cent). The eight per cent figure suggests that Algeria received about half-a-million tourists, a figure virtually impossible even if one assumes that those visiting were emigrants visiting relatives in Algeria.[18] This is unthinkable in view of the warnings that have been issued by many European countries and the United States, advising their citizens not to travel in Algeria. In fact, on 31 January 1997, the United States recommended that 'Americans who choose to be in Algeria despite this warning have substantial armed protection while travelling overland, on their work sites or their accommodations; other Americans in Algeria should depart'.[19] Although no American citizens have been assassinated, the United States government issued travel warnings to Algeria because the Armed Islamic Groups (GIA) killed more than one hundred foreigners working or visiting Algeria. In 1993–94, the GIA issued several warnings to foreigners demanding that they leave the country or they would be killed. Although not all foreigners left Algeria or stopped coming to visit relatives or do business, they kept a very low profile and ventured only into highly secured areas. Foreign workers, especially in the hydrocarbon sector in the southern part of the country, are well protected. Undoubtedly, the figure provided by the World Tourism Organization for 1997 is unbelievable; in 1993, for instance, more than 300,000 Moroccans had visited Algeria. However, the borders between the two countries were closed in 1994 following a deadly terrorist attack carried out in Marrakech against Spanish tourists. The Moroccan authorities accused the Algerian secret services of involvement. Surely, the figure that the World Tourism Organization provided has no basis. The Algerian Office National des Statistiques (ONS) provides figures that contradict those reported by the World Tourism Organization, although the figures do not relate to the same year. According to ONS, in 1992, 121,514 foreign tourists visited Algeria.[20] This figure excludes those who came from the Maghrebi countries (Morocco, Tunisia, Libya and Mauritania); their number amounted in 1991 to 521,170.[21] Algerian official statistics reports, published by the government in 1997, indicate very clearly that the number of tourists decreased dramatically between 1990 and 1994. The number of foreign visitors, Maghrebis and other Africans excluded, represented a mere 73,091.[22] There is no reason to believe that the number of tourists suddenly increased, although security had improved considerably in the period 1996 onward.

The situation in 1993–94 aggravated the already-bad image that Algeria held within the tourist industry. In 1992, a high Algerian official in the Ministry of Tourism admitted candidly that 'for the tourist, Algeria is a black hole, rampant with Islamic fundamentalism, situated between Morocco and Tunisia' (Rowland 1992: 4). The government, which still sought to promote tourism, attempted to increase earnings from that sector. Reports indicated that Algeria's earnings from tourism amounted to a mere $85 million, whereas tiny Tunisia collected US$1 billion (Rowland 1992: 4). The problem, however, remained that even though it wished to promote that sector, the government did little to facilitate the task of private tour operators who had difficulty not only in obtaining bank loans, but when they did, the interest rates were prohibitively high. But by May 1992, the authorities decided to change course and to help the private sector to expand in tourism. In May 1992, the secretary of state for tourism announced the state's decision to discontinue all direct involvement in tourism and to privatize its assets. He declared that 'henceforth, the state totally disengages itself from investment and management of tourism, to concentrate on its true role which consists of stimulating national and foreign private investment with guarantees and incentive measure'.[23] In 1993, Italian, French, and German companies seemed quite interested in investing in tourism; the Chinese had already begun building a hotel near the Algiers (Houari Boumedienne) Airport, whereas the Koreans had started feasibility studies.[24]

Undoubtedly, the insecurity prevailing in the country in 1993–94, coupled with the death threats against foreigners, led to a sharp drop in the number of tourists visiting Algeria. The south, which was relatively spared from the violence, saw a 100 per cent drop due not only to the political violence, but also due to the banditry related to 'illegal African emigration [from Mali and Niger] to Tamanrasset'.[25] The violence and its impact did not seem to discourage the Algerian authorities, who decided in June 1994 to create a new venture, Sodextour, made up of several state companies, to promote tourism.[26] Clearly, in spite of the violence, the process of economic liberalization and privatization has proceeded unceasingly. The government has been positioning that sector for better times. As put by Tourism and Handicrafts Minister Mohamed Bensalem, 'In this transitional period, tourism can play an important role by quickly improving its products and yielding perceptible results. But we must be realistic; we are now working to prepare the after-security crisis take-off.'[27] The Mokdad Sifi government made the opening of tourism to local and private investments one of its priorities. The interest in this sector stemmed from the capacity of tourism to generate capital and to create jobs reasonably quickly. The government, encouraged by Legislative Decree No. 93–12 of 5 October 1993, which greatly facilitated foreign and local private investments, sought to completely disengage the state from management. According to a publication of the Ministry of Tourism and Handicrafts, the objectives of the privatization programme in tourism were:

- total privatization of management;
- total or partial privatization of capital;
- total privatization of future investments.

Furthermore, the stated long-term goal was the establishment of 'a tourist industry largely dominated by private capital'.[28] The only conditions that the authorities insisted upon were concern for the environment, especially in the fragile Tassili area in the Deep South, and that tourism be a dynamic, competitive, and high-class industry which will yield sizeable revenues. They also fear 'reckless' development which would lead to high-rises such as exist in coastal Spain. The preference was undoubtedly to encourage investments in the southern regions; those regions have traditionally depended on tourism as a source of revenue, but also because of the southern region's familiarity with that industry.

The process has affected the tourist industry, especially the acquisition of hotels by foreign investors. For instance, in 1995, Flamingo Hotels signed a contract with the Algerians to manage two 150-bed, four-star hotels (the Riyadh resort at Sidi Frej, and Les Zianides in the western city Tlemcen).[29] That same year, the Algerian government had issued yet another decree on privatization which covered, among other things, hotel business and tourism.[30] In June 1996, the state decided to sell more hotels. It invited bids from investors for the procurement of 13 hotels.[31] Clearly, one might expect that this process would continue, especially when Algeria succeeds in bringing civil peace and stability. This observation is corroborated by the fact that, in April 1998, Algeria hosted an important five-day international conference on tourism (SITEV 98, April 13–17, 1998), whose main objective was to attract foreign investors.[32] Although Algeria has the potential to become an important tourist attraction, the question remains, of course, whether the country can provide the necessary guarantees for foreign direct investments and whether the government authorities can convince the population of the benefits that tourism can bring to the country.

Conclusion

Unquestionably, the prospects for tourism in Algeria are tremendous. The natural setting makes Algeria one of the best candidates for tourism in the Arab and Islamic worlds. The beautiful virgin coastal areas, the mountains, the considerable health spa potential, and the huge desert make Algeria an ideal place for tourism. As stated earlier, the Algerian government has attempted to promote this sector forward because of the need to generate revenue and to create jobs. But the success for tourism in Algeria depends on a variety of factors. Obviously, security has been the primary concern since the early 1990s. Although it is true that the situation in 1997–98 is not as unstable as it was in 1993–95, the magnitude of the massacres and their horrifying barbarity have made Algeria an unlikely candidate for foreign tourism. The fake roadblocks, the bombing of trains, and the threats to foreign visitors from Islamist extremists and bandits are important considerations. Despite tangible progress, the state security forces are still not in full control of the vast territory.

However, problems of security are not the only obstacles to the development of tourism in Algeria. As the head of the National Tourist Agency, ONT, admitted in Naples, Italy, in late 1996:

> There are 18 hotel agencies totalling about 40,000 beds, only 7,000 of which are in conformity with international norms. In the private sector, there are more than 620 hotel establishments with a capacity of 20,000 beds, and around 250 travel and tourist agencies.[33]

But as the CEO of ONT correctly pointed out, this 'deficit in infrastructure can also represent real possibilities for investment, either in new constructions or in the renovation of hotel structures'.[34] There is no reason to doubt that this might well be a plausible development if foreign investors decided that Algeria would indeed be an attractive place for investment in the tourism sector. But this also depends on yet another factor: can Algeria develop a 'tourist culture?' The negative attitudes that the proud and nationalistic Algerians have toward waiting on tables or catering to tourists with a smile in order to merit tips are legend. The problem is how to change those negative attitudes. The other major cultural question relates to the puritanical character of Algerians. Not only do the Islamists object to a certain kind of behaviour, but so do most Algerians in general. The authorities, though eager to encourage tourism, are still reluctant to allow its liberal development as it has in neighbouring Tunisia. But perhaps economic necessities and education in inculcating a tourist culture among people will succeed in changing those attitudes.

Recently, Algeria has witnessed some interesting developments, which might have positive repercussions for tourism. The policy of 'Civil Concord and National Reconciliation' that President Abdelaziz Bouteflika launched after his election in April 1999 resulted in substantial improvement of the security situation. The regime has recognised that foreign direct investments in the non-hydrocarbon sectors would not be forthcoming unless security in the country was restored. Furthermore, the authorities have accelerated privatisation and economic liberalisation. There is consensus that the growth of the private sector is essential for the development of the country. Not surprisingly, tourism has been identified as one of the sectors that need strong impetus. The Algerian newspaper *Liberté* reported on 17 August 2000 that improvement in security has already resulted in an influx of foreign tourists; Algeria now appears as one of the suggested destinations in European tourist brochures. The first trimester 2000 has witnessed an increase of 52 per cent in the number of tourists compared to the same period in 1999. This trend will likely continue, albeit slowly, should the security conditions continue to improve. The end of political instability and careful economic liberalisation could potentially attract European tourists waiting to discover one of the most beautiful Mediterranean countries.

Notes

1. Permanent Mission of Algeria to the United Nations: Travels and Hotels, http.www.undp.org/missions/algeria/p61.htm.
2. Etude Synthèse, 'Evolution du secteur touristique et perspectives nouvelles pour le développement d'un tourisme de masse: le modéle algérien,' *Revue de l'Institut des Sciences Economiques d'Oran* (April 1982): 247.
3. On Algeria's industrialization, see Bennoune, pp. 114ff.
4. Democratic and Popular Algerian Republic. *National Charter* (Algeria: Ministry of Culture and Information/SNED 1981), p. 160.
5. Ibid., p. 161.

6. Interview of Yahia H. Zoubir with high official in the Ministry of Tourism, Algiers, June 1997.
7. *National Charter*, op. cit., p. 161.
8. Ibid.
9. Ibid.
10. Ibid.
11. Ibid.
12. Ibid.
13. *Algérie-Guide économique et sociale* (Algiers: Agence Nationale d'Edition et de Publicité 1989), p. 390.
14. These figures were taken from *Algérie-Guide économique et sociale*, op. cit., p. 386. In 1986, the number of tourists visiting Algeria was 347,745 compared with 407,355 in 1985. See *Reuter*, May 12 1987.
15. *Reuters North European Service*, January 9 1987.
16. *Reuter Textline, Gas Daily Risk Monitor*, January 14 1987.
17. *Travel Trade Gazette*, February 12, 1987 p. 51.
18. *North Africa Journal* (http://www.north-africa.com/archives/0_102497.htm).
19. US Department of State, Travel Warning, January 31 1997; see also http:eurogate.iit.nl/travel/algeria/algiers/all_zz01.htm.
20. *Annuaire Statistique* (Algiers: Office National des Statistiques 1996), pp. 308. Note that the number of Moroccan tourists in 1991 reached 338,158.
21. Ibid., p. 305.
22. *Annuaire Statistique* (Algiers: ONS, 1997), p. 289. The number of Maghrebi tourists fell from 417,653 in 1990 to 257,222 in 1994. The number of Moroccan tourists fell to 180,673; ibid., p. 290.
23. 'Algeria: Tourism Sector to be Privatized', *Middle East Economic Digest*, May 22 1992, p. 16.
24. 'Algeria: Tourism – The Undiscovered Sector', *Euromoney Trade Finance and Banker International*, April 1 1993, p. 23.
25. 'Insecurity Scares Tourists Away from Sahara', *The Reuter Business Report*, January 25 1994.
26. 'Algeria: New Venture Created to Promote Tourism', *Middle East Economic Digest*, June 27 1994.
27. 'Algeria: Supplement on Algeria – The Attraction of the Sahara', *Euromoney Trade Finance and Banker International*, September 30 1994.
28. Ministry of Tourism and Handicraft. *Privatization and Investments in Tourism in Algeria* (Algiers: National Tourism Office, n.d.), p. 9. The law is extremely liberal and provides appealing incentives, for foreign investors in particular.
29. 'Spain's Flamingo Goes into Algeria', *Travel Trade Gazette Europa*, September 14 1995, p. 6.
30. 'Algeria Publishes Privatization Decree', *United Press International*, October 4 1995.
31. 'Algeria: Second Attempt to Privatization Process States with Sale of Several Hotels', *Middle East Economic Digest*, June 17 1996.
32. See *La Tribune* (Algiers), April 20 1998; *La Tribune*, April 14 1998; *El Watan* (Algiers), April 15 1998. See also the web page of SITEV 98, http://www.algeria-tourism.org/index.htm.
33. 'Le Tourisme en Algérie – Situation d'aujourd'hui: Communication du DG de l'ONT á la semaine italo-arabe á Naples', *Algérie*.
34. Ibid.

15 Moroccan tourism

Evolution and cultural adaptation

Jacques Barbier

Three types of tourism, successive and later superimposed

Over the past 50 years, Morocco has experienced three radically different types of tourism: the tourism of curiosity and discovery, seaside resort tourism and, lastly, interior tourism often an imitation of the first two, with each provoking different reactions. Obviously these three alternatives have not merely succeeded each other but have been mingled and superimposed. However, for clarity's sake, they will be dealt with successively in this chapter, in the order of their emergence.

Tourism of 'curiosity' and discovery

For over a thousand years, Morocco has been seen as a 'bridge' between Europe and Africa, but far closer to the first than to the second. An old tradition of cultural and commercial exchanges has always been accompanied by visitors: the curious, the learned, the artists (painters, musicians). Later on, the Protectorate and the strong French presence brought lively interest in anthropological, ethnographic, geographic, historic and linguistic research. With the increasing transport facilities, this tendency has expanded to form a genuine touristic movement of discovery. The early travellers of our century approached this great kingdom with respect. In order to win acceptance for their presence and capture the real identity of the country they travelled through, modifying it as little as possible, they adopted the same living conditions, sometimes the same clothes, even the same language when they could. Later on, these forms of mimicry became naturally attenuated but without disappearing altogether: the hotels of our epoch imitate, more or less successfully, the traditional palaces or the shaded houses of the medinas and anachronic horse-carriages still carry visitors to Marrakech on tours of the walls, in streets invaded by diesel engines. In the South, it is not unusual to see visitors wearing turbans and djellabas or caftans descending from the 4 × 4s of the travel agents.

The tourist was seen as a guest, even as a friend, come to see the marvels of Morocco. His presence and curiosity comforted the dignity and pride of the Moroccans, and he was willingly invited to share in the celebrations of important events in the social and family life.

Finally, and probably most important in this type of tourism, the visitors were generally people of means, inspired to leave their houses in Europe only by curiosity, sometimes by cold winters and, above all, by the desire to see the true Morocco, with a fairly strong dose of old-fashioned nostalgia.

This touristic movement was mainly directed towards the 'Imperial Cities' and the South. Tangiers and Marrakech, thanks to their climate and their gardens, soon succeeded in retaining the visitors for long sojourns, particularly in the cool season (October to May).

The invention of the Mediterranean seaside resort and its emergence in Morocco

After the 1939–45 war and the subsequent reconstruction, a radically different touristic motivation emerged in Europe. The strong urbanization movement, the mediocrity of living conditions in hastily reconstructed towns, the raising of the standard of living, the beginning of family motorization and, no doubt, the need to make up for years of misery and frustration, led to the invention of 'vacations'. The leisurely and unpredictable journeys of the well-to-do classes of the beginning of the century were replaced by a new product: the short period, almost obligatory, programmed for nearly all the summer months, with a strong desire to escape from the cities, to regain nature, the sun, to relax and recover health and to escape temporarily from yet another year of greyness and constraints. This new need, born of new singularities and frustrations, has brought the manifestation of new touristic products, including the 'Mediterranean seaside resort' type, a frenetic search for sun, sea, sand and *dolce farniente*, rather quickly followed by freedom of behaviour and exaltation of the body. This unwinding began by colonizing, developing and degrading the Mediterranean sea fronts of southern Europe (since redeveloped and rehabilitated), then went in search of other shores able to fulfil the required conditions, colonizing them in turn. Once air transport became accessible to the middle classes, the countries of North Africa were targeted. Tunisia has become rapidly integrated into this market. Algeria, reticent at first, made tourism a matter of state policy towards the 1970s, quickly aborted and abandoned. Morocco, prepared by cultural tourism, entered this market timidly, up until the time of the reconstruction of Agadir, after the earthquake caused the majority of its tourism in this sector to teeter.

The appearance of seaside resort tourism of this type in the Moroccan landscape has completely modified cultural impacts. Morocco, like most of the Mediterranean coast, is not a country of sailors. Its tradition is continental. Prior to the twentieth century, the kingdom's capitals had almost always been interior cities. The coast is often mountainous, lacking in good agricultural land and sweet water, and the sea was considered as the refuge of pirates, who swept down from coastal locations to raid the population, and a permanent source of dangers for the coastal inhabitants, hostile to sea bathing and few of whom could swim. Fishing was only a marginal activity until the progress in freezing made it possible to enter the big markets. So, the astounded inhabitants observed, uncomprehendingly, those tourists come from Europe to expose themselves under the scorching summer sun on arid beaches, half-naked and living in straw huts, whereas the Moroccan imagination is nourished on cool, shady houses, flowering, luxuriant gardens, streams and fountains.

Just as tourism for culture and curiosity was a factor for integration and exchange, so seaside resort tourism began under the sign of separation, segregation and incomprehension. Studies launched at the beginning of 1980, by foreign enterprises for the development of the coast, are witnesses to this change in values. In one of them, the touristic potential is evaluated according to the length of the beach, multiplied by a coefficient allowing calculation of its capacity in tourists, with no other criteria than accessibility from the generating markets. The population, its customs, its way of life, were only considered as secondary variables, liable at best to provide the manpower needed for running the resorts. These were, at best, paradoxically described as 'integrated', even though within their perimeters they offered all necessary facilities and services, sparing the tourist any temptation to venture out and visit the country itself.

In this second wave, which dominated the decade of 1980–90, the leading product has without doubt been Agadir, which still today offers 31 per cent (1992) of all the country's tourist beds, and whose initial success under very special conditions had aroused expectations never to be realized.

Quantitatively, seaside resort tourism will rapidly forge ahead of cultural tourism and today it still represents the main motivation for around 16 per cent (1999) of tourists.

Now, the essential characteristic of this tourism is that the tourists no longer come to see the country for itself, but that suitable sites are developed to become what the market expects.[1]

Tourism becomes a separate activity, without a real contact with local life. It is very often confined to holiday villages or clubs, which are rarely left by the clients. It provides them with distractions, sports, company, meals and entertainment, plus the possibility, against payment, of making some excursions with fellow captives to the best known sites in the proximity. This almost caricatural regime has at least had the effect of limiting cultural shocks during the early years and of avoiding attitudes of rejection or certain perverse effects suffered by other countries. But it has been fundamentally different from the early category of tourism, due to the almost total lack of interest concerning the host country.

The emergence of interior tourism

Fortunately, Morocco is also going through a process of economic growth and urbanization, provoking the appearance of an urban middle class. This class also suffers the constraints (poor housing, unhealthy environment, pollution of all types, overcrowded housing conditions) which give rise to frustrations and needs to compensate. This explains the relatively rapid growth of interior tourism, imitating, but not altogether adopting, the dominant model of international tourism and, primarily, the seaside resort. The beaches near the large towns are invaded on Sundays by thousands of people. The beach is 'reinvented', to use A. Corbin's expression. It becomes the territory where nautical games are not forbidden to children, adolescents and young men, but in which women do not participate.[2] The young people adopt beach games and sports fairly rapidly, whereas the older ones remain dressed and observe these activities with an impassivity sometimes tinged with disapproval. During vacations the families rent modest apartments at the beach resorts. 'The holiday reunions of families from the towns, summer vacationing mainly in resorts of strongly urban character, help to strengthen the coherence of the family, while contributing to appreciation of the diversity within the family itself ... allowing a play of behavior patterns varying around a common base' (Miossec 1994). In summer the youngest and the less well-to-do herd together in camping areas improvised on beaches lacking infrastructure, transformed into dumping grounds in autumn and cleaned by the high tides and storms of winter.

This type of leisure is parallel rather than convergent with international seaside resort tourism. The cohabitation of the two forms of occupation in the same spaces is often a source of potential conflicts: the behaviour of the foreigners is considered by many to be offensive to decency, and that of the nationals contrary to norms of hygiene and, above all, to the ecology.

In fact, the spaces are easily distinguished. For instance, the real Mediterranean coast, between the Strait of Gibraltar and the Algerian frontier has progressively been overtaken by interior or Algerian tourism (before the closing of the frontier), whereas the big resorts in the south, especially Agadir, remain largely dominated by the international model. On a smaller scale, around urban agglomerations, the beaches frequented by foreigners and 'Europeanized' Moroccans are not the same as those of mass interior tourism.

Interference between tourism and socioeconomic evolution

Tourism has played a big role in the expansion of the Moroccan economy, directly and indirectly. It is a great provider of foreign currency and direct employment. In 1994, the total income from tourism was estimated at 25 billion dirhams (around 2.8 billion US dollars), and the value-added tax for the sector at 16.5 billion dirhams (1.3 billion US dollars, namely 6.7 per cent of the GNP). Jobs are evaluated at over 500,000, direct and indirect, including all types of tourism (six per cent of the total employment). The foreign currency accounts show a surplus income of 9 billion dirhams (one billion dollars) and represent 18.2 per cent of the balance of payments. After transfers from Moroccans abroad, the sector is the country's second biggest source of foreign currency and represents 41 per cent of the total commercial balance, 35 per cent of the total of goods and services and 1.9 times the total of the current account.[3] It can be fairly said that, in spite of difficulties encountered at the beginning of the 1990s, now overcome, the sector has remained one of the pillars of the Moroccan economy.

Indirectly, tourism has modified the habits of Moroccans as consumers and helped exports by obtaining recognition of Moroccan products and by creating a positive image of the country for consumers. Also, while not yet becoming 'the California of Europa', as somewhat prematurely claimed by some, the touristic potential favours localization for industrial and service-related activities. The relocation of activities to Morocco is helped by its reputation with Europeans as a pleasant place of residence but also, and more concretely, by easy access and the density of its air transport service. The latter could not be maintained without the contribution of its tourists. These favourable factors should not obscure certain contradictions. On the one hand, an important element has been the economic and social development, the rise in incomes, and the availability at all social levels of ordinary consumer products, greatly reducing the capacity of Moroccan towns to seem exotic. The medinas, which 50 years ago were a magic world for the visitor, have lost much of their charm since refrigerators, plastic babouches and basins are sold there and craftsmanship is reduced to a folkloric role, concentrated in bazaars that only foreigners visit. These same medinas, that were by tradition the town centres inhabited by the well-to-do (as still witnessed by the big houses and ruined palaces), are ill-adapted to the functions of modem life. So the classes with means have migrated to the 'new towns' and have been replaced in the historical centres by the poorer immigrants who share the old, now dilapidated houses. The rich merchants now only keep their shops or workshops there and live in villas accessible by car. By contrast, the least qualified workers of the industrial zones live in the medina. In this way, there is a double swing of the pendulum, well described by A. Adam, who tells of his discovery of Fez in 1934, 'during many far journeys, we have never experienced such a shock. . . . No town in the Arab world – except perhaps Aleppo – has ever given us so much as this city. . . . The impression of plunging into the East, and into an almost immutable East, just as it must have been several centuries ago.' He compares it to a visit in 1971 when, 'I experienced the same impression as in the old medina of

Casablanca. . . . It was a stream of workers, still half peasants, returning to their homes, often in old bourgeois houses, rented by floors and sublet by room.'[4] We ourselves, who discovered Fez in 1976 and who, in 1991, worked on the project to preserve the medina, can bear witness to an impression of similar denaturation, and there is now open talk of the degradation into slums of these old and venerable town centres.

The above example reveals an evident but rarely evoked fact. In Morocco, as elsewhere but far more intensely, due to the large size of the country and the force of its contrasts, tourism is inserted into a space that is, in a way, diametrically opposed to that occupied by the local population. In the old towns, tourism is interested in the ancient medinas, abandoned by the population, in the countryside it avoids the overcrowded zone and the polluted borders of the conurbation of Casablanca-Rabat-Kenitra, where the economic potential, the industry, is concentrated and towards which the population from the countryside flows. It besieges the south coast and the pre-Saharan region, abandoned by their population for the Casablanca coast. It will soon conquer the mountain and the desert, main sources of the rural exodus. These contradictory tendencies contribute to increasing the gulf between the population and the tourists and considerably diminish the pleasure of discovery. In places where the tourism of discovery and curiosity draws the tourists, they find landscapes and ruins but fewer and fewer people or traditional societies. The tourists visit the medinas during the daytime, when they are full of other tourists and shoppers from the city. By evening, the streets are deserted and the visitors, virtually confined to the hotels of the new town, bore themselves watching television.

This phenomenon is one of the results of a socially desirable evolution and we are not disputing it. However, it creates 'new frustrations' which must be urgently attended to with 'new offers'.

This 'wearing out' of traditional touristic products can also be seen in other areas. The seaside resort is taking a hard blow from competition offering more distant, more exotic destinations (because they are in less developed countries), which meet the demand better for reasons of climate or less severe customs. Morocco is a Muslim country, known for tolerance and liberalism, especially with regard to its guests. As long as it was confronted with the competition from countries such as those of the south and east coasts of the Mediterranean, with that of Spain, Corsica or Sardinia, 25 years ago, it could hold its own. Today, caught between the very liberal, even permissive European countries and far-off destinations with easy-going ways of life, Morocco is in a more difficult position with regard to seaside resort tourism. Its star resort, Agadir, has aged considerably and does not offer compensation for an obvious lack of animation. At a time when the demand veers clearly towards more active vacations and towards 'having fun', the Moroccan clientele stagnates and grows old. The touristic environments hesitate between the desire to stay with this ageing clientele (which offers a certain security in the short term) and the risk of catering to a younger clientele and disturbing their traditional visitors. This contradiction, which would solve itself in an expanding market, is an important obstacle at a time when the attraction of the Mediterranean, in general, is decreasing in favour of tropical countries and when the countries of the north coast, which are both the main reservoirs of clients for Moroccan tourism and their main competitors for seaside resorts, have recently renewed and modernized their offer, adapting it to the new demands: 'green' tourism, ecology, hygiene, etc.

The difficulties of the sector, oriented towards seaside resorts, and the progressive retreat of the international clientele, encourage the taking over of less competitive installations by interior tourism. This substitution makes the modernization of the product even more difficult because even though the national demand gradually approaches the international demand, it is still very different and blocks the efforts to modernize. The family gatherings look for apartments to rent rather than hotels, since the preparation of meals is one of the fundamental rites of Moroccan conviviality. With the exception of a small fringe of young people from the 'upper-middle class', who practice sports and imitate the behaviour of the international tourists, the Moroccan holiday-makers have not yet the financial means of access to games, sports and entertainment appreciated by the foreign clientele and, therefore, in no way contribute to their economic feasibility. As for the rich classes, truth demands it be said that they incline to spend their vacation abroad, often on the very nearby Spanish 'Costa del Sol', with its reputation for being more lively than the Moroccan coast which faces it a few kilometres away.

The 'new products', initiation of a return to the sources

Many sociologists of tourism think that the era of 'vacations' is reaching its end and that tomorrow's tourism will want something other than the simple 'rebuilding of working energy'. Observation also reveals substantial modifications in the division of leisure time over the year and, also, during life itself. On one hand, the annual periods of rest are increasingly split up. Instead of the four-week holiday taken in summer, an increasing number of the population and employers prefer shorter, more numerous holidays, better spread over the year. They make it possible to undertake different activities, often oriented towards action and adventure (voyages, discoveries, treks, cruises, etc.). The more sedentary activities and limited physical expenditure required by work turn the demand towards sports and discovery rather than towards the *dolce farniente*, or the return to the countryside. Whereas, ten years ago, it was still considered that a seaside holiday should last at least three weeks (three days for arrival and installation, three days to 'destress', 12 days of real vacation and three days for the return), the preference today, especially for destinations reached almost always by plane and in a few hours, is for much shorter stays (in Morocco, in 1995, the average trip lasted 10.4 days (1998)). On the other hand, the decrease in the global time worked and the advancing of the retirement age frees a potential clientele, with less money (perhaps) but with far more time to spend, who care more about travelling and doing than about resting. In the Moroccan context, a relative decrease in demand for the traditional seaside resort is to be expected. (It will no doubt continue to be important, modernized, especially in the southern zone where the climate favours exploitation throughout the year). Also to be expected is a strong increase in discovery-related touristic products, in adventure, mountain and desert, observation of nature and, also, sports, particularly those which can be practiced for a long period of a person's life (for example, golf, tennis, fishing, sailing, excursions, mountain biking). This naturally applies to tourism with a specifically cultural aim, but renewed to compensate for the popularization of its present settings.

How will this transformation be culturally accepted and integrated? This is an essential question in a country which has emerged from under-development and which now has the possibility of choosing its means of growth.

Tourism and social and cultural values

Tourism has never provoked a reaction of rejection in Morocco. Its traditional role of passage, its close proximity to Europe and a pronounced open attitude towards others on the part of the Moroccan people explain this easy acceptance. But we must add two other elements, already mentioned in the introduction. The tourism for culture, coming within a historical continuity, was essentially urban (that is to say, concentrated in the very places where the social change takes place), and it was a tourism that integrated more than it shocked. It was often felt as a valorization of the Moroccan heritage, urban first and then traditional. The interest shown by tourists from afar, and from 'important' and developed countries, in the customs of daily life gave these new dignity. Bennani Chraibi reports that 'during a local religious feast, the presence of foreign tourists, just as that of the Governor, appeared as a recognition, an added value to this ceremony'.[5]

The seaside resort tourism, far stranger to Moroccan society, provoked less favourable judgements. By authorizing, in certain parts of the national territory, unfamiliar behaviour, it runs the risk of rejection based on refusal of its values – and, particularly with regard to the 'aspects related to morals and sexuality',[6] and a certain bitterness, 'they [the tourists] come for the sun ... they would have liked Morocco without the Moroccans ... they consider Moroccans like monkeys to photograph.'[7] Since these types of leisure are imitated by the younger people, they are also judged as encouraging behaviour forbidden to Moroccans (consumption of alcohol, audacious clothes, prostitution) and a source of destruction of the fundamental values of the country. But, as we have seen, this type of activity is concentrated in only a few places and remains removed from Moroccan society.

The new types of tourism expected, which will partly happen in the remote zones with more hermetic, less 'modern' societies (mountain, desert), risk coming up against more determined reactions. Already the development of 'trekking' in the Berber Atlas is viewed with reserve, not only because of the maladjusted attitudes of the tourists (clothes judged to be indecent, groups reuniting young people of both sexes, photography of inhabitants with zoom lens and without permission) but perhaps, above all, through fear that the visitors will create in the young people aspirations that are impossible to satisfy in these traditional societies, inciting them to emigrate. Among the Berbers of the Atlas, where the traveller was considered as being sent by Allah, because he too was poor and needed help, the tourist, even when he behaves discreetly and tries to integrate, is regarded with suspicion for fear that his easy circumstances, his material possessions or his apparent freedom may give birth to impossible dreams in the young people. Such reactions are doomed to spread if the penetration of tourism in remote areas is not accompanied by rapid economic and social progress, which the single touristic activity alone cannot guarantee.

Corresponding to these fears, there are also opportunities for development. The traditional societies are also, in many ways, archaic societies, poorly situated for facing the modern world. The withdrawal into self and isolation will not protect them from destruction nor from progress. Very modern techniques (solar electricity, for example) are introduced in villages by emigrants returning from the towns or through the initiative of residents abroad. The closed societies must open up to adapt and the tourism of discovery and ecotourism are also vehicles of progress and of preparation in this respect. They encourage instruction, for example and create ties which are inevitably established between the guests and the host environment in this type of activity. The training centre for mountain skills, installed in the valleys of the Aït Bougmès with the support of France, is a good example of this diffusion of knowledge. Founded originally to train guides and escorts for trekking groups, for spring-time skiing, and mountaineering; it has rapidly extended its training to other jobs connected with the reception of visitors, including the craftsmen to improve the refuges and the reception in the homes of inhabitants. In this respect, a well-adapted touristic activity can play an important role in the preparation of groups exposed to changes, rendering them more apt to benefit from other opportunities. Social progress is not possible without change and conservative attitudes are rarely the best. The cultural exchange, the processes of imitation and adaptation initiated by tourism constitute, when not too violent, is one of the least traumatic ways to initiate the change. In this respect the new types of tourism will certainly be more efficient than that of beach resorts.

Tourism and emigration: an interesting synergy

It is not really possible to evaluate the sociocultural impact of tourism without reference to the large colony of Moroccans residing in foreign countries. Several Mediterranean countries, including Morocco,[8] have millions of nationals resident in countries that are principal generators of tourists. This population stimulates part of the touristic movement. In Morocco, the Moroccan nationals from abroad (RMEs) who rejoin their country for the big religious feasts and, above all, for their summer vacations, contribute around 10 per cent of touristic expenditure. Above all, these compatriots from abroad have partially adopted the leisure habits of the countries in which they live and they introduce them in their regions of origin. They are also often small investors in the sector of touristic installations, especially since their position allows them to 'draw' a (modest) clientele from among friends and acquaintances, and because their knowledge of a foreign language helps them to win clients and, perhaps, prepare an activity for when they finally return to stay. Since emigration is particularly strong in remote and poor regions, the emigrants become intermediaries who encourage the new tendencies of tourism (nature, mountain, discovery) and a better integration of tourism in the 'microlocal' economy, by directly associating their families with their 'receptive' activities. This phenomenon is very noticeable on the Mediterranean Riff coast, going so far as to build houses which the emigrants rent out until they finally return to occupy them. In this way, they initiate a type of residence (long term, on a rental basis) which is well-suited to the new clientele of retired visitors, who are still almost non-existent in 'formal' tourism. Another very striking example is that of the Dades valley, south of the Atlas, which has seen the opening of a multitude of inns and restaurants in a setting that was, up to then, confined to hasty visits 'without leaving the bus'. These emigrants, who are familiar with the habits and even the manias of the European clientele (obsession with cleanliness, for example, or distaste for bargaining) make effective agents in helping their compatriots to understand this behaviour, and for softening the cultural shock. They should be taken into account and associated to the maximum in the development of lasting tourism. In other countries on the northern shores (Italy, Spain and Portugal) observation reveals that the emigrants have played a big role in the phase of transforming vacation and short-trip tourism towards a formula of long-term residence, purchase of houses and direct investment by clients in a second residence.

Conclusions

The example of Morocco is characteristic of the process of cultural adaptation and integration which normally follows a phase of mass tourism oriented exclusively by the demand for 'beach resort products', at first strange to the local community, then progressively assimilated. During the period of such assimilation, the international demand, satisfied in beach resorts (partly by competing countries emerging on the market) turns once again towards a search for identity. Well, Morocco itself is seeking its identity. A profoundly Muslim country, but far from any integrism, it hesitates between its fidelity to the Arab world and its millenary adherence to the 'Andalusian' block which it founded on both sides of the Strait of Gibraltar and which now spreads to the Maghreb (in the South) and to the European Community (in the North). A country of tradition, it witnesses the development of its country and the modernization of its way of life, to the detriment of the picturesque. Some sour spirits will think that these changes are all synonymous with the loss of values. We believe, to the contrary, that historically the Mediterranean only maintains its equilibrium when undergoing change, like an airplane, failing when it ceases to advance. Tourism plays an essential role in these changes and is no doubt directed, after a period of reciprocal surprises, towards a fusion, giving birth to new societies.

Inch'Allah!

Notes

1. This is not peculiar to Morocco, nor to seaside resort tourism. The other products of this generation of tourism, dominated by market demand, have undergone the same drift: the development of the French and Swiss mountains, with integrated resorts, remodelling of slopes to adapt them to the average skiers' technical levels, the installation of mechanical ski-lifts, are all forms of the same phenomenon, which is not to be criticized in itself but which reduces the contact with the 'natives' merely to needs involving services.
2. J. M. Miossec, 'Tourisme et loisirs de proximité dans le monde arabe', in Maghreb-Machrek 1994, p. 149.
3. URBAPLAN, 'Stratégie national damènagement touristique', Rabot 1996, Vol. 5, pp. 86–7.
4. A. Adam, 'Urbanisation et changement culturel au Maghreb', in Ducmas R. ed., 'Viller et sociétes zu Maghreb', CNRS Paris 1974, p. 217.
5. M. Bennani-Chraibi, 'Soumis et rebelles, les jeunes du Maroc', Le Fennes. Casablanca, p. 73.
6. Ibid. p. 76.
7. Ibid. p. 80.
8. This is also true of Turkey, Greece, and Tunisia.

Part IV

Spatial reorganization of Mediterranean tourism

Part VI

16 Tourism, sustainable development, and the environment

Philippos Loukissas and Pantelis Skayannis

Introduction

Tourism has been, and will continue to be, one of the most important economic and social development activities in the Mediterranean coastal regions. History and the natural environment have played an important role in this direction. However, rapid and uncontrolled tourism growth have contributed to the deterioration of the coastal environment, especially in the more developed regions of Spain and Italy. There is a growing recognition that the natural and cultural environment is an important economic resource worth preserving.

It would rather be trivial to limit the problems tourism faces in the fields of tourism infrastructures, etc., as conventional approaches usually do. Such approaches, employed for decades, have attempted to face tourism's problems, yet reality shows that problems have been augmented, and accentuated. Fragmentation and partial understanding would once more lead to the past results. The Mediterranean regions cannot afford any more to experience an environmental degradation, a form of tourism that is totally alienated from the wider cultural environment in which it develops, infrastructure bottlenecks and environmentally traumatic solutions to these organizational and institutional inadequacies, as well as social and spatial differentiation leading to the exacerbation of developmental cleavages.

The problem tourism poses at the local level (sub-national) is, in fact, one of its positioning among the factors of development. Yet development should be understood as sustainable, i.e. as targeting simultaneously the areas of economic growth, sociocultural development and environmental development. Tourism up to now has focused on the first part of this triangle, i.e. solely on economic growth. The other two parts were sacrificed and the results of this sacrifice have shown their negative impacts.

Sustainable tourism development adopted as goals of modern tourism the importance of environmental, as well as the need for conservation of social, cultural and economic resources (Inskeep 1991; Lorch and Bausch 1995; Papayannis 1994; UN 1992; Van den Berg and Nijkamp 1994; World

Tourism Organization 1980). Responsible planning for tourism development must recognize the capacity of the existing natural environment to
absorb tourist activity. Long-term effects on ecological, sociological and
economic functions must be taken into account.

Sustainable development is a guided process which envisions global
management of resources so as to ensure their viability, thus enabling our
natural and cultural capital to be preserved. Tourism should contribute to
sustainable development and be integrated with the natural, cultural and
human environment; it must respect the fragile balances that characterize
many tourist destinations. In order for tourism to contribute to the
process of sustainable development, it is necessary to ensure the participation of all actors, both public and private, in the tourism arena. The goal
of sustainable tourism can be achieved if the resources available to
support human activity in a particular place can be assessed and not
exceeded (Manning and Dougherty 1995).

This chapter argues that some of the problems that have arisen because
of the current pattern of tourism development can be overcome by the
adoption of strategic management actions relating to tourism (SMART).[1]
The chapter will draw from the SMART project as a whole, yet special ref-
erences will be made to the Greek case, as it was in Mt Pelion that the

Figure 16.1 Greece.

pilot projects were fully developed, hence relevant conclusions can be derived. The SMART project's goal was to explore ways to improve the management of tourism through the protection and utilization of the natural and cultural environment and to promote sustainable local development. The project's specific aims were to promote alternative forms of tourism and transportation, redistribution of tourist flows through time and space from popular coastal areas to less developed mountain areas in the hinterland and to sensitize the local public and private agents regarding the need for tourism management and preservation of the environment through training. In order to accomplish these goals a number of pilot actions which have potential for transferability were planned and tested in the three research areas in Greece, Italy and Spain.

In the second section of the chapter, we state the fundamental problems that tourism faces today in the Mediterranean regions. We try to distinguish between the various kinds of tourism. A typology of tourism examples are examined in brief and certain conclusions as to their suitability for the Mediterranean regions are drawn. In the third part there is a description of the SMART project, and particularly its pilot actions in Pelion, Greece, while in the last section certain proposals in the form of guidelines are put forward for Mediterranean areas in general and for the case of Pelion in particular.

Typology of tourism models: advantages and disadvantages

In fact, this brief reference to tourist accommodation construction and planning shows nothing but the tendencies of the tourist market itself: today's tourist preferences favour smaller scale and diversity, in combination with higher quality standards. This issue can be related to the question of mass- and non-mass tourism, but also to the broader economic developments as well. In fact, the story of the transition from Fordist to post-Fordist modes of economic activity, hence paradigms, start to be present in tourism as well.

The appearance of post-Fordist models in production and consumption presupposes the collapse of mass markets, of standardized products, etc., favouring or expressing market fragmentation, diversification of consumer tastes and quality aspects. Driven by demand, this development is now threatening traditional tourism modes. The rigidities of the supply structures, including the strong economic interests behind the curtains (e.g., the tour operators) have not yet fully matched demand in this respect. The hesitant steps so far taken by the operators to provide variety and diversification do not seem to satisfy consumer needs to the desired extent, thus leaving space of manoeuvre to the proponents of alternative ways of tourism development.

Related to this are the issues of tourist accommodation and of the environment. Similarly to the case of mass tourism, the era of vast hotel complexes may not have passed, yet it is being challenged by a number of tourists whose preferences lie more to smaller establishments, and their holiday aspirations lead to a form of tourism closer to nature. However, the most prominent development of the last decade is the turn to quality, a request put upon all forms of tourism, massive or not.

At the same time, preferences of tourists worldwide tend to undergo changes. Although the classical pattern of mass tourism (tourist packages organized either as visits to places or as summer holidays by the sea) is not seriously threatened, forms of tourism existing yet not dominant in the past, tend to gather pace and to claim a share of the tourist market. These are either forms of non-mass tourism or of alternative tourism. There have also been examples of particular mixes of mass with non-mass, or even alternative, tourist forms, such as the one in the Caribbean, where entrepreneurial spirit has elaborated flexible packages appealing to various tourist tastes.

In order to deal with the question of tourism in the Mediterranean areas, one must first analyse the types, or styles of tourism operating in the area. Tourism in the Mediterranean is primarily oriented to recreation related to the sea, but in this respect the tourism stereotype is no different than it is in several very popular tourist sea-oriented sites worldwide. (The tourist jargon has named this pattern as the four 'S', namely sea, sand, sun, sex.) This stereotype is most prominent in Spain, where there is a mass tourism industry. However, Greece and Italy (as well as Tunisia and Turkey) have followed similar trajectories, albeit not in exactly the same massive fashion. However, the diversity of the Mediterranean landscape and culture provides the opportunity for diversification of the interests of tourists. This, accompanied with periods of crisis that certain areas go through at certain time intervals, as well as the need for a new kind of protection of the cultural and natural environment, raises the question of whether this stereotype can be challenged and, if yes, in which way. What has frequently been proposed is alternative tourism. Yet the term is not clarified properly and may cause confusion. Therefore, in this section, which in some ways is methodological, we shall try to clarify this term, putting it into the appropriate context, which is related to the issue of mass and non-mass tourism.

The styles of tourism can be divided according to the criteria of conventionality-alternatives, or 'massiveness', as is shown in the Table 16.1.

Mass tourism is the kind of tourism organized through the package

Table 16.1 Styles of tourism

	Conventional	Alternative
Mass	1 Benidorm	2
Non-mass	3	4 Pelion

system, and at the same time involving large numbers of people, while non-mass tourism refers to either individual tourism or package tourism of small numbers.

The term 'alternative' means that the tourist engages in an activity which is not conventional. What is conventional depends on the tourist habits of a certain period of time in a certain place. For instance, underwater fishing in Greece is not an alternative activity, as it is estimated that a very large proportion of the population does this sort of sport on a regular basis. Alternative tourism may also be organized in packages or involve certain numbers of people, though in this case packages are smaller in size and different in quality.

The advantages of mass tourism mainly consist in the reduction of risk, the stable income, and the security it offers for the local entrepreneurial community, as well as of the pre-organizational advantages it offers to the tourists. Benidorm is the prime example of mass tourism. The main disadvantages are the necessity for the creation of large infrastructures, the alteration of the life styles of the local population, the dependence on tour operators (on which many of the advantages depend), the lack of flexibility and usually the short high-season, as most of the mass-tourists are working people who take their holidays in July or August. Although there are exceptions to this (see the example of Benidorm where the tourist season has been extenuated to all year around). The question of the environment also comes into the picture here. Mass tourism is in some ways disadvantageous for the environment and in some other ways not. If it involves large infrastructures and big hotels there is a high possibility that all such structures are, at least from the aesthetic point of view, environmentally damaging. Of course, this depends on architecture and the aesthetics of the designers. In addition, one can argue that large numbers of people produce more waste, and that the projects themselves give a new shape to the environment which is not necessarily the one desired. It is quite obvious as well that the ecosystems are altered and threatened. However, mass tourism if properly organized might not harm the environment to the extent believed, mainly from the scope of waste, since special infrastructure (technical and organizational) projects can prevent this.

Places which are suitable for mass tourism should be endowed with the appropriate physical infrastructures, have quantitatively and qualitatively sufficient human resources and personnel, and have a natural environment able to satisfy large amounts of people doing similar things at the same time (e.g., large beaches). Mass tourism tends to involve low-middle-income groups.

In contrast, non-mass tourism has the advantage of involving smaller groups of tourists, tourism can be spread time-wise and needs lighter infrastructures, can utilize traditional skills and a varied physical environment. It has the disadvantages of temporal instability of profits for the local entrepreneurial community and of the requirement of a variety of infrastructures. Places suitable for non-mass tourism should have a multi-faceted natural environment and a very interesting human-made environment. This kind of tourism usually attracts either low- or high-income tourists. The question of the environment is also important here. Non-mass tourism is not organized, so there is a high probability that suitable infrastructures, however light, may not be adequately preventing environmental damages. In this sense, non-mass tourism, from the environmental point of view, is a double-edged sword. Small groups or individual special purpose tourists may prove the best beneficiaries of the environment, or the worse destroyers (especially because of the various waste), yet not on a very large scale.

The most difficult combination is the one of non-mass alternative tourism (cell 4 of Table 16.1). This is because it represents a very demanding situation for a locality, as the place should be organized in a way to offer possibilities for alternative tourism, while at the same time be prepared to run risks and accommodate a variety of small groups and individual tourists having different tastes and needs. Yet this kind of tourism, because of its special audience, who are sensitized in many ways, has a high probability of good environmental behaviour. Mt Pelion is representative of this type of tourism.

According to the criteria stated above, most Mediterranean areas have a good chance of becoming places suitable for alternative tourism. The Mediterranean areas can combine the sea with the mountains and a particularly rich cultural environment (historical and religious monuments, interesting traditions, exceptional food habits – 'Mediterranean diet', – cultural events, ecosystems in good condition and so on). These are some of the reasons and criteria by which the areas of the SMART project were selected.

The case of Mt Pelion

To illustrate some of the above problems and opportunities, one could take the case of tourism accommodation in Greece. Tourist resort areas in this country developed during the post-war period, primarily targeting northern European and North American markets. In the post-1990 era, the country has been progressively attracting additional tourist flows from East European countries, especially from the Balkans and from the Independent States. (In this respect, Greece has widened the scope of tourist attractions and opportunities it may have to offer to this wider 'audience'.) In 1995, tourism accounted for about 7.5 per cent of GNP, while its multiplier effects raised this number to 9.3 per cent (Kyriazi 1996).

In the field of tourist accommodation, Greece experiences a duality: on the one hand there are large hotels located in the big cities (Athens, Thessaloniki and so forth) and in the most popular holiday resorts (in particular Rhodes, Corfu, Halkidiki, and Crete), while on the other, there are small hotels and an unknown,[2] yet certainly very large, number of 'rooms to let', especially on the islands (of both the Aegean and the Ionian seas). Of course, between these two extremes are plenty of small hotels and alternative form of accommodation such as camping, agro-tourism units, etc. Yet all these have come about solely as a result of the market forces. While it might be argued that there is nothing wrong with this, the weak state regulation and virtually absent planning has resulted in an undesirable outcome: tourism accommodation is concentrated in particular coastal zones. In most cases this accommodation is violating the law in the sense that both large hotels and 'rooms' (but second – vacation – housing as well) are built in part illegally, or are violating other regulations, most commonly by obstructing free access to the sea which, in Greece, is guaranteed by the law (Centre for Planning and Economic Research 1994 Vol. II p. IIA.1.2)

Up to 1973, the main concern of the state tourist policy was to increase the foreign exchange transfers in order to balance the trade deficit through increasing the supply of hotel beds in the major urban centres and subsequently in smaller centres which, however, were lacking the appropriate infrastructure. Tourist development policy was targeted towards promotion of regional development of less developed areas, and mostly during the years of the dictatorship, i.e., 1967–74, the construction of large hotel complexes was favoured via special permits and loans provided in this direction. This was the period when some large hotel complexes were developed, such as the ones in Halkidiki and Rhodes. This policy continued after 1980. The incentives were proven to be insufficient to stop the concentration in already developed areas and the development of illegal and uncontrolled building, which led to environmental degradation.

This situation exists partly due to lack of planning or due to lack of those mechanisms which would enforce planning regulations. In fact, the purely indicative planning of tourist accommodation in Greece has undergone various stages. From 1983 until 1987, the five-year 'Economic and Social Development Plan' prepared by the Ministry of National Economy (1985) favoured smaller and environmentally more acceptable tourist enterprises and accommodation of a scale compatible with the host settlements. Financial incentives promoted the development of smaller enterprises in touristic destinations that were already in place. This was the period where the first ideas for agro-tourism were introduced in Greece.

However, this direction, being indicative, did not result in any genuine small-scale enterprises in the spirit of the direction indicated. Instead, the 'rooms to let' mushroomed, lacking qualitative facilities, violating land-use regulations, evading tax payments and so forth. In the post-1990 period, tourist accommodation construction continued to follow more or less the rules of the free market. However, the tendency to build large complexes has slowed down, while the first attempts to regulate land uses in the coastal areas have appeared, and the general public seems to be more sensitized by the issue of the environment and of sustainable development. (Centre for Planning and Economic Research 1994 Vol. II A2 p. 23)

While planning, as far as accommodation is concerned, underwent the controversial trajectory depicted above, spatial planning was entirely absent until the early 1990s when the first special physical planning master plans were assigned in the frame of the European Programme ENVIREG. A recently completed regional plan for the Mt Pelion area sets the protection of the built and natural environments as its main goal (Ministry of the Environment, Regional Planning and Public Works on-going 1997). It proposes a zoning strategy which includes an absolute protection zone for areas of sensitive ecosystems, as well as protection of rich agricultural lands. The built environment in traditional communities has been protected by a special building code since 1980. The proposal has already found opposition among those whose property rights are affected. Land use planning decisions rest exclusively in the hands of central government. Local property owners are limited to appeals and the local government has only an advisory role. This reveals nothing but the other extreme of overplanning; a typical case of planning carried out by bureaucrats which, because of its unrealistic nature, will undoubtedly be violated. In this case, what was needed was a more flexible and complex approach based not only on mandates but also on economic incentives which would allow a variety of places and possibilities of economic activity to materialize.

For Greece, flexible tourist markets, the request of quality and the demand for alternative and non-massive tourism forms are anticipated to play an important role in the formulation of the supply of tourist services.

Lessons and guidelines for strategic management actions relating to tourism for Mediterranean areas

The areas selected for the SMART project have a lot of characteristics in common: coastal tourism and a mountainous hinterland more or less not utilized for tourist purposes, mountainous communities that are, to a little extent, engaged in tourism and that need technical and other support to perform in a qualitative way. The aim of the project was to propose ways in which two fundamental problems could be tackled:

1   the problem of the concentration of tourist activity in the coastal areas, and
2   the problem of the limited duration of the tourist period.

These two problems had to be faced in the context of sustainable development. This means that the proposed actions should be compatible with the sociocultural context of the areas concerned and should not violate basic rules related to environmental protection. This sort of reasoning has been applied in other scientific areas. For instance, transport telematics aim at providing solutions for the relevant bottlenecks utilizing soft methods rather than hard infrastructure projects. In such a way the proponents of transport telematics claim that appropriate management of traffic flows can, in many cases, substitute hard projects, thus not traumatizing the urban tissues and not generating new spatial exclusions.

In a similar fashion, the SMART project was limited to application of soft actions, rather than hard infrastructure projects. Emphasis was placed on actions such as redistribution of tourist flows through time and space by enticing tourists who concentrate in coastal areas to visit the mountain areas and by luring them to visit the area during the off season, as well as by promoting alternative modes of transport. Another goal was to sensitize and educate local agents regarding the need for management of tourism and preservation of the environment.

In this section, a review of the most important findings of the project is attempted. First, a comparative analysis of the areas is presented. Second, the major lessons from the project are discussed. The analysis and evaluation of the problems and opportunities has led to the development of strategic management guidelines for tourist development in the three study regions, and although it is difficult to draw conclusions from a sample of a few cases, we will also attempt to develop some general policy guidelines for other similar areas in the Mediterranean basin. We start with a discussion of common elements and differences in the three areas and some conclusions are drawn regarding factors which hinder or influence creation of tourist products.

Similarities and differences between the three areas

The three research areas, Mt Pelion in Magnesia, Etna Region in Catania and Marina Baixa in Valencia, were chosen because of their similar geomorphology, climate and sociocultural characteristics. They possess a combination of coastal and mountain regions and the same pattern of uneven development, with a concentration of high development on the coast along with high density of tourism development (Skiathos, Taormina and Benidorm) and slow growth in the inland mountain areas (Milies, Randazzo and Guadalest) respectively.

The three areas under study seem to have quite a lot of features in common. Their physical aspects are similar, each one including a narrow coastal strip and a mountainous inland, which in the case of Etna has a maximum altitude of 3323 m, in the case of Pelion 1610 m and in the case of Marina Baixa, 1508 m. All the areas are easily accessible by all means, except Pelion that is insufficiently serviced by plane. Zones under protection included in the study areas are the Regional Park of Etna, the Marine

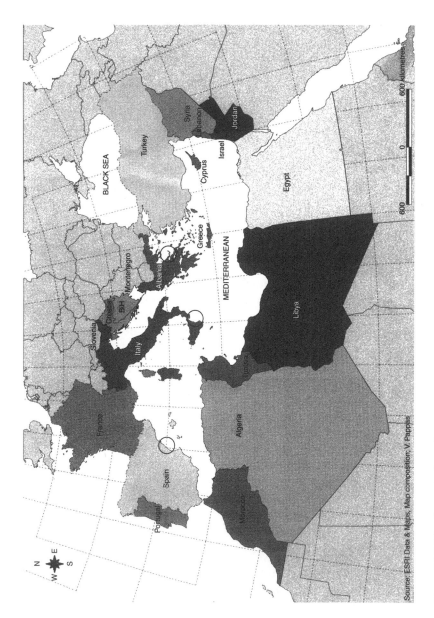

Figure 16.2 The three study regions.

Park in the Sporades Islands and the proposed protection zone on Mt Pelion in the Prefecture of Magnesia.

The three areas are very unequal with regard to their population. The Valencia Community has almost 4,000,000 inhabitants and its capital, Valencia, has 764,293 inhabitants, while the Province of Catania has 1,000,000 inhabitants and the city of Catania has 333,000 inhabitants. The Department of Magnesia has only 200,000 inhabitants and the town of Volos has 116,000 inhabitants. Both the cases of Marina Baixa and Pelion present an unequal distribution of population between coastal and inland municipalities. In Marina Baixa the density of population ranges from 1227 inhabitants/km² in the coastal area to seven inhabitants/km² in the mountainous municipalities. In the case of Pelion, the population of the coastal communities keeps developing while the population of the inland areas keeps decreasing, although the gap is not as large as in the case of Marina Baixa.

Agriculture still plays an important role in the economic activity in all three areas. In Valencia it is a part-time activity in the coastal municipalities. In Catania it represents a basic economic activity with which local industry and commerce are linked, although the Etna region is suffering a crisis in the primary sector. In Pelion, agriculture is still the most important sector, although it suffers from the general problems of Greek rural areas (Agricultural University of Athens and University of Thessaly 1998). The industrial sector is quite important in Catania, held by small firms and focused mainly on traditional sectors and handicrafts, though suffering the general delay of the South of Italy, compared to the North. Magnesia used to be one of the most industrialized departments of Greece but underwent a severe crisis since the beginning of the 1980s that doesn't seem to be slowing down.

All three areas have much tourist potential and are quite popular vacation places. Marina Baixa is the most developed, with 32,000 hotel beds, while the whole Department of Magnesia has 15,445 beds and the whole Province of Catania has 7628 beds. In Benidorm alone, the most important tourist place of Marina Baixa, 9,000,000 visitor nights were recorded in 1994, while the respective numbers for Magnesia and Catania are about one million.

Despite the differences of scale and of the broader socioeconomic frame of the three areas, a similar tourist policy is practicable. The common goal is the diffusion of tourist flows from the coastal zones to the inland sites, within the areas under study in the cases of Marina Baixa and Pelion and, in the case of Catania, from the popular neighbouring coastal settlements, such as Taormina.

The differences among these areas are in terms of size of population, the density of development activity and the level of institutional organization. The Etna region, for example has a much higher density than Pelion. It is part of a less developed region of a more industrially developed country, with all the advantages which come from this relationship. As a result, for example, its road network is far more developed than that of Mt Pelion. The Marina Baixa district has one of the highest concentrations of tourist development worldwide. In the following section we will limit ourselves to the description of only the Greek pilot cases. Similar attempts were undertaken in the other two case studies.

The pilot actions in the Mt Pelion area

The SMART project offered the opportunity to test new ways for the management of the tourist flows and considering the aspects of quality in the supply of tourist products. Three pilot actions were organized, two of which are briefly presented below. The third included a training seminar for Greek and Italian persons involved in the tourist sector regarding creation of new products, quality management in small hotel businesses and their promotion. The seminar was developed and administered by the Fundacion Cavanilles de Altos Estudios Turisticos and took place in Sicily.

a) *The day trip action.* The concept behind the day trip action was to help to acquaint tourists with the Pelion hinterland, so that the idea of spending some time in the mountainous villages becomes more widespread. If this becomes an established pattern, then tourism may gradually be distributed in a more balanced way over space and time, as the mountainous villages are very suitable for periods of the year other than the summer. In parallel, as long as the trip proposes certain modes of transport, such as tracking, riding a traditional train and using a minibus contributes to environmental protection. It needs to be noted that the current traffic in Pelion is dependent on large tourist buses and private cars. In the mountainous area in which the road system comprises of narrow roads and where parking places are scarce, these create a multitude of problems of carrying capacity. The utilization of alternative means of transport was considered to be essential. The day trip aimed at demonstrating that the utilization of more environmentally friendly means is possible and that a tourist activity built on such premises had a good chance of becoming profitable.

The particular day trip targeted the utilization of the hinterland of Pelion as a focus of tourist activity, at the experimentation in a new style of tourist activity and the provision of new services, at the investigation of the views of the tourists for such a kind of activity and at the understanding of the organizational demands of such activities. This day trip, as a tourist activity, had the prospect of being utilized by the local entrepreneurial community as well as by various local, public or quasi-public agents.

The actual activity included a tour of four villages. The distance between the villages was covered partly on foot, and partly in a minibus, while the final return was made on the traditional train. Breakfast and lunch was offered, prepared by local people with traditional recipes and local products. The tour of the villages included visiting special points of attraction such as museums, an old library, churches, etc.

The pilot action happened exactly as scheduled and was very successful, as the answers to questionnaires distributed to the participants revealed. The trip took place several times during the summer of 1996 and once during the spring of 1997. The two phases involved different groups of people. Despite the fact that the groups were very different in many ways, several things in common came out from their answers. Participants in both groups were satisfied with the day trip.

As a result of the detailed observations made above, one can derive the conclusion that if alternative tourism is to become a pattern for tourism in Pelion, certain points have to be taken into consideration. First, if the type of alternative tourism is one including physical effort, the best period is certainly not the high season of July or August when temperatures are high. Spring and possibly autumn are better for such tourism. The two high season months may prove more suitable for cultural events during the evenings, or for other types of special-purpose tourism. Any type of tourism involving physical effort will have to take special consideration of three different target groups, the young people, the 35–55 age group and the older. Ideally, packages should differ according to the target age of the group. However, all age groups demand the element of environmental quality which Pelion has to offer.

The lesson from this pilot action was that Pelion is a suitable place for alternative tourism. This tourism can be organized during off-season months and can attract young people, as the area can provide an environment suitable for various activities. This will prolong the tourist period and may create an additional income for the local population. The future creation of such a tourist stream may alter the current pattern of tourist flows and spread tourism more evenly in terms of time and space. Evidence of the pilot action's success is that a local tourist agent started offering the trip as a tourist package on a regular basis.

b) The cultural–ecological event. To organize such a pilot action aiming at getting more people to become acquainted with the Pelion hinterland, several additional ideas were examined. One idea that the team concluded to have potential was to organize a seminar targeted to groups of school-age pupils. The programme would include lectures on the natural and cultural environment of the area, dance demonstrations, etc., offered by local experts.

The main aims of the proposed seminar were to design a new tourist product focusing on the Pelion hinterland, to transfer organizational technology to members of the local community, to sensitize the local community and to utilize and train local human resources while taking advantage of the rich cultural heritage of the locality. The experience from the seminar could be utilized by the local community and other agents, while it can transform itself into a regular event if permanent structures were established for this purpose.

The seminar took place twice during winter and spring. The purpose of the first seminar was to test the organizational structure in the village, and the purpose of the second was to test the programme and its contents. The first seminar was attended by 15 pupils from the City of Larisa (a group which already had experience from similar events), and the second by a group of 40 pupils from an International School in Athens. Despite minor problems that occurred, both seminars were assessed to have been very successful.

Lessons and transferability of results

Local conditions and attitudes can vary from place to place and need to be taken into consideration. What holds for one place and one period of time is not necessarily transferable. Local elected officials in Pelion have expressed fears about the negative impacts of mass tourism and the loss of control associated with it. On the other hand, in Benidorm local authorities appear to be satisfied with the assurance of repeat tourists and the guaranteeing of high levels of occupancy that only mass tourism can provide. Experience in the Etna region place it somewhere in the middle. All three areas are trying to differentiate in the tourist product offered by providing alternative attractions. Trips to the inland attractions are common. In Benidorm, there is a plan to bring in a major theme park as a means of maintaining the region's competitiveness. Such an idea would not even be in the realm of possibilities for Pelion at this stage.

There are still many areas in Pelion where tourism is not considered as the prime economic activity and there is still a strong feeling among many workers that tourism does not provide quality jobs in the way that the manufacturing sector does. Agriculture remains the prime industry. This multiplicity of employment is something important to be preserved and should be reinforced in the future. That is why agro-tourism appears to be such an appropriate approach in which to specialize. Representatives from local authorities in Pelion have expressed the concern that agro-tourism has not consistently produced successful results so far. There are still bureaucratic obstacles in the development of alternative forms of tourism, which fit the Pelion area so well. In parts of Pelion, seasonal homes are more prominent than hotel development.

Infrastructure projects like road and parking construction are at the top of the list of most urgent projects for local officials. Of course there are those who believe that tourism has the potential to provide work for those unemployed from manufacturing jobs.

There is a prominent feeling that national governments in general do not pay sufficient attention to tourism. The National Tourist Organisation of Greece (NTOG) is a centralized rigid bureaucratic institution which, in the past, has suffered from a lack of policy continuity. (There have been changes in the top administration, on average, every six months.) On the other hand, the Organisation has control of most decisions that have to do with promotion, marketing, development, licensing and regulation of hotel establishments and other tourist infrastructures. Although there are numerous laws and regulations, they are frequently changing and are not properly enforced.

The decentralized Italian and Spanish models of tourist organization appear to have worked well. The short experience from the innovative activities already tried in the Provincial Administration of Magnesia in the field of tourism promotion in partnership with the local Association of Hotel Owners have shown positive results. The Prefecture of Magnesia's Committee for tourist promotion has established a new system of tourist promotion through a public/private venture and have initiated a process of promoting quality of services.

Factors which influence the creation of new tourist products

The quality of the natural and cultural environment is essentially the primary tourist attraction in these areas. Waste and water pollution, noise and traffic congestion, and a mixture of inappropriate uses are only some of the results of uncontrolled development. The new products must be designed to be less polluting, more efficient and socially and culturally appropriate.

A very important factor related to the design of the new tourist products is the one of complementarity. It has become clear from accumulated experience, and was also proven in the SMART project, that complementarity among various agencies and products is essential for the final output-product offered in the tourist market. This notion of complementarity, though well worked out internally in the case of the tourist packages whereby the operator is in control of all facets of the package, in the case of non-mass or alternative tourism is threatened. This is due to two reasons. First, the current trend towards flexibility tends to divide the supply into many different suppliers of a variety of complementary services and products. This needs co-ordination and common standards, which are difficult to achieve. The second reason is applicable to the mass tourism model as well and is related to exogenous factors which are not related to the direct responsibility of the tourist industry. Such factors are the administrative system of the country under consideration, the quality of general infrastructure and services, the attitudes of the population, etc.

For instance, the Greek administrative system is characterized by high centrality and rigidity and a lack of effectiveness in implementing policies. Numerous laws are instituted, but there is an inability to effectively enforce them. On the other hand, the current fragmented system of local governments presents an additional obstacle to the successful creation of new tourist products. Small communities lack the critical mass and resources and are experiencing difficulties in managing programmes on their own. The recent restructuring of municipal boundaries was aimed at addressing exactly that issue. Successful privately-motivated activities, like the cultural festival in the town of Milies during the early 1990s could not continue due to lack of support by the rest of the community. On the other hand, the effectiveness of the Spanish and Italian systems can be attributed to the strong autonomous regional administrative systems in these countries.

Another example would be the co-ordination between the operators and the state services, such as banks, railways, etc. The SMART project had the chance to witness a lack of co-ordination between the Hellenic Railways Organization and the day trip pilot action. However, this problem has deeper roots. A hypothesis could be that the attitudes to tourism are different among the various state agencies, and that the role of tourism is not yet fully understood by all parties concerned. This fact can be related to a variety of psychological issues linked to xenophobia, ethnic self-esteem, and so on, but to the economic and business cultures of countries as well. In Greece, where for certain periods demand has surpassed supply in the case of tourism, provision of low quality services had not been an obstacle to the tourist industry. Today, in the era of globalization, tourist markets have expanded and competition has accentuated. Therefore the issues of quality have become important issues on the agenda and drastic improvements related to changes of attitudes are urgently required.

Implications for tourism strategic planning

In all three regions, local authorities have realized the importance of tourism for the stability and growth of the local economy and have sought to plan and implement policies that a) improve the quality of services, and local infrastructure, b) attempt to differentiate local services and products, and c) attempt to deal with the negative aspects of mass tourism and prevent the adverse consequences for the environment. Current development is characterized by high seasonality and concentration on coastal areas, while interior mountainous areas do not have a fair share of the benefits accruing to the region from tourism. There are environmental problems due to overuse of the private car (old, narrow and not properly maintained road network, lack of parking in small mountain villages, air and noise pollution). Policy decisions must avoid the use of only mandatory and normative measures and start providing incentives which would promote alternative forms of governance, in the form of public–private partnerships. The values and needs of the local population must be respected. According to Lakshmanan (1983), the key to Greek development lies in improving the human capital and the organizational and institutional infrastructure, through investment in education and innovations in the design of institutions that take advantage of emerging new technologies and service industries.

a) Management of tourist flows to achieve a more even seasonal and spatial distribution through product differentiation. Exploitation of the opportunities that the natural and built environments have to offer and better marketing practices of the natural and historic assets of the area by promoting eco-, agro-, cultural and athletic tourism.

b) Management of transportation on mountain villages through more ecologically sensitive design of infrastructure facilities and promotion of traditional and alternative modes of transportation which are appropriate for the local topography. Traditional mountain paths, mountain bikes, animals, steam trains and boats are among the recommended modes. The emphasis on traditional modes of transportation should not preclude the applications of new technology in solving problems, i.e. the use of minibuses, even a funicular railway on which cars are pulled up and lowered by cables, if that proves to be the most cost effective and environmentally sensitive solution (see Loukissas coord. 1999).

In both the cases stated above, various 'soft measures' related to information may prove to be important. The tourist industry, or some of its components, has already developed informational networks for transport – especially air transport, hotel bookings and money exchange. However, in Greece, this does not apply at the local level, as far as the organization of the host local authorities are concerned. The informational infrastructure to fine tune tourist flows could prove essential for the best possible management of these flows even under the constraint of the plans of the large tour operators. However, this is in part related to the issue of planning, building up the competence and promote co-operation between authorities.

c) Environmental and cultural protection and promotion of alternative forms of tourism. There is a need to upgrade local human resources, sensitize local agents for the need to protect the environment and promote intermunicipal co-operation by creating networks. The new law regarding redistricting – the union of small municipalities into larger administrative units – has the potential to benefit, in spite of the reactions it has raised due to its compulsory nature. It provides an opportunity to save money through economies of scale and to produce a more effective administrative unit. There are still institutional barriers which hinder the implementation of alternative forms of tourism which need to be overcome. Infrastructure improvements are necessary but not sufficient to create new tourist products. There is a need to involve and upgrade local human resources and promote public–private partnerships in the management of tourism.

Greece is one of the few countries where regional planning and the preservation of the natural and historic environment are protected in the constitution. However, it lacks the capacity to implement plans and effectively enforce the laws. Some of the most important reasons for the current situation are considered to be the heavy reliance on a centralized and bureaucratic administrative system which is not prepared to accept the political cost of such decisions, coupled with weak protection of individual rights and the absence of significant local and private initiatives. Solutions imposed from above, no matter how sound they may be, cannot be enforced without taking into consideration the local needs and individual property rights.

Taking into consideration all the above, policy decisions must be based on the following goals and principles.

- The natural and cultural resources must be utilized and protected within the framework of sustainable development.
- The values and needs of the local population must be respected and, at the same time, must be educated about the value of preservation.
- Policy planning must avoid the use of only normative measures and start providing incentives which would promote alternative preservation and rehabilitation initiatives by local and private sectors. At the same time, emphasis should be placed on mechanisms such as fair compensation of property owners, as well as transfer of development rights for those negatively affected by public policies.
- The costs of preservation should be equitably shared among beneficiaries. The institutional framework must be enabling, flexible, decentralized and efficient to assure effective plan implementation.

Notes

1. SMART was a research project supported by the E.C. D.G.XXIII – Tourism Unit. The partners included the Department of Planning and Regional Development at the University of Thessaly, in co-operation with the Prefecture of Magnesia, The National Tourist Organization of Greece, the Universita di Catania in Italy and the Fundacion Cavanilles de Altos Estudios Turisticos in Spain.
2. Only a fraction of them are officially declared and approved by the Greek authorities. It has been estimated that in 1990 about 77 per cent of the beds in rental rooms in Magnesia operated illegally without a license and the actual tourist nights are three times as many as those officially reported (Centre for Planning and Economic Research 1994).

17 From tourists to migrants

Residential tourism and 'littoralization'

Lila Leontidou and Emmanuel Marmaras

Changing cultures, consumption patterns and new forms of leisure over the last two centuries have affected tourism profoundly. As a form of consumption, it spread from a small minority of affluent populations to broader social groups, as income and leisure time have increased in Western Europe and the USA. Its types have also diversified and expanded from a small elite to become massive: sun-and-beach tourism, cultural tourism, alpine tourism, and, more recently, eco-tourism and international residential tourism. The latter is still limited, but contributes in a qualitative change in tourism along the European Mediterranean coasts. Here, tourism tends to combine with migration as a phenomenon and a research topic. In Mediterranean African and Near Eastern destinations, international residential tourism is still too limited to attract our attention here, though, as discussed below, some Southern Mediterranean countries have seen relevant phenomena at a limited scale.

The very concept of international residential tourism is so new, that it is not included in most books on tourism (e.g., Apostolopoulos, Leivadi, and Yiannakis 1996; Montanari and Williams 1995; Williams and Shaw 1998). It is defined as the seasonal relocation of people to foreign resorts for long periods of time (several months rather than weeks; Barke 1991: 20), usually in owner-occupied houses. With this definition, caravans, boats and holiday cottages rented for one holiday are excluded (Shucksmith 1983: 174). International residential tourism is a phenomenon evolving from domestic seasonal homes, and combines semi-permanent migration and second-home holding in a foreign country. It may be theoretically relevant with time-sharing, but a closer analysis indicates that the latter is much closer to traditional forms of tourism in practice.

International residential tourism was rare in the past. What was more usual, was *domestic* residential tourism: the phenomenon of second homes within the home country, used for leisure and recreational purposes. More affluent populations in continental Europe still keep a summer home in the countryside, usually around metropolitan areas or across the seafront and in island villages, while they live within the city for the largest part of the year. This is much more pronounced in the Mediterranean, where values for central living keep affluent populations in the compact city, combined with a rural second home for days of leisure (CPER *et al.* 1998; Leontidou 1990, 1997). However, summer homes are not restricted to affluent populations; they are often located at the place of origin of migrants to cities, which a lot of the Mediterranean urban populations are. Second-home access throughout continental Europe has been important in the creation of compact cities: 8–20 per cent of urban residents had access to a rural second home in the early 1980s (White 1984: 163–4), and the rate has increased since then, with the spread of residential North–South tourism lining the Mediterranean shores (King *et al.* 2000).

There is an important difference between domestic residential tourism in second homes, and *international* residential tourism: the latter causes the thinning of the line between tourism and international migration. This line has actually been trespassed in the 1990s in the European Union, as Northerners have increasingly populated Mediterranean coasts and islands for longer and longer periods of time. Evidence appears of their numbers increasing since 1993, because of European legislation permitting European citizens to buy property anywhere within the European Union (Marmaras 1996: 18). Mediterranean Africa and the Middle East have not yet participated in this phenomenon in any significant degree; hence, the focus of this chapter on Mediterranean Europe and on the 1990s, when residential tourism escalated after the removal of institutional barriers within EU countries and the new freedom of movement of the population. The development of transport technology is an additional channel facilitating frequent movement between work and leisure, even across boundaries.

Residential tourism is growing to be an important aspect of postmodern lifestyles in the EU, though not yet on a massive scale. It concerns Mediterranean coasts and islands and is related to climatic conditions. The traditional tourism industry does not welcome this trend, which causes strains in the hotel sector. Several types of movers can be discerned in the current research findings:

- families splitting their time between a job in the city and leisure at the countryside abroad (often involves mixed marriages);
- entrepreneurs who combine leisure with income earning in the countryside;
- artists, intellectuals and people who seek alternative lifestyles; and
- retirement migrants moving to the Mediterranean coasts and islands.

The last category is the most sizeable and the best-studied in current research (Cribier 1982; Oberg, Scheele and Sundstrom 1993; King, Warnes and Williams 2000). Aspects of the other three categories often coincide into one, especially the first and the second, but also the second and the third. These variations present a challenge to the widespread idea that residential tourism is a non-economic matter; in fact, it is consumption-oriented but has interesting linkages with the production sphere (Leontidou coord. 1997).

Finally, besides international and domestic residential tourism, there is a third variant of movement which borders between international and domestic migration on the one hand, and often relates to tourism, on the other: *return migration.* This important variant has escalated in Southern Europe after the mid-1970s, and its relationship with residential tourism has been reported for Italy (King 1984) and Portugal (Mendonsa 1983). The seasonal relocation of the diaspora is very important in Greek islands and coastal villages during the summer. Throughout Southern Europe, emigrants returning from the North, or *retornados* from Africa, set up their businesses in tourist resorts, besides towns and cities. Returning migrants often opted for the coast and opened up tourism-oriented businesses, using little capital investment. The opening to indigenous capital their activities involved was welcomed in the relevant reports. Repatriated capital was enhanced, shortly afterwards, with capital from new emigrants: the international residential tourists.

Migration networks and interaction with the local society

The villas lining the coasts of Nice bear witness to the roots of international residential tourism in the early twentieth century and even before, when English elites used to live seasonally in the French countryside. The French, for their part, used to reside in their colonies in North Africa. In early twentieth-century Tunisia, the picturesque Andalusian-style village of Sidi Bou Said owes its fame to three young painters who stayed there after 1914: Paul Klee, August Macke and Louis Moillet (Baedecker 1992: 239–40). However, it is difficult to speak of international residential tourism in this case, as well as other cases of diaspora communities in colonies: the French communities in Algeria, Italian ones in Libya, Egypt, Erytrea, the Greek communities in Egypt are not properly 'touristic'.

Those were rare phenomena, or particular instances related with imperialism. At present, by contrast, international residential tourism is embedded within an overall radical change in migration waves and the dynamics of migration in Europe. At its outset, though limited, this movement was interesting from a theoretical viewpoint: it has tended to shift the migration dynamics from the production to the consumption sphere. The tendency for investment to tourism-oriented facilities was an in-between activity related with tertiary 'production'. At present, trickles of these tourism/migration waves are increasingly composed by economically-active populations, besides the pensioners or leisure migrants who were predominant during the beginning of the 1990s.

However, the relationship with the production sphere is incomparably weaker than it was after the Second World War, when labour migration caused major South–North population relocations. This economic migration trend, which has been characterized as Fordist (Fielding 1993), subsided in Europe by the 1970s (King 1993a, 1993b; Leontidou 1990).

Since the 1980s, migration waves in the European Union actually became much more complex, composed of several streams: repatriation from North to South and, in the case of Portugal, from the African colonies since the mid-1970s; the arrival of African migrants to Southern Europe, which has caused the southward shift of the parallel axis dividing North/South in migration waves, to the North of Africa (King and Rybaczuk 1993); post-socialist restructuring creating a 'new South' in the East of Europe and the Balkans, which sends migrants and refugees to the West, including Greece and Italy; job-related migrations by highly-skilled professionals within EU countries (Salt 1996); the stepping up of Third World migration from the Middle but also the Far East (King and Rybaczuk 1993); highly skilled migration related with the brain drain; and North–South residential tourism (Leontidou 1997; Leontidou coord. 1997).

There are several novel trends here, for both Southern and Northern Europe. The passage from emigration to immigration is the most important aspect for the South (King and Rybaczuk 1993), whether in the case of labour migrants from the Third World, repatriation, or residential tourists from the rest of Europe. The two latter trends are combined in the phenomenon of returnees toward tourist resorts, islands and coastal localities, who arrive to open up small businesses or to spend part of the year, and are often encouraged to this effect. In Portugal, *retornados* opted for the urban periphery as well as tourist resorts. The Greek government has created tourist settlements for diaspora Greeks since the mid-1980s, in the prefectures of their origin, especially in the Peloponnese: the 'Arcadian village' was thus created by DEPOS (Public Corporation for Urban Development and Planning), and other such developments are planned for Laconia, Helia, Thessaly and Lesvos. Settlements have been also built by regional authorities, such as the one under construction in Samos. These cater for seasonal or permanent relocations at diaspora.

Besides returnees, retirement tourists are encouraged by certain countries, because they ease out seasonality tensions. In Spain, legislation has been passed to help them use their pensions locally. For Northern Europe, the novel trend is that people from nationalities who did not take much part in international migrations within Europe in the past, such as British, Germans and Dutch, are now involved in them (Zlotnik 1992). The in-between case of France is also interesting: its large immigrant population was city-bound and labourer-oriented (Ogden 1989), until its Mediterranean coast and countryside recently started to attract international residential tourists. Its migration inflows (but not yet its total immigrant population stock) were dominated by North Europeans by 1990, many of whom were not economically active (Buller and Hoggart 1994: 2–3).

Researchers of Mediterranean coasts have found that mild climate and warm weather are now prime reasons for migration – effective 'pull' factors, to use a dated terminology from the years when migration was employment-oriented. Valero Escandell (1992: 67) has reported this for 49 per cent of his respondents on the coasts of Alicante, Spain, with the low cost of living as the second important factor. This is echoed by 51 per cent of those interviewed among British holiday-makers, who said they arrived in Alicante to 'take the sun' (Rodriguez Martinez 1991: 58). This has traditionally been an important reason for holiday-making in the area, of course, but the actual British population of Alicante was only 2875 in 1981 and rose to 11,535 in 1986 and 17,682 in 1988 (Valero Escandell 1992). According to the survey in Greek islands and coasts in 1995, 40 per cent of residential tourists were found to be British (80 per cent in the Ionian islands) and 18 per cent Dutch (40 per cent in the Northern Aegean). Only a few had moved prior to their settlement: 53 per cent had previously lived only in their own country, and 26 per cent in another European country besides Greece (Marmaras 1996).

The operation of migration networks here is very interesting. Chain migration in the years of employment-related labour migration, especially the 1960s, has been repeatedly documented. Networks of families and friends directed the migration of Southerners to Northern Europe (Boisvert 1987). Now, similar migration networks seem to operate for Northerners arriving on Mediterranean coasts. In a comparison between Spain and Greece, chain migration has been shown in the latter, on the basis of respondents' views and ways in which they have found their homes. In Greece, homes found through recommendations of Greek friends (37 per cent) or marriage (22 per cent) predominate, while the role of local agents (15 per cent) or foreign advertisements (4 per cent) is restricted. In Spain, by contrast, it is through local estate agents (50 per cent) that most homes are found (Marmaras 1996). This reflects the different levels of development of real estate markets and advertisement in the two countries. It also provides evidence about the different type of residential tourists emerging in Greece and Spain, as discussed in the following section.

Residential tourists appear committed to their new homes and more general environment, despite reservations with regard to certain aspects of the host societies. Foreigners in Greece feel welcome by the local, insular societies, but at the same time unable to penetrate the native social structures, apart from in cases of mixed marriages. Complaints are related to the Greek bureaucracy and substandard public services and infrastructure. Several of these residential tourists have their friends dispersed in Europe (25 per cent) rather than in their home town or country (22 per cent), but many have also developed networks in the Greek settlement (19.5 per cent). A comparison with Spain in this respect indicates the cosmopolitanism of residential tourists in Greece: those living in Spain show a more marked disposition towards friends from their home town or country (35 per cent) or those in the Spanish settlement (23 per cent), rather than a dispersal of their social networks in Europe (15 per cent, see Marmaras 1996). In both countries, few residential tourists have indicated any networks in other localities of the host society (11 per cent and 12 per cent respectively).

Destinations, residential patterns and profiles

The Mediterranean has always known strong population pressures during the summer months, but these have had a short duration. Seasonality has always been a major concern. In the early 1990s, Mediterranean Europe had received about 113 million package tourists a year (Perry and Ashton 1994) and in 1990 it accounted for 34 per cent of all international tourists in the world (Lanquar 1995). Tourists to Spain, the top destination, increased tenfold between 1960 and 1994, from 6 million to over 61 million, of whom 88.8 per cent were European (European Commission 1997; Lardies Bosque 1997; Secretariat General de Tourismo [SGT] 1995). The activities of tour operators created a spatial concentration of both demand and supply (Valenzuela 1991): top destinations were Catalunya, which received 28 per cent of Spanish tourism in 1994, and the Baleares with 16 per cent (Lardies Bosque 1997; SGT 1995).

This 'seasonal counter-urbanization' of Europe is related with a much less seasonal phenomenon: among the above tourists, more enduring residential communities have emerged. Domestic residential tourism has been better covered in research, especially since the 1970s. Second homes have been springing up around large cities and have lined the Mediterranean coasts. They have populated the French countryside, where timesharing among family members, friends and purchasers in general is customary. 'Weekend homes' as a reality and an aspiration encircled Spanish cities (Canto 1983; Valenzuela 1991: 47) and tourist regions (Miranda 1985). However, until recently, little research has appeared on return migration, with an overwhelming concentration on the case of Portugal; only anecdotal evidence exists about the impact of Northern European migrants/tourists and their tourism-based initiatives on Mediterranean economies.

Each Southern European country has its own geography of foreigners. In France, rural areas are more attractive than coastal ones (Tuppen 1991): a comparison between Spain and France found 64 per cent of migrants in Catalunya living in coastal municipalities, contrasting with only 19 per cent of migrants in Lanquedoc. Another interesting aspect of residential patterns is the tendency of foreigners to conglomerate by nationalities in the various regions of Europe. In Languedoc, 65 per cent of non-French EU residents are Spanish (Lardies Bosque 1997). In Italy there were about 40,000 Germans and 30,000 British citizens in 1991

(Leontidou coord. 1997). Many were scattered in the Sylvan landscapes, renovating abandoned houses of Tuscany and Umbria, to the disadvantage of locals wishing to buy property (King 1991: 78; Pedrini 1984).

Spain seems to have been the prime destination of international residential tourists attracted by the sea and sun (Valero Escandell 1992). Large-scale settlement of foreigners started in the 1970s in the Canary Islands and the Costa Blanca (Gaviria 1977b), and toward the Costa del Sol (Jurdao 1979). New types of property acquisition, such as 'lease-back' and 'time-sharing' appeared in areas where residential tourism threatened to over-run the capacity of municipal services. This was especially so in the 'leisure towns' of Torremolinos, Benidorm and Lloret de Mar, conceived in the 1960s (Gaviria 1977a: 24–31; Williams 1997): Benidorm had a permanent population of 33,700 in the mid-1980s, but received 3 million tourists annually. Seasonality receded as elderly residents lived along the coast in autumn and spring, while younger residents arrived during the hotter summer months. In the mid-1980s, just before Spain's accession to Europe, there were more than 1 million foreign-owned dwellings on the Spanish coasts, and foreigners used to buy 50,000 dwellings a year (Valenzuela 1991: 48–9). Permanent UK residents increased by 33 per cent in 1987–89, from 55,318 to 73,535; German residents numbered 44,220 in 1989. By 1991, there were dense concentrations of foreigners in the Baleares, Canarias, Castilla la Nueva and Valencia (Marmaras 1996). Changes in EU legislation after 1993 created yet more opportunities. Contrary to Greece, concentrations of residential tourists in Spain did not appear mainly in insular areas.

In Greece, the regional distribution of second homes can be indirectly deducted from empty houses during population censuses, which are the most numerous around large agglomerations and in the Aegean islands closest to the coasts (Leontidou 1991: 97–8). Research in the mid-1990s identified a marked density of foreigners (but not necessarily residential tourists) in Attiki and the Southern Aegean islands. An important variation between the years 1981 and 1991 was observed in the Cyclades islands. In the former year, foreigners concentrated in the standard tourist resorts, such as Myconos and Santorini; in 1991, by contrast, they were rather concentrated in smaller islands from the viewpoint of both population and area, such as Antiparos, Donousa and Ios (Marmaras 1997). Preferences for isolated islands are indicative of tourists' profiles. These match the figure of the individual seeking an alternative lifestyle near nature, as is also portrayed in the stories of Houliaras (1998). The image clashes with the affluent mass tourist residing on the coasts of Spain.

In several large islands in Greece (e.g., Lesvos, Chios) and France (Corsica), the international residential tourism phenomenon has not taken root. Diverse economic structures including the establishment of new universities and the lack of industrial activities there, has had the positive effect of maintaining a good level of infrastructure and a clean environment, but this has not attracted any large numbers of tourists. Lesvos and Chios have not developed accommodation facilities in an organized way, while other Greek resorts tend to plan facilities (Loukissas 2000). Corsicans, by contrast, had initially seen the arrival of mass tourism with some hostility, but now consider tourism as a suitable activity for socioeconomic revival (Leontidou coord. 1997). None of the three islands have valorized the opportunity of seasonal combination between the student population and tourism.

On the basis of questionnaires posted to 14 per cent of foreigners in Greek islands, the profile of the international residential tourists is as follows: retirement tourists over 65 years old are a minority (16 per cent), most residents have a university degree (72 per cent) and they originate from Britain (38 per cent), Germany (19 per cent), the Netherlands (14 per cent, see Marmaras 1996). Gender differences are not very important, but foreign women tend to predominate among the residential tourists who come to stay after marrying a native: 25 per cent of foreigners in Greek islands fall into that category.

The comparison with foreigners on the coasts and in the islands of Spain and the in-depth analysis of the causes and effects of residential tourism in selected localities of the two countries through institutional interviews, questionnaires to residential tourists and other fieldwork, has revealed different profiles of foreigners in the two countries. Spain has 25 per cent of its residents over 65 years old, and only 45 per cent have a university degree. Residential tourists attracted to Spain are part of a mass-tourism process linked with retirement tourism and organized by agents in the home countries. Those choosing Greece, by contrast, move individually and act through networks of friends rather than though organized agents (Marmaras 1996). Many seek isolation and alternative lifestyles (Houliaras 1998).

Activity and employment patterns

International residential tourism has been studied almost exclusively as a consumption-related phenomenon (Buller and Hoggart 1994). In fact, it can be distinguished, as a migration experience, from labour migration during the Fordist stage, because consumption-led migration predominates. Economic inactivity reaches as high as 67 per cent in Catalunya and 73 per cent in Languedoc among EU residents in 1990 (Lardies Bosque 1997) and is not restricted to the elderly retirement tourists (Perry, Dean and Brown 1986).

Consumption-led migration predominates even at present, when the number of economically-active residents rises. Increasingly, certain productive activities are identified among foreigners in the Mediterranean. The decision to migrate to the coasts and tourist resorts is not related with these activities: employment follows the move, and helps the comfortable and more permanent settlement of residential tourists (Lardies Bosque 1997). In the past, the most usual type of small-scale economic activity practised by domestic residents and residential tourists was the use of their property for income, where property included small workshops besides the house for themselves and their family. In few cases, broader types of domestic tourist businesses have been studied (Williams, Shaw and Greenwood 1989). Recently, however, it has been found that many international tourists and return migrants are actively involved in entrepreneurship and productive initiatives (Lardies Bosque 1997). Small-scale enterprises spring up and are encouraged by policy-makers, as large corporate involvement in the tourist industry has come under increasing strain in Mediterranean Europe.

In France, Spain and Portugal many tourist-related enterprises, run by tourists, spring up. Tourism projects are based on the initiatives of international residential tourists and returning migrants. A comparison of the LEADER initiative and tourism in rural Spain and Portugal has led to the conclusion that rural tourism is a viable alternative for localities hit by seasonality in the past, and has highlighted the structure and impact of tourist projects in Andalucia and rural Portugal. It has also stressed the role of immigrants and returning migrants in them (Fernandez Martinez in Leontidou coord. 1997).

North European immigrants have created tourist enterprises in Catalunya and Languedoc, two very different regions, one of mass tourism and the other less dependent on tourism, respectively. Questionnaires to foreign entrepreneurs in the two regions have highlighted their migration histories, their motivations and management techniques, their role and impact of their enterprises in the localities studied. Differences arose between the two regions, but an important similarity has led to further reflections: namely, the predominance of non-economic decision making among North European tourism entrepreneurs in Spain and France. According to the survey, personal lifestyle reasons rather than economic considerations have attracted foreign entrepreneurs to these Mediterranean coasts. Naturally, this poses the problem of the viability of their enterprises (Lardies Bosque 1997). Such findings underline the persistence of consumption-led migration in the phenomenon of residential tourism, contrasting with production-led migration during the 1960s.

Research in Greece has indicated that foreigners on coasts and islands are not all idle populations or pensioners. The majority have been found to be self-employed (53 per cent) rather than retired (18 per cent), and some are employees (18 per cent) and housewives (12 per cent). These may be married women with Greek husbands in rural activities. In Spain, a limited sample of foreign population was researched and was found to be divided between employees (those living permanently) and retired persons (those using the house as a second home, see Marmaras 1996).

The question has been whether they practice their economic activities seasonally in the cities of origin, or also in the resort where they move during the warmer part of the year. A rising number of residential tourists have been found to choose the second option, but the opening of enterprises does not follow economic considerations: it was found that the main attraction for entrepreneurs to move to Costa del Sol was the pleasant climate (60 per cent) and the lower cost of living (Eaton 1995; Lardies Bosque 1997). Again, it is not economic decision making which motivates residential tourists.

International counter-urbanization and the 'littoralization'

Trickles of international residential tourists and domestic second-home movers have started to affect the 'postmodern' European migration trends. As shown above, urban residents of Southern Europe, but also several Northern urbanites, have traditionally kept a rural second home for days of leisure within their own country. This trend is now becoming international, and spreading to Europe as a whole. The settlement network is affected by southward population shifts and movements to the coast. Processes of 'littoralization' are presently at work (Leontidou *et al.* 1998).

As stressed previously, international residential tourists settle as residents and occasionally entrepreneurs, after visiting the areas and enjoying short vacations there. Their most outstanding economic impact on the Mediterranean coasts and islands relates to the rising demand for holiday houses rather than with their entrepreneurial activities, which are still limited. The property market boom in the warmer part of Europe contrasts with the depression of the real estate market in Northern Europe during the last 15 years. Rising property prices in the North have had an impact, pushing populations to other property markets. The term 'international counter-urbanization' (Buller and Hoggart 1994; Perry *et al.* 1986) indicates a movement of population and real estate activity from the larger Northern cities to the Southern coasts and islands. This follows the summertime counter-urbanization of the whole of Europe.

The present stage of diffuse urbanization in Mediterranean Europe is combined with this parallel coastal and island population increase (Leontidou 1990). Sales of plots of land and of old buildings have escalated during the last years in the Aegean and Ionian seas. Private building societies have entered the market, constructing big complexes of houses and apartments in Cyclades, Crete (Greece), Corsica (France), Majorca, Ibiza (Spain). This was actively sought through state policy and place marketing. The latter is suggested by the traditional nomenclature of the Spanish coasts: Costa del Sol, Blanca, Dorada.

These impacts and the entrepreneurial investment tendencies outlined in the previous section, relate with cultural change. International residential tourism tends to surpass the 'leisure' category and acquire the character of a postmodern lifestyle, where spatial fixity in one home is in question (Leontidou 1997). Studies of elderly migration in North America have found that seasonal residence in warm locations has become a way of life. Many occupy their summer residence for longer periods and for long blocks of time rather than several short stays, so that it is their winter residence that has come to be a second home (McHugh 1990; Sullivan 1985).

In postmodern Europe, the increase of seasonal residential mobility reflects material affluence and changing lifestyle preferences, combined with the opening of European borders and the improvement of communication technology. This new type of *seasonal/semi-permanent residential mobility* spread from pensioners to more people seeking a more varied lifestyle, where autumn and winter, spring and summer, are spatially as well as temporally differentiated. The corresponding adaptation of the job market to follow this trend may be just a matter of time, as we move further into the new millennium.

18 Spatial dimensions of marine tourism

Outlook and prospects

Kostas Krantonellis

Introduction

The main external factors affecting tourist development in recent years include: demographic and ecological pressures, social change, adversities such as recession, unemployment and political instability, the initiatives of the European Union aiming at an Euro-Mediterranean partnership, and the massive technical infrastructure in the form of a Trans-European transportation network. A result of the co-operation of all these factors is the redistribution of tourist activities at an international level and the gradual restructuring of Europe's spatial pattern, in particular concerning the traditional Mediterranean pleasure peripheries.

An essential social change, which accelerates tourist reformation, is the growing awareness of the public in favour of natural and cultural environments. In this sense tourist demand is increasingly determined by qualitative criteria while tourist supply designed by the niche market tour operators promotes environmentally-sound forms of tourism. The tourism-generating countries regard these forms, including marine tourism, as an alternative to the prevailing resort tourism concept. This new setting establishes competitive market conditions for marine tourism in the Mediterranean, whether it is yachting or cruise ship travelling. A necessary prerequisite is the protection and even the future improvement of the marine environment in the area by means of co-ordinated collective action. First priority ought to be the sustainable management of natural and cultural resources.

This chapter deals with the marine tourism traffic in the Mediterranean according to the spatial distribution of resources. More detailed reference is made to the essential features of yachting and cruise ship tourism, the relevant problems and the development prospects of the industry.

The extent of the subject and the scarcity of comparable statistical data on marine tourism in the Mediterranean countries prescribe a rather general approach. Nevertheless, the overall picture presented here could be regarded as reflecting the existing situation in the area.

Cruise ship tourism

Cruise ship tourism constitutes a new form of holiday tourism and, indeed, a special category – marine tourism. Until now its contribution to international tourism has been small, due to the restricted number of tourists who chose to holiday in this fashion. Up until now there has been little demand for this, which has been due to a limited market and lack of publicity. Although, especially as far as the North American public are concerned, this lack in demand has been due to an inactive tourist drive characteristic of a conservative market.

Cruise ship tourism, making up part of holiday tourism, follows the rules of supply and demand, which apply to customary package tours. At the economic and political level it is influenced by the same externally-created factors of international economy and politics.

Apart from these similarities, the basic difference rests on the fact that, in contrast to resort tourism, cruise ship tourism does not appear to be in decline but is relatively increasing. This is due to the European tourist who, having become saturated with mass tourism, is branching out, where as the American tourist, even though a little belatedly, is entering the tourist scene having been tempted by this new form of tourism which combines snob appeal and fairly reasonable prices.

The aggressive advertising campaigns in the United States must also not be overlooked. As competition intensifies, so advertising expenditure increases while, on the other hand, there is a potential of attracting financially adequate North American clientele. In this way cruise ship tourism can easily be characterised, at the moment, as a small but dynamic form of tourism.

Even though cruise ship tourism, as a form of mass holiday tourism, has only been in existence for 20 years or so, it has already passed through the phase of readjustment towards the new requirements of the international tourist market. These requirements refer firstly to the duration of the cruise, which is gradually being restricted to three or four days. These short holidays on the sea coupled with the mixed system of fly-cruise package holiday assures a cheap and quick arrival back to the port of departure. A second requirement concerns the choice of entertainment on board the cruise ship, amongst which include a casino, thermal baths, etc.

In this way today, compared to the past, cruises offer the tourist a more complex product as far as the duration, destination, the fittings of the cruise ship, the activities aboard and trips are concerned. In this way they are able to meet every demand, and cruise ship tourism is in a completely successful competitive position on the international tourist market.

However favourable these prospects are, they are based on internally-created factors which affect the development of the industry. For the immediate future, particularly in the Mediterranean, growth will depend mainly on externally generated factors. That is to say the relationship between the economic and the political situation, which naturally directly influences the tourism-generating countries as much as the tourism-receiving countries.

The history of modern cruises covers the period between 1966 and 1974. It started with the entrance of the Norwegian Cruise Lines onto the market in 1966, followed by the Royal Caribbean in 1970 and ended with Carnival in 1974. The common factor linking these three pioneer companies was a week's cruise in the Eastern Caribbean, attracting middle-class customers through low pricing.

Until the 1980s there was an even distribution of the potential of cruise lines where the larger ones served mass tourism and the smaller ones a more selective clientele. As a rule these companies belonged either to families or to people who had founded them, and they were run on traditional business lines. In contrast to this, the years between 1986 and 1991 saw a reorganisation of these companies, which resulted in the consolidation of the lines through mergers, acquisitions and strategic alliances, surpassing anything in the past. The reasons for the above were mainly to try to reduce the running costs and to apply the principle of 'strength (and a variety of cruises offered) in unity'.

Today there are several major cruise lines that embrace other smaller ones and together they control the largest part of the market. They control not only mass tourism but also the special interest groups through the smaller lines that these major lines have acquired. Characteristic of the fleets of these major cruise lines is not only the number but also the relative newness of the ships and their enormous size. Grand Princess, which is the largest cruise ship in the world, is capable of carrying 2500 passengers (Steavenson 1997).

After the developments of the last five years, cruise companies can be classified into three categories. First, those of the major cruise lines; second those belonging to the small specialist lines and thirdly those belonging to the medium-sized lines offering 'traditional' cruises. In the recent past similar groupings have occurred in other industries with the main aim being economic partitioning. As far as the cruise ship industry is concerned it is possible that at both extremes, that is to say the large and specialist cruise lines, they will survive easier than the traditional ones. An essential presupposition, not only for this, but for every form of tourism, is the competitive differentiation between supply and demand which, for cruise ship tourism, means new markets and new destinations.

In many economies and private businesses the steady slowing down of the increase of European and worldwide tourism creates a climate of uncertainty for the future. Along with this, a continual degrading to the natural environment has been observed, as well as a lowering of the cultural environment, due to the unbalanced development, up until now, of mass tourism. The existing infrastructure is not able to meet the increasing demand on quality for this different type of tourism.

Given that the major role of Europe in tourism, where most economic interests are concentrated, is gradually being reduced, the effort to face the above problems, by most European countries, relies not only on the increase in inter-European tourism and the attraction of overseas tourists, but also to include new social groups in the process of tourist consumption. At the same time the aim to protect the environment also makes it necessary to find new forms of alternative tourism.

The international perspective of the development of tourism depends on new social values and the accompanying consumer prototype that will emerge as society moves towards the post-industrial age. Certain tendencies in the near future will include the turn towards for quality, a reduction in the duration of holidays, and a decrease in expenditure. There will also be an increase in out-of-season city tourism as the established seaside resorts are environmentally degraded.

The presumption for verification of the mid-term forecast of the World Tourism Organisation that is related to the industry in general, including marine tourism, is a healthy population basis in the tourism-generating countries, adequate publicity for the consumers and an increased accessibility of destinations. At the same time obvious factors needed are economic recovery in many countries, political stability and a clean natural environment.

Cruise ship tourism, as a special form of tourism, will follow the general development of the industry. As such, it must be pointed out that even though not very large, it is characterised by a particular strength compared to many other forms of tourism. This is confirmed, firstly, by the average annual growth rate of 7.6 per cent since 1980 and, secondly, by the indisputable fact of the continual renewal of the products offered.

The known opponents to this are over availability of bed-days offered due to the amount of new ships that have invaded the market, in conjunction with a drop in demand due to the recession and, finally, discounted prices. At the same time there is an increase in expenditure, mainly in advertising the product, and a decrease in income, due to the reasons already mentioned, which are reducing the profit margin.

In spite of the negative occurrences of the past few years, it is certain that the share of cruise ship tourism in the worldwide tourist industry will increase considerably. Also the fact must not be ignored that the expected 10 million passengers yearly around 2000 are likely to exhaust the greatest possibility of the American market. Thus the question must be asked as to how far in the future will the cruise ship industry continue to be profitable.

The importance of the Mediterranean as a cruise region is due to the short distances from European countries that generate tourism and the variety of tourist resources that it contains. It must also be noted that, especially for the Eastern Mediterranean area, there has been a problem in recent years in promoting this tourist product on the North American

Table 18.1 Mediterranean passengers 1998

Ports	Cruise ships	Other ships
Ancona	5730	991,423
Balearic Islands	499,755	1,114,754
Barcelona	373,799	507,731**
Bari	120,527	734,942
Cagliari	12,878	483,315
Cannes	56,466	–
Corsica	102,896	839,340
Cyprus	608,376	–
Genoa	364,647	1,846,122
Gibraltar	98,760	–
Haifa	411,614	–
Livorno	195,743	1,206,311
Malaga	105,965	174,263
Malta	147,484	337,048
Messina	72,269**	11,280,718**
Monaco	51,609	–
Napoli	334,685	620,844**
Nice	218,198	619,985
Palermo	83,171	920,122
Piraeus	432,456	5,530,515
Savona	102,755	–
Toulon	39,026	101,274
Trieste	12,706	166,237
Tunis	201,691	309,755
Valencia	255,953	1,774,576
Venice	335,483	365,207
Volos	13,805	346,668

* 1996
** 1997

Source: Medcruise Association

market due to the general instability of the area and the fear of terrorist attacks. Moreover, the sensitivity of American tourists towards their personal safety is well known.

It must be mentioned that the passengers of Mediterranean cruises especially appreciate the cultural sights of the area. This means that measures must be taken to reduce every sort of difficulty that can occur at ports, such as during disembarkation, local trips and visits and especially the return of passengers to the cruise ship. Already many Mediterranean countries have commenced major port alterations, such as those at Genoa, Marseilles, Malta and Piraeus, so as to improve conditions and services offered to passengers of the cruise ships.

A problem of the long-term development of the Mediterranean is mainly the restricted internal market which, at the moment, is not in a position to preserve a fairly independent demand for cruises in the extent of a drop in demand outside the Mediterranean. An added obstruction is the high price of air transport that is making Mediterranean fly-cruises non-competitive compared to cruises towards the Caribbean. This has resulted, in many cases, in British and German passengers tending to prefer a non-European destination.

The shape of the future demand of cruise ship tourism is influenced by various social, economical and political factors, which either directly or indirectly influence demand. As is well known, the determining factor that influences demand is mainly the family income in the tourism-generating countries and the prices of the services offered in the tourism-receiving countries. In the particular case of Mediterranean cruise ship tourism, critical factors influencing demand will be the price of the cruise, inclusion of air transport, as well as the price of services not included in the holiday package both on the cruise ship and at the ports of call.

As has already been mentioned, cruise ship tourism is only one form of holiday tourism. It has been established that holiday tourism has shown itself especially durable up to now to negative occurrences. This is a thing that will benefit the industry in general. This has already been shown in the elasticity of tourist demand in relation to prices and income. From these above facts, it is clear that the short-term development of cruise ship tourism will improve. In 1996 leading European markets were Northern Europe with 7.2 per cent and the Mediterranean with 8.4 per cent of the total traffic.

Yachting

The long and rich history of the Mediterranean Sea, its natural beauty and mild climate, both as far as air temperature and prevailing winds are concerned, make it ideal for yachting from April to October. Other important features of this sea, interesting to both yachtsmen and harbour engineers, are the absence of tides and the considerable depths close to the coast. This last feature enables easy landing of boats on the shore.

Yachts are boats with on-board accommodation facilities, and they are usually longer than 6–7 metres. There are motor yachts without sails or with auxiliary sails, sailing boats with auxiliary motors and bare boats that do not require a crew. When we talk about yachting we mean pleasure cruising that can be done with motorised or sail boats. The word 'yacht' is Dutch in origin but the British have given it worldwide recognition.

The nineteenth century and the beginning of the twentieth century saw yachting in the Mediterranean as a pastime of rich people. Their motivation was exploring the historically-rich seas with ships as large as they were luxurious. Later on, in the years between the First World War and the Second World War, it became fashionable to combine yachting with hunting, for example deer hunting in Albania, the hunting of birds in Greece and wild boar in Turkey. Directly after the Second World War, a few small boats started sailing in the Mediterranean and, as late as the 1960s, yachting in the Eastern Mediterranean basin was considered a perilous journey into the unknown.

At the beginning of 1960s descriptions from travellers using sailboats gave interesting information about the prevalent situations of the area – in particular about bureaucratic formalities, which on occasions could be time-consuming, organised yacht-clubs at many large harbours, as well as the few berth facilities at smaller harbours. There was also information about the frequent difficulties encountered when renewing supplies and the mutual friendship that developed between yachtsmen and the local, hospitable inhabitants. From the point of territorial distribution at this time, the most popular routes were crossing the Mediterranean as part of a longer voyage, sailing in specific regions such as the Central Mediterranean between Italy and Greece and local trips around the archipelagos of the specific area. Visits to archaeological sites and interesting sights of the hinterland were also typical of these sea voyages.

With the general development of mass tourism, chartered boats with or without crews began to appear. During the 1970s, the British Yacht Cruising Association began to import into Greece the first flotillas, that is to say small fleets of sailboats. Holidays with flotillas soon spread to the expensive tourist destinations of the Mediterranean, which soon became established as the pleasure periphery of Northern and Western Europe. At the same time an increase in private boats was noted, due to the gradual improvement in the standard of living of the inhabitants of Mediterranean countries that were members of the European Union. Today the ratio of those who own some type of boat to the rest of the population is 1:64 in France, 1:71 in Italy, 1:105 in Greece, 1:300 in Spain and 1:400 in Portugal (Adie 1984).

The first, and most important, construction built for yachting in the Mediterranean has been the tourist development of the Languedoc-Rousillon coast in the south of France, that started at the beginning of the 1960s. Even though the short-term political plan of the French Government was the socioeconomic improvement of a problematical border area, as well as trying to keep French tourists in their own country, as far as possible, the benefit to yachting in the western end of the Mediterranean is obvious. The area is divided into six regions, each with the same number of marinas and all the supplementary activities that go to make up an entire tourist development based on yachting. The final size of the complex was to include hotel accommodation and second homes for 250,000 people and a total of 20 harbours with different capacities – but this has not been realised so far.

The approximate distance between each harbour was ten miles so that it would be easy for amateur yachtsmen to sail from one to the other. Finally there was to be the provision of 10,000 berths and 30,000 inland storage in shore berths. In the case of Langueder-Rousillon, tourism was used as a vehicle for regional development and the French coast from the mouth of the Rhône to the foothills of the Pyrenees was promoted as the first pleasure periphery, in Europe, for mass marine tourism.

A few years later, at the other end of the Mediterranean, the National Tourist Organisation of Greece, as the official state authority, started laying down the long-term tourist development programme for the country. It shaped a new approach, which was aimed at developing yachting at a national level. Up until then the Greek practice had been to build marinas on the edge of a large urban area or on the boundary of a harbour area of a commercial port or a fishing port, for pleasure boats. With this rationale in mind the first marina was opened in 1965 at Vouliagmeni, 30 km east of Athens.

This new approach is based on the typical geographical characteristic of the Greek seas with their many islands, and on the idea of pleasure sailing zones as a subsystem of a national system of marinas. In this way sailing zones signify an organised, working unit, which includes installations for the service and safe stay not only of the boats but of the crews. This can be characterised by weather conditions that do not fluctuate very much and by a variety of natural and cultural resources. In a properly-delineated zone, all the practical and recreational needs for yachting must be found within this zone.

In practice, the national maritime territory divided into eight regions, each one with the capability of sailing 200–300 miles in ten to fifteen days. This is about the length of time a foreign tourist stays in Greece. In each sailing zone there is a total of harbour units in a hierarchy of three levels. At the first level, one marina of 500–1500 berths with all the facilities is foreseen. At the second level, a number of smaller marinas with 50–100 berths for basic renewing of supplies after a day's sailing. And thirdly, the last level takes in natural shelters in protected areas only to be used in an emergency.

The two examples mentioned above, although of a different level and scale, go back to the concept of the integrated tourist plan as a responsibility of the central or regional administration, strongly supported by state subsidies.

At most marinas, the main inhabitants are mostly made up of clients of either long or short stays. They may be amateur fishermen, yachtsmen, users of speedboats, holiday-makers and day-trippers. For all these different categories of clients, who give a certain colour to the pleasure area near to the sea, their foreseeable consumer needs must be met at competitive prices. Especially for the private yachtsmen, the essential qualities of the marina needed are easy access from both land and sea, security and the possibility to renew all required supplies. On the other hand, for the boats that just pass through, reception installations and sightseeing in the surrounding area are more important.

The depth of the sea, the currents and the winds determine, to a great extent, the type of boats that the marina can service and, of course, this will affect the expected income. Adverse weather conditions reduce the life expectancy of the harbour facilities and make their repair more costly. For a marina to be rendered an economical success, it has to work for 12 months a year. This presupposes it neighbouring onto an urban centre and that it is connected to a tourist network, as is the case with many island marinas. Up until recently there was often an incorrect estimation of the berth capacity and the types of extra facilities in demand. Similar mistakes in planning led many marinas into an economic deadlock. Gradually it was realised that marinas were not units that could be produced according to a set of rules that would result in their correct functioning. As a rule the opposite was found to be true; their success was based on a number of external factors and these had to be pinpointed if the venture was to be successful.

The coastal marinas of the Mediterranean countries belonging to the European Union are distributed, depending on their demand, between the most popular tourist destinations. Thus first comes Spain, Italy and France with 201, 150 and 134 marinas respectively. Portugal has 14 marinas and Greece has 13.

In Portugal, from the beginning of the 1980s, the marina of Vilamoura on the Algarve has been in use. The marina has 1000 berths and it has been strategically placed because it is on a sea passage. It also has a favourable microclimate and the local area boasts interesting sights to see. It appears that the marina of Vilamoura will gradually take over Gibraltar as the first stop on the way into the Mediterranean sea for those coming from the Atlantic. At the same time there are a series of smaller marinas that enable coastal yachting with the possibility of making stops inbetween. The aim of this is firstly to create organised sailing zones with marinas that are basically used by wintering boats under foreign flags or as business centres for yacht brokers.

Gibraltar, having two marinas, can also claim a place in the international market of marine tourism as most of the tourism comes from daytrippers. In the last few years tourists coming with pleasure boats were only 0.3 to 0.4 per cent of the total of visitors and their expenditure was only 2.5 to 3.0 per cent. Therefore, as with Portugal, the construction of marinas in conjunction with second homes and other complementary facilities should be a priority for Gibraltar.

In Spain the privately-owned boats are displacing foreign ones which are going to the Portuguese marinas. The majority of these marinas are to be found along the 750 miles of coast on the mainland and some on the Balearic Isles. There are also moorings at fishing ports where the picturesque scenery compensates for the lack of organised facilities. Due to the mild climate, overwintering at Spanish marinas and especially these of the Balearic Isles is very popular, given the fact that out-of-season flights to the rest of Europe are frequent and fairly cheap.

In terms of tourist geography France, compared to it neighbouring Mediterranean countries, is the one with the least amount of foreign tourists, with only 15 per cent of them visiting the French Mediterranean coast. For Spain and Italy the percentages are 73 per cent and 80 per cent respectively. Nevertheless, French marinas are showing a high level of occupancy not only for those along the Mediterranean coast but also those along the Atlantic coast too. This is due to the local production of sailing boats; the French market absorbs 50 per cent of this production. The high level of occupancy is also due to the lack of berths outside of the marinas in comparison to Spain, for example, where there are numerous moorings and anchorages at commercial ports. However, yachtsmen agree that the French marinas, including that of Monaco, offer the best services in the whole of the Mediterranean.

In Italy the marinas are mainly to be found along the West Coast of the peninsula, being more concentrated towards the north than the south. The exception to this being the length of coast between Venice and Trieste, where due to its geographical position in relation to the European hinterland, many foreign boats over-winter on a permanent basis. There is also a chain of marinas along the length of the Italian Riviera. Although there are not many marinas on Sardinia and Sicily, boats can be serviced at fishing ports. Italian marinas exhibit to a great extent the problem of seasonality, coming to a peak in the months of July and August. During this time occupancy can, as a rule, exceed 100 per cent. This is due to the local summer holidays and also to the wave of foreign tourists coming from neighbouring France.

Approximately half of the Greek marinas are to be found in Attica, servicing yachtsmen who live in the built-up areas of Athens and Piraeus and also the thousands of foreign tourists who, with either private or chartered boats, start out each spring for the Greek seas. The rest of the marinas are to be found on the larger islands and in Northern Greece. Apart from there being marinas at nearly every port, there are also stations for the renewing of supplies. Greece is classed among the countries with an excellent network of services provided and this is due to the small distances between ports.

The extensive Yugoslavian coast is made up of three chains of islands, which compose a sailing zone of attractive natural features. For tourist yachting, the most important part is to be found along the coast between Dubrovnik and Split. As far as facilities are concerned there are about 30 marinas and numerous anchorage or shelters in natural ports. The boats in this area belong mainly to Austrian and German yachtsmen. As far as Albania is concerned, once political stability is established in this area, it could enter the international market for yachting, especially the southern area by Sarandes.

Turkey has eight marinas with 7500 berths, as well as 1500 moorings at fishing ports. The coasts of the Aegean Sea are of great interest for tourists

due to the fact that they have not only the most, but also the best organized marinas. Other countries of the Eastern Mediterranean basin, namely
Syria and Lebanon, show little interest in yachting, although they have a
historically-rich hinterland. For this reason, yachting infra-structure is
practically non-existent except for Lebanon, where there is a marina near
Beirut. Israel, on the other hand, has three marinas, the most important
being at Tel Aviv. For the time being, as in Albania, the political instability
in this area of the Eastern Mediterranean is not suitable for either tourism
or yachting.

In the Republic of Cyprus there are two large and well-organized
marinas at Larnaca and Limassol in contrast to the occupied part of the
island where services for boats are very poor indeed. The marina at
Larnaca has been established for many years as an over-wintering port
before crossing the Mediterranean, for all those approaching this sea via
the Suez Canal. This is in accord with the Comprehensive Tourism Development Plan for Cyprus of 1988 – the strategy followed by those responsible for the plan was to target the profitable management of natural and
cultural resources of the country. This strategy co-exists with the encouragement of politically-motivated investments, among which yachting is
included as one of the most economically-viable forms of marine tourism.

Morocco and Algeria have only one marina each and boats are serviced
mainly at existing commercial ports and fishing ports. Yachting as a form
of marine tourism does not show any particular interest as far as the
natural environment is concerned. Tunisia is found on the dividing line
between the Eastern and Western Mediterranean and, as it is on the sea
passage for Malta, it has three marinas. Malta, due to its central position in
the Mediterranean, has historically become very attractive to international
yachting. For all this though, it has became better known for over-
wintering rather than as a sailing zone. There are two marinas that are in
operation and one other is to be constructed soon. The Maltese Islands
Tourism Development Plan of 1989 foresaw the construction of marinas
as part of the effort to enrich tourism on the island.

Although the Egyptian archaeological sites of the hinterland are well
known worldwide, Egypt does not exhibit any interest in sailing. The
harbour infrastructure at Port Said and at Alexandria, as well as the old,
well-established nautical clubs, serve as stations for the renewing of supplies for boats crossing the Mediterranean. Libya, by its own choice,
remains outside the tourist market. Given the fact that the government is
planning to construct marinas for local use only, the level of services to be
offered is still unknown.

The most extensive areas of the Mediterranean in which yachting can
take place belong to two basic categories characterised by natural geographical features. These are coastal and island regions. The coastal
regions include the north-western coasts of the Mediterranean, better
known as the French and Italian Riviera, the western coast of the Italian

peninsula and the Yugoslavian coast of the Adriatic. The island regions include the Balearic Isles, the large islands of the Central Mediterranean area including Malta, the Greek archipelago, and Cyprus.

Contrary to the type-casting of coastal yachting, the islands of the Mediterranean, for all their differences from the point of view of existing facilities, are able to offer a great variety of sea routes within a high quality natural and cultural environment. Within the Mediterranean maritime regions, extreme conditions exist, not only from the point of view of socioeconomic development but also tourism. As such, it is well known that the countries of the Western Mediterranean area take precedence over those of the eastern end and that the northern coast of the Mediterranean attract more tourists than those of the south. These differences can be seen in yachting, which makes up a type of marine tourism especially for those with a high income. About 90 per cent of marinas in the Mediterranean are to be found in tourist countries of the European Union: Spain, Italy, France, Portugal and Greece.

Recognizing this new situation, tourist countries in the Eastern Mediterranean are adjusting their conditions to take in yachting. Therefore, it is soon anticipated that there will be a doubling of berths in Greece and Turkey and the construction of three marinas in Cyprus, one of which will include a tourist village. Given the fact that marinas on the north-western coast of the Mediterranean will become more expensive due to increase in demand and lack of supply, the marinas in the Eastern Mediterranean will remain competitive. The expected increase of incoming tourism, due to the 2004 Athens Olympic Games, will positively affect overall tourist traffic and in particular, the future of marine tourism which already has a comparative advantage in that area.

Concluding remarks

Marine tourism in the Mediterranean is certain to continue to grow. The physical and cultural environments present significant attraction factors in the area. Marine tourism is moreover an alternative in form. In fact, it relocates the fundamental man-to-sea relationship, for modern tourism, from the beach to the open sea.

During the last few years, there has been a growing marine tourism demand moving down from the higher to the upper-middle incomes. This means that the market is expanding but at the same time it is getting more vulnerable to economic vibrations. Although there is an apparent popularization trend, marine tourism will remain a rather expensive recreation, restricted to relatively small numbers.

A marina is no longer the commonly-known boat-carriage of the 1960s. During the last decades marinas rather have tended to convert to integrated resorts, containing a number of interconnected facilities. In this sense the so-called third generation marinas or residential marinas are space-consuming complexes looking for extended coastal areas to settle in. Taking into account the rarity of the coastline, which is apparent in many European maritime regions, new marinas are expected to develop in areas of cheap available land, shifting the centre of gravity of future yachting towards the Eastern Mediterranean.

Concerning cruise ship tourism in the Mediterranean, there is statistical evidence of an increasing demand. Earlier this demand was exclusively limited to the British consumer, but during the last few years there is a spectacular market expansion towards the countries of Southern Europe. At the same time the given interest of the American market for new destinations is stimulated by the strongly promoted fly-cruise packages. Consequently it is expected that there will be a permanent deployment of some vessels now operating in the overcrowded Caribbean down to the Mediterranean area for the summer season. Many ports in the Central and Eastern Mediterranean are already improving their harbour equipment in order to upgrade the services offered. Rapid developments in new technologies of high-speed boats, as are manifested in the recent orders of major shipping companies sailing in the Adriatic Sea between Greece and Italy, point to the trends for the future. By April 2000, some of the fastest vessels in the world are expected to enter the market between Patras and Ancona. Their specifications are 173 m long, they can carry 1600 passengers and their speed exceeds 28 miles per hour (Anagnostou 1999).

As already mentioned, marine tourism is a little sector of the total tourist traffic. In any case, the possibility to concentrate even small numbers in a restricted and fragile area as, for instance the archipelagos in the Central and Eastern Mediterranean basin, could cause environmental problems. It is important to strictly observe all the regulations concerning marine pollution, to establish carrying capacity indices for the same traditional settlements and to effectively protect the coastal ecosystems.

There are factors with a negative impact on almost any tourist activity and, in particular, on marine tourism which, by its nature, is rather sensitive to risks. Such negative factors are the lasting environmental pressures all over the area and the political destabilization in many parts of the region during the recent years. The political interest of the European Union and the initiatives of the international organizations aiming at stability and the protection of the environment are the necessary conditions for a sustainable development of marine tourism in the Mediterranean.

New directions in Mediterranean tourism

19 Restructuring and strategic alliances in Mediterranean tourism

Yorghos Apostolopoulos and Sevil Sonmez

Introduction

This volume presents itself as the first comprehensive effort to examine Mediterranean tourism as a coherent unit of analysis. The influx of over a quarter of a million travellers[1] (not including domestic tourists) to Mediterranean shores has brought to the fore the need to study the nature and effects of tourist mobility and industry on the region's socio-economic and cultural spheres. Issues such as the political economy of tourism, planning, management, policy, sustainability and restructuring have held fundamental significance for the book's analysis of Mediterranean tourism. The contributors have extended tourism literature by firmly placing Mediterranean tourism in the discourse of political economy, socioeconomic development and cultural change.

Although the focus of the volume has been centred on 14 distinct, national tourist industries with their respective unique features, the next phase of Mediterranean tourism, within the context of globalization, will definitely have a more unified form. A 'new' restructured Mediterranean tourism – by forming co-operative alliances to replace the dead-ended fierce competition of the past between and among Mediterranean countries – holds greater potential for success in the world arena. The critical examination of the organization, structure, and ramifications of the 14 countries' tourism sectors provides the basis for an understanding of the commonalities of their national tourist products and it also functions as the guiding light for formulating strategies to strengthen the position of Mediterranean tourism in the transnational arena in the twenty-first century. The discussion below suggests ways in which this book's contributors shed light on the dynamics and future prospects of Mediterranean tourism.

Restructuring and sustainability in Mediterranean tourism

In its conceptualization, this volume was built on two indisputable premises:

1 a comprehensive analysis of Mediterranean tourism requires the adoption of a 'system' approach, which accepts the diverse nature of

the region's tourist industry, while it treats Mediterranean tourism as a coherent unit of analysis; and

2   in an era of international reshuffling, the predominantly coastal nature of Mediterranean tourism has gone through a prolonged 'crisis' phase and is now in a state of flux, due to a multitude of exogenous and endogenous factors.

These factors include:

a   the emergence of new geographic regions as competing tourist destinations, and subsequent traveller shifts;

b   consumer need and the search for alternatives (i.e., ecotourism, agrotourism, cultural tourism, water-based tourism) despite the enduring popularity of the traditional resort tourism model;

c   the Mediterranean's slow, unco-ordinated, and often hesitant development of new tourist products;

d   seasonality – over 40 per cent of tourist activity takes place between June and September (WTO 1999); and

e   sporadic regional political conflicts and terrorism (i.e., such as the ones in the Middle East, North Africa, and the former Yugoslavia).

The foregoing factors have constituted some of the chief reasons for the stagnation of the Mediterranean's share of the traditional market (northern European travellers) for the past decade (WTO 1999). Consequently, the region has turned to long-haul travellers (from the USA and Japan), new markets (Eastern Europe and intra-Mediterranean travel), and niche market areas (i.e., cruise liner vacations) (Apostolopoulos and Sonmez 2000; Kradonellis 1999) to increase tourist arrivals.

In addition to these efforts, Mediterranean countries need to overcome the aforementioned challenges to guarantee future success. In this framework, the 18 previous chapters, in their quest to examine Mediterranean tourism, have brought to the fore two overarching themes:

a   the imperative sociopolitical need for balanced tourism development in the framework of physical and human resource sustainability; and

b   the expert-supported mandate for restructuring the Mediterranean tourism system as a whole – which is not only necessary but almost inevitable for its sustained welfare (Apostolopoulos and Sonmez 2000).

These two topics deserve further elaboration here.

First, the growing concern about environmental pollution during the 1960s and 1970s has accelerated in the 1980s and 1990s, while focusing on balancing inter-relationships between physical and social environments. Questions about the acceptability of the objectives, strategies and policies of conventional tourism growth have entered public debate, and the concept of 'sustainable development' has become well known due to the 1987 World Commission on Environment and Development report (also known as the Brundtland Report, named after its president). Sustainable development, by 'meeting the needs of the present without compromising the ability of future generations to meet their own needs' (WCED 1987) provides the framework for the integration of environmental policies within economic development strategies.

Mediterranean tourism, therefore, must go beyond mere macroeconomic indicators; it must be a positive force for both environmental and human resources. Toward this direction, the fact that Mediterranean tourism is predominantly organized by third-country-based operators functions in a counteractive manner, as the priorities are often different. Therefore, the citizens of the region's countries should be actively involved in the tourism planning and decision-making processes at a meaningful level. Tourism development alone, without active grass-roots participation in the framework of sustainability, would not be able to alleviate the chronic imbalances between the northern, southern and eastern shores. And, even the development of collaborative alliances among the nations of the region would be ineffective and literally irrelevant, if the ecosystem is damaged and human social patterns are alienated by inappropriate tourism development. Tourism policies should incorporate sustainability and enhance culture as well.

Second, from the 1950s and until the early 1970s, there were relatively stable relationships in the broader contexts of the production and consumption – constituting what is known as the Fordist regime of accumulation (Montanari and Williams 1995). But in the early 1970s, several inherent crises of the Fordist regime (exacerbated by the oil crisis) triggered major inflationary and recessionary problems, resulting in a revision of forms of accumulation. Further, globalization trends linked to market and technological changes also affected the nature of production and consumption.

In the context of tourism, until the 1960s the dominant mode of production was the traditional model of mass tourism (representing a form of mass consumption). Mass tourism has been characterized by a standardized product and market domination by producers. In this context, there was the emergence of a state-supported sector along with the oligopolistic dominance of a few companies (primarily tour operators and airlines) and price (rather than the particularities of places) competition. But gradual shifts in consumer preferences and in production requirements led to a transition from Fordism to post-Fordism models of production and consumption – that is, to flexible systems of production and organization – just as the way in which services are consumed changed (with rapidly-shifting traveller tastes and the emergence of niche markets).

In the Mediterranean context, the region should adhere to these broader changes in production and consumption of tourism services and provide the consumer of the tourist experience with innovative new products. Toward this direction, diversity in traveller preferences and proliferation of alternative destinations, diversification of vacation types, development of new destinations, growth of 'softer' tourism forms, and interweaving tourism-related activities into regionally identifiable and differentiated tourist products, constitute necessary dimensions for regional and local development.

Finally, adherence to the principles of sustainability and restructuring will contribute to a positive direction for the Mediterranean tourist product, tourist industry, and tourist market. Subsequently, these efforts can boost employment, increase revenues, minimize social tensions, and ultimately lead to broader peace and stability in the region. Furthermore, such efforts will create the most appropriate background for achieving the formation of broad collaborations between and among the Mediterranean nations.

Co-operation and alliances in Mediterranean tourism

Fragmentation, within both the private and public sectors, has constituted one of the major problems facing the global travel and tourist industry. Tourism is rarely accorded the status it deserves within national governments[2] and because it constitutes a 'horizontal' economic activity, which crosses competencies of several governmental and public sector bodies, most governments have failed to view tourism as a valuable export. Governments which run national tourism boards, offices, organizations, or ministries rarely demonstrate the private sector's marketing expertise; nor do they possess adequate resources to carry out effective marketing strategies to give direction to their countries' tourism sectors.

Traditionally, each Mediterranean state has viewed its tourist product(s) as competing with that of neighbouring states (i.e., Greece with Turkey; Spain with Italy). For this reason, it will require dedicated effort on the part of Mediterranean countries to regard their traditional competitors as partners in regional collaboration. Regional co-operation is not impossible – it is an endeavour requiring strong political will and common sense from tourism entrepreneurs. For example, co-operative promotion for the Cypriot, Greek and Turkish product can stress aggregate tourism offerings and function as an immense competitive advantage, not only *vis-à-vis* other Mediterranean states, but also *vis-à-vis* other competing regions outside the Mediterranean (Sonmez and Apostolopoulos 1999). In light of lower airfares, greater tourist affluence and travellers' quest for more exotic destinations and exciting vacation experiences, any country or region with a warm climate offering beach resort vacations (i.e., the Caribbean, South Pacific) is in a position to compete with the Mediterranean, intensifying its declining share of world tourist arrivals (WTO

1999). Due to the marked socioeconomic improvement of various Mediterranean countries, the region is no longer a low-cost destination for its traditional market of northern Europeans. Considerable research will be necessary to attract new markets (i.e., from the USA, Canada, Australia, Japan), primarily to better gauge travellers' destination choices and decision-making behaviour. Especially for long-haul arrivals from Japan and the USA, travellers' perceptions of service quality and standards, price, destination image, as well as safety and security, need to be well understood.

It has been widely recognized that the Mediterranean's market share will decline even faster if it continues to focus on mass tourism. The region must clearly explore niche markets and practice focused market segmentation to maintain its strong position in world tourism. For a targeted marketing strategy, the Mediterranean tourist product will need to be re-evaluated in terms of its price competitiveness, destination image, market position, added value, and product quality within the context of customer sensitivity and perceptions (Apostolopoulos and Sonmez 2000). Although the Mediterranean tourist product is unique and not subject to substitutability, it is vulnerable to the major tour operators' demonstrated power to sway travellers to other regions, based on their own and their customers' perceptions. For this reason, effective marketing research is an essential prerequisite for neoteristic marketing strategies, which involve marketing and promoting the Mediterranean region as a 'single' destination or clusters of destinations (Apostolopoulos and Sonmez 2000). Quite clearly, the entire Mediterranean region can not compete with other destinations. Rather, sub-regions within the Mediterranean representing clusters of countries with similar characteristics and offering specific types of vacation experiences (i.e., Cyprus/Greece/Turkey; Algeria/Morocco/Tunisia; Egypt/Israel/Lebanon) can compete with other destinations outside the Mediterranean. A superficial approach to promoting the Mediterranean would not produce the desired results – it would, in fact, be detrimental and waste collective efforts and funds. Various forms of promotion are effective and important; however, advertising alone cannot rectify a poor tourist product's deficiencies. For example, tour operators – the so-called 'export/marketing arms' of nations for their tourist products – have found in their research that hotel comfort and service standards are paramount to a destination's attractiveness (Brackenbury 1996). It is recommended that, in order to be successful, co-operative marketing efforts for the Mediterranean include the exploration of a spectrum of factors such as urging the public and private sectors into co-operating on environmental protection and encouraging the public sector to upgrade infrastructure (Kopp 1996).

Marketing the Mediterranean as a unified destination might be attractive to overseas tour operators as well; however, such marketing must also strive to maintain the uniqueness and brand image of each nation-state.

Numerous possibilities for multi-country tourism itineraries exist for the Mediterranean region. To carry out these functions, a Mediterranean marketing body could be modelled after the Caribbean Tourism Association, with an added advantage of a common heritage beyond simple geographic and natural links. Furthermore, it is feasible for the region to communicate a 'Mediterranean brand' to potential travellers, while co-operating to assure synergy and cohesion in other marketing functions (i.e., distribution, pricing, positioning). A brand strategy can inspire the private sector to develop comprehensive itineraries and as a result, re-package Mediterranean vacation products. A good example of co-operative marketing is that of the European Union's (EU) strategy for its 15 European countries (European Union 1999), which suggests that a generic image can be conveyed in marketing a rich and multi-country region to an overseas market. The EU example is also useful in pinpointing problems of co-operative tourism marketing. In Europe, the public sector-initiated co-operative marketing efforts received lukewarm support from the private sector. The lack of support by the tourist industry and insufficient funding were identified as reasons for not achieving the expected benefits of co-operative marketing. One potential way to avoid repeating the same mistake in the case of the Mediterranean would be to tap into the EU's MEDA programme – established after the 1995 Euro-Mediterranean conference – for funds.

On the other hand, the private sector has expressed its concerns about marketing a generic Mediterranean product based on the premise that a geographic region would be difficult to brand (Zoreda 1996). This concern may appear more serious in light of the fact that tourism marketers are increasingly focusing their strategies on niche markets involving special interest tourism for specifically-targeted audiences. Fortunately for the Mediterranean region, numerous market niches can be addressed, such as religious, historical, sports, and festival tourism, as well as countless architectural and archaeological attractions.

Co-operative marketing is a complex process, involving long-term strategic as well as tactical plans (Hill and Shaw 1995). Numerous decisions are involved, such as the nature of the advertising campaigns, the development and communication of a unified image, the nature of the co-operative marketing effort, the name and logo of the marketing initiative, and appropriate positioning in the world tourism market. To convey one image and one message, a representative name and symbolic logo would be required for the Mediterranean region. The choice of this name may be influenced by the existence of regional clusters, such as the northern Mediterranean shores (i.e., Spain, France, Italy), eastern shores (i.e., Turkey, Syria, Israel) and the southern shores (i.e., Egypt, Libya, Algeria). Examples of successful joint and regional marketing initiatives include tourist attractions/destinations shared by several countries, such as the Alpine states, the Maya route in the central American states, the (proposed) Silk Road, the Amazonia, and the Andes. Other examples of successful co-operative marketing efforts include regional organizations such as the Middle Eastern and Mediterranean Travel and Tourism Association and the PROALA in Latin America.

The simple agreement by Mediterranean countries to participate in co-operative marketing would not guarantee a competitive edge for the region. A primary goal for the Mediterranean region would be to first identify, then convey its competitive advantages over competing destinations. It is also very important for co-operative marketing objectives to be clearly delineated and representative of the needs of the participating countries as well as of the region as a whole (Apostolopoulos and Sonmez 2000). Will the emphasis of co-operative marketing be to increase tourist aggregates, draw higher income visitors, or target new markets? Will efforts focus on traditional mass markets or promote niche vacations? Will co-operation constitute a political initiative to counterbalance regional inequalities?

Concluding remarks

Travel and tourism have been a major force in shaping the Mediterranean for several centuries. More than any other economic activity, travel and tourism have always favoured peripheral regions – even if limited mobility at the time of the Grand Tour focused travel activity in a rather restricted belt stretching from England to Italy. Over the years, travel and tourism have been transformed into the world's largest economic sector with enormous ramifications for both the physical and human spheres of industrialized and developing nations alike.

In the Mediterranean, where diversity and commonality coexist, tourism as one of the leading economic sectors of the region awaits the approach of the third millennium, at the crossroads of important decisions. In order to maintain and improve its status in the world tourism arena, the Mediterranean region needs to effectively reposition its tourism sector in the continually shifting world arena. Toward this direction, the restructuring of its tourist product, industry and market within the framework of sustainable development and through the establishment of strategic alliances (i.e., co-operative marketing, joint ventures) is crucial to the future prosperity of the Mediterranean. For any type of restructuring to have enduring success, the Mediterranean region would also have to overcome issues of political instability, terrorism and religious fundamentalism, which have hurt Mediterranean tourism in the past. A challenging but exciting road lies before the Mediterranean countries in their quest for a strong and enduring place in world tourism.

Notes

1. This represents the most conservative forecast. According to Grenon and Batisse (1989), total arrivals in the 21 Mediterranean countries will be between 268 and 409 million, by the end of this century.
2. The characteristic example of Greece is definitely not an exception. While tourism constitutes the most vital economic sector, its administration has been erratic and spasmodic, exemplified by the fact that the 'life expectancy' of each Secretary General of the Greek National Tourism Organization (responsible for the sector's strategic planning/marketing) is less than eight months.

Bibliography

Acherman, A. 1995. *Schengen Agreement and its Consequences: the Removal of Border Controls in Europe.* Bern: Stampfli.

Achituv, G. 1973. 'The European Potential', pp. 67–74 in *The Second Million – Israel Tourist Industry Past-Present-Future,* C. H. Klicn (ed.). Jerusalem: Amir Publishing.

Adam, A. 1974. 'Urbanisation et Changement Culturel au Maghreb', in *Villes et Societes au Maghreb,* al R. Duchac. CNRS. Paris.

Adie, D. 1984. *Marinas: A Working Guide to Their Development and Design,* 2nd edn, London: The Architectural Press Ltd.

Africa Research Bulletin (Economic Series) (1990). 28.

Afshar, H. 1991. *Women, Development and Survival in the Third World.* London: Longman.

Agelis, Y. and Falirea, L. 1996. 'Tourism and the Greek Economy.' *Tourism and Economy,* 214: 53–79 (in Greek).

Agricultural University of Athens and University of Thessaly. 1998. *A Study of the Greek Rural Areas.* Ministry of the Environment Regional Planning and Public Works (in Greek).

Aktas, A. 1995. 'Turkish Tourism in the Planned Periods (1963–1994): Tourism Investment Incentives and Foreign Capital Evolution.' *Proceedings of the 1st International Conference on Investments and Financing in the Tourism Industry,* 13–126. Jerusalem: Israel Ministry of Tourism and WTO.

Al-Ahnaf, M., Botiveau, B. and Frégosi, F. 1991. *L'Algérie Par Ses Islamistes.* Paris: Khartala.

Allman, T. D. 1998. 'Special Report, Egypt, the Agony and the Ecstasy.' *Condé Nast Traveler,* March, pp. 110–25 and 187–201.

Anagnostou, A. 1999. *Apogevmatini,* Business Section, 15 August 1999 (in Greek).

Anderson, L. 1986. *The State and Social Transformation in Tunisia and Libya, 1830–1980.* Princeton: Princeton University Press.

Andronikou, A. 1979. 'Tourism in Cyprus', pp. 237–63 in *Tourism: Passport to Development?,* E. de Kadt (ed.). New York: Oxford University Press.

Andronikou, A. 1987. *Development of Tourism in Cyprus: Harmonization of Tourism with the Environment.* Nicosia: Cosmos.

Andronikou, A. 1993. 'The Prospects for Cyprus Tourism in 1993.' *The Cyprus Weekly* (March 12–18): 16–17.

Apostolopoulos, Y. 1996. *The Effects of Tourism Expansion in the Greek Islands: Regional Tourism Planning and Policy Lessons for Sustainable Development.* Athens: Papazissis (in Greek).

Apostolopoulos, Y., Leivadi, S. and Yiannakis, A. (eds). 1996. *The Sociology of Tourism.* London: Routledge.

Apostolopoulos, Y. and Sonmez, S. 2000. 'Mediterranean Tourism: Restructuring and Strategic Marketing Alliances.' *Thunderbird International Business Review* (forthcoming).

ArabNet. 1998. Egypt, Business (www.arab.net/egypt).

ArabNet. 1998. Egypt, Economics, Business (www.arab.net/egypt).

Azariahu, M. 1993. *Eilat: the Ambiguity of a Frontier Situation.* Unpublished paper presented at the International Conference on Regional Development: the Challenge of the Frontier Beer Sheva.

Azariahu, M. 1993. The Beach at the End of the World: Eilat in Israeli Popular Culture. Paper presented at a conference on 'Regional Development: the Challenge of the Frontier' and organized by Ben Gurion University, The Dead Sea, December 27–30, 1993.

Baedecker 1992. *Tunisia.* Stuttgart: Automobile Association.

Bailly, A., Jensen-Butler, C. and Leontidou, L. 1996. 'Changing Cities: Restructuring, Marginality and Policies in Urban Europe.' *European Urban and Regional Studies,* 3(2): 161–76.

Baker *et al.* 1997.

Barke, M. 1991. 'The Growth and Changing Pattern of Second Homes in Spain in the 1970s.' *Scottish Geographical Magazine,* 107(1): 12–21.

Bastin, R. 1984. 'Small Island Tourism: Development or Dependency?', *Development Policy Review* 2: 79–90.

Beltrán 1995. 'La Experiencia Española en la Promocion *Spain: Towards Sustained Tourist Competitiveness,* Josep Ivars Baidal Jose Union Europea: el Programa Leader,' Papers de Turisme, no. 17, Valencia, Institut Turístic Valenciw, pp. 27–31.

Bennani-Chraibi, M. 1994. *Soumis et Rebelles, les Jeunes au Maroc.* Le Fennec: Casablanca.

Bennoune, M. 1988. *The Making of Contemporary Algeria, 1830–1987.* Cambridge: Cambridge University Press.

Berberoglu, B. 1981. *Turkey in Crisis – From State Capitalism to Neo Colonialism.* London: Zed Press.

Berriane, M. 1988. *Une Nouvelle Fonction du Centre – Ville Moderne Marocain: Loisir et Convivialite.* Element sur les Centres Villes dans le Monde Arabe. URBAMA, 19.

Berriane, M. 1992. *Tourisme National et Migrations de Loisirs au Maroc.* Rabat.

Best, M., Murray, R. and Pezzini, M. 1989. 'Industrial Consortia and the Third Italy', in *Report on Fourth Stage of the Cyprus Industrial Strategy,* R. Murray (ed.). Brighton, Sussex: Institute of Development Studies.

Bicak, H. A. and Altinary, M. 1996. 'Economic Impact of the Israeli Tourists on North Cyprus.' *Annals of Tourism Research,* 23(4): 928–30.

Blacksell, M. and Williams, A. (eds). 1994. *The European Challenge: Geography and Development in the European Community.* New York: Oxford University Press.

Blanco. 1997. *'El Proyecto España Verde': Ideas e Instrumentos para el Desarrollo Sostenible del Turismo.* Proceedings of the conference, Turismo y desarrollo sostenible en la España Verde, Bilbao, pp. 2–24.

Blizovsky, Y. 1973. 'The Role of Tourism in the Economy', in *The Second Million – Israel Tourist Industry Past-Present-Future,* C. H. Klien (ed.). Jerusalem: Amir Publishing, pp. 107–29.

Boisvert, C. C. 1987. 'Working-class Portuguese Families in a French Provincial Town', pp. 61–76 in *Migrants in Europe*, H. C. Buechler and J.-M Buechler (eds). Westport: Greenwood.

Borojević, G. 1995. 'Turistički Trendovi u Svijetu i u Europi s Osvrtom na Hrvatsku.' *Turizam 43* (11–12): 212–18.

Bote, V. and Marchena, M. 1996. 'Pontica Turística', in *Introduccion a la Economía del Turismo en España*, A. Pedreño and M. Monfort (eds). Madrid: Civitas.

Boyer, M. 1968. *Le Tourisme dans le Sud-est Méditerranéen Français*. Comité des travaux historiques, Imprimerie Nationale.

Boyer, M. 1972. *Le Tourisme*. Paris: Seuil.

Boyer, M. 1996. *L'invention du Tourisme*. Paris: Gallimard, collection Découvertes.

Boyer, M. 1997. *L'invention du Tourisme dans le Sud-est de la France, XVIème-fin XIXème*, en 21 fascicules (disponibles Université de Lille III).

Boyer, M. 1998. *Le Tourisme en l'an 2000*. Presses Universitaires de Lyon.

Brackenbury, M. 1996. 'New Opportunities for the Mediterranean.' Paper presented at the inter-ministerial tourism conference 'Marketing the Mediterranean as a Region.' Sliema, Malta, November 7–9.

Braudel, F. 1977. *La Mediterranee: L'Espace et l'Histoire*. Paris: Arts et Metiers Graphiques.

Briassoulis, E. 1993. 'Tourism in Greece', in *Tourism in Europe: Structures and Developments*, W. Pompl and P. Lavery (eds). Oxford: CAB International.

Briassoulis, H. 1992. 'Environmental Impacts of Tourism: a Framework for Analysis and Evaluation', pp. 11–22 in *Tourism and the Environment*, H. Briassoulis, and van der Straaten, J (eds). Boston: Kluwer.

Briassoulis, H. and van der Straaten, J. (eds). 1992. *Tourism and the Environment*. Boston: Kluwer.

Britton, R. A. 1978. International Tourism and Indigenous Development Objectives: A Study with Special Reference to the West Indies. Unpublished Ph.D. dissertation, University of Minnesota, Minneapolis, MN.

Britton, S. G. 1982. 'International Tourism and Multinational Corporations in the Pacific: the Case of Fiji', pp. 252–74 in *The Geography of Multinationals*, M. Taylor and N. Thrift (eds). London: Croom Helm.

Britton, S. G. and Clark, W. C. (eds). 1987. *Ambiguous Alternative: Tourism in Small Developing Countries*. Suva, Fiji: The University of the South Pacific.

Brohman, J. 1996. *Popular Development: Rethinking the Theory and Practice of Development*. Cambridge, MA: Blackwell.

Bryden, J. M. 1973. *Tourism and Development: a Case Study of the Commonwealth Caribbean*. London: Cambridge University Press.

Buckley, P. J. and Papadopoulos, S. I. 1986. 'Marketing Greek Tourism – The Planning Process', *Tourism Management*, 86–100.

Buechler, H. C. and Buechler, J. M. (eds). 1987. *Migrants in Europe*. Westport: Greenwood.

Buhalis, D. 1998. *Tourism in Greece: Strategic Analysis and Challenges for the New Millennium*. Aix en Provence: ICRST.

Buller, H. and Hoggart, K. 1994. *International Counterurbanization: British Migrants in Rural France*. Aldershot: Avebury.

Burns, P. 1995. 'Sustaining Tourism Under Political Adversity: the Case of Fiji', pp. 259–72 in *Island Tourism: Management Principles and Practice*, M. Conlin and T. Baum (eds). London: Wiley.

Butler, R. W. 1980. 'The Concept of a Tourist Area Cycle of Evolution: Implications for Management of Resources.' *Canadian Geographer*, 24: 5–12.

Butler, R. W. 1993. 'Tourism Development in Small Islands: Past Influences and Future Directions', pp. 71–91 in *The Development Process in Small Island Nations*, D. G. Lockhart, D. Drakakis-Smith and J. Schembri (eds). London: Routledge.

Butler, R. W. 1997. 'The Concept of Carrying Capacity for Tourism Destinations: Dead or Merely Buried?', pp. 11–22 in *Tourism Development – Environmental and Community Issues*, C. Cooper and S. Wanhill (eds). Chichester: John Wiley & Sons.

Canto, C. 1983. 'Presente y Futuro de las Residencias Secundarias en Espana.' *Annales de Geografia de la Universidad Computense de Madrid*, 3: 83–103.

Central Bank of Greece. 1997. *Report of the Governor for the Year 1996*. Athens:. Central Bank of Greece Printing Works.

Central Bank of Greece. 1998. *Report of the Governor for the Year 1997*. Athens: Central Bank of Greece Printing Works.

Central Intelligence Agency. 1999. *The World Factbook*. Washington: CIA.

Centre for Planning and Economic Research. 1990. *The Development of Greece: Past, Present and Policy Proposals*. Athens: CPER (in Greek).

Centre for Planning and Economic Research. 1994. *Preliminary National Economic and Spatial Plan for Tourism*. Athens: GNTO and CPER (in Greek).

Centre for Planning and Economic Research and National Tourist Organization of Greece. 1994. *Preliminary National Economic Regional Plan for Tourism*. Athens (January) (in Greek).

Centre for Planning and Economic Research, University of Thessaly. 1998. *Seasonal Homes and Housing Development in Greece*, Ministry of the Environment Regional Planning and Public Works, Athens.

Charlton, S. M. 1984. *Women in Third World Development*. Boulder: Westview.

Charmes, J. 1994. *Visible et Invisible: Le Secteur Informel dans l'économie Urbaine du Monde Arabe*. Colloque International sur la Société Urbaine dans le Monde Arabe: Transformations, Enjeux, Perspectives.

Chilcote, R. H. 1984. *Theories of Development and Underdevelopment*. Boulder: Westview.

Christodoulou, D. 1992. *Inside the Cyprus Miracle*. Minneapolis: University of Minnesota Press.

Cohen, E. 1982. 'Thai Girls and Farang Men: the Edge of Ambiguity.' *Annals of Tourism Research*, 9: 403–28.

Cohen, E. 1988. 'Tourism and Aids in Thailand.' *Annals of Tourism Research*, 15: 467–86.

Coleman, D. (ed.). 1996. *Europe's Population in the 1990s*. New York: Oxford University Press.

Compendium of Tourism Statistics 1991–1995, Edition 1997. Madrid: WTO.

Compton's Interactive Encyclopedia. 1995. Compton's News Media, Inc.

Consejo Superior de Investigaciones Científicas (CSIC). 1995. *La Demanda Turística Española en Espacio Rural o de Interior: Situacion Actual y Potencial*. Unpublished research, Convenio de Colaboracion entre el CSIC, Secretaría General de Turismo-Instituto de Turismo de España, Empresa Publica de Turismo de Aridalucía, y Fundacion Cavanilles de Altos Estudios Turísticos.

Conselleria de Agricultura y Medio Ambiente. 1995. *Plan de Gestion de las Instalaciones Recreativas de los Montes de la Comunidad Valenciana*. Generalitat Valenciana.

Constantinides, G. 1991. 'Planning and Coastal Development'. *Phileleftheros* (in Greek), March 14–15.

Cooper, C. P. (ed.). 1990. *Progress in Tourism, Recreation and Hospitality Management*, Vol. 2. Chichester: John Wiley & Sons.

Cribier, F. 1982. 'Aspects of Retirement Migration from Paris: an Essay in Social and Cultural Geography', pp. 111–37 in *Geographical Perspectives on the Elderly*, A. M. Warnes (ed.). London: Wiley.

Croatian Almanac. 1997. Zagreb: Hina.

Cruise Lines International Association. 1996. *Five Year Cruise Industry Capacity Outlook*, Executive Summary, First Quarter 1996.

Cruise Lines International Association. 1997. *The Cruise Industry – An Overview.* Marketing Edition, January 1997.

Crush, J. S. and Wellings, P. A. 1983. 'The Southern African Pleasure Periphery, 1966–83.' *The Journal of Modern African Studies*, 21: 673–98.

Cyprus Tourism Organization. 1976. *Annual Report.* Nicosia: CTO.

Cyprus Tourism Organization. 1990. *New Tourist Policy.* Nicosia: CTO.

Cyprus Tourism Organization. 1992. *Annual Report.* Nicosia: CTO.

Cyprus Tourism Organization. 1993. *Annual Report.* Nicosia: CTO.

Cyprus Tourism Organization. 1994. *Annual Report.* Nicosia: CTO.

Cyprus Tourism Organization. 1994. *Annual Tourist Survey.* Nicosia: CTO.

Cyprus Tourism Organization. 1995. *Annual Report.* Nicosia: CTO.

Cyprus Tourism Organization. 1996. *A Guide to Hotels, Travel Agencies and Other Tourist Services.* Nicosia: CTO.

Cyprus Tourism Organization. 1998. *Annual Report.* Nicosia. CTO.

Debbage, K. 1990. 'Oligopoly and the Resort Cycle in the Bahamas.' *Annals of Tourism Research*, 17(4): 513–27.

Department of Town Planning and Housing. 1990. *Town and Country Planning Law.* Nicosia: Cyprus.

Dicken, P. 1992. *Global Shift: the Internationalization of Economic Activity.* London: Chapman.

Din, K. H. 1989. 'Islam and Tourism: Patterns, Issues and Options'. *Annals of Tourism Research*, 16: 542–63.

Doxey, G. V. 1975. 'A Causation Theory of Visitor-resident Irritants, Methodology and Research Inferences', in *The Impact of Tourism*, proceedings of the Travel Research Association, 6th Conference, San Diego, CA: 195–8.

DPT. 1963.

Dunford, M. and Kafkalas, G. (eds). 1992. *Cities and Regions in the New Europe.* London: Belhaven Press.

Eaton, M. 1995. 'British Expatriate Service Provision in Spain's Costa del Sol.' *The Service Industries Journal*, 15(2): 251–66.

Economic and Tourism Intelligence. 1997. *International Tourism Reports*, No.1.

Economist. 1993. 'Last Chance Sisyphus: a Survey of Greece.' *The Economist*, 1–18.

El Beltagui, M. 1995. *Inauguration Speech on Occasion of the Fourth Congress of the International Academy of Tourism*, Cairo, June 1–5.

El-Hayawan, H. 1998. *Highlights on Privatization in Egypt.* Cairo: American Chamber of Commerce, Cairo, 1–2 (www.amcham.org.eg).

Enloe, C. 1990. *Bananas, Beaches and Bases: Making Feminist Sense of International Politics.* Berkeley: University of California.

Ephratt, A. 1993. 'The Master Plan', pp. 146–8 in *The Second Million – Israel Tourist Industry Past-present-future*, C. H. Klien (ed.). Jerusalem: Amir Publishing.

Esteban, A. 1996. 'El Marketing Turístico', in *Introduccion a la Economía del Turismo en España*, A. Pedreño and M. Monfort (eds). Madrid: Civitas.

Esteban, A. A. and Pedreño. 1987. *Renta de los Municipios de la Comunidad Valenciana*. Alicante: Caja de Ahorros de Alicante y Murcia.

European Commission. 1995. *Europe 2000+: Cooperation for Spatial Development in Europe*. Luxembourg: Office for Official Publications of the EC.

European Commission. 1995. *Green Paper on Tourism*. Brussels: Author.

European Commission. 1997. *Yield Management in Small and Medium-sized Enterprises in the Tourism Industry: General Report*. Luxembourg: Office for Official Publications of the EC.

European Community. 1994. *Europe 2000+*. Brussels: EC.

European Union. 1999. *Tourism and the European Union*. Brussels: EC.

Eurostat. 1995. *Tourism in Europe*. Brussels: European Commission.

Ferchiou, R. 1991. 'The Social Pressure on Economic Development in Tunisia', pp. 101–8 in *Tunisia: The Political Economy of Reform*, I. William Zartman (ed.). Boulder: Lynne Riener.

Fielding, A. J. 1993. 'Mass Migration and Economic Restructuring', pp. 7–18 in *Mass Migrations in Europe: The Legacy and the Future*, R. L. King (ed.). London: Belhaven Press.

Fleischer, A. and Mansfeld, Y. 1995. *Urban-Rural Linkages in Tourism in Israel* (in Hebrew). Rehovot: Development Study Center.

Forestier, S. W. 1950. 'Through the Greek Archipelago.' *Yachting World*, March 1950, pp. 106–9.

Fourneau, F. and Marchena, M. (eds). 1991. *Ordenacion y Desarrollo del Turismo en Espana y en Francia*. Madrid: Ministerio de Obras Publicas y Transportes.

Garcia-Lizana, A. (ed.). 1992. *Espana-Africa*. Malaga: Edinford.

Gaviria, M. 1977a. *Benidorm, Ciudad Nueva*. Madrid: Editora Nacional.

Gaviria, M. 1977b. *El Turismo de Invierno y el Asentamiento de Extranjeros en la Provincia de Alicante*. Alicante: Instituto de Estudios Alicantinos.

Gearing, C. Swart, W. W. and Var, T. 1976. *Planning for Tourism Development: Quantitative Approaches*. New York: Praeger Publishers.

Gillmor, D. A. 1989. 'Recent Tourism Development in Cyprus.' *Geography*, 74: 262–5.

Ginosar, O. and Mansfeld, Y. 1993. 'The Application of Personal Construct Theory in Studies of Locals' Tourism Development Perceptions.' *Horizons in Geography*, 37–8: 35–50 (in Hebrew).

Goldenberg, Y. 1996. *National Master Plan for Tourism Enterprises and Recreational Areas* (#12). Jerusalem: Ministry of Tourism, Ministry of the Interior and Israel Land Authority.

Goldenberg, Y. 1996. *National Master Plan for Tourism and Recreational Enterprises*. Jerusalem: Israel, Ministry of Interior, Ministry of Tourism and Israel Land Administration.

Goodman, D. and Redclift, M. 1991. *Environment and Development in Latin America*. Manchester: Manchester University Press.

Government of Israel. 1994. *Development Options for Regional Co-operation*. Jerusalem: Ministry of Foreign Affairs and Ministry of Finance.

Government of Turkey. 1998. *Web Page*.

Graburn, N. H. H. 1983. 'Tourism and Prostitution.' *Annals of Tourism Research*, 10: 437–43.

Gray, M. 1997. 'The Political Economy of Tourism in Syria: State, Society, and Economic Liberalization.' *Arab Studies Quarterly*, 19: 2.

Greek Association of Tourist Enterprises. 1992. *The Greek Tourism Industry.* Athens: GATE (in Greek).

Greek Association of Tourist Enterprises. 1993. *The Greek Tourist Industry: Problems and Prospects.* Athens: GATE (in Greek).

Greek Association of Tourist Enterprises. 1995. *Plan of Action for a Qualitative Upgrade of Greek Tourism.* Athens: Horwath Consulting (in Greek).

Greek National Tourism Organization. 1997. *Tourism Statistics.* Athens: GNTO (in Greek).

Greek National Tourism Organization. 1998. *Tourism Statistics.* Athens: GNTO (in Greek).

Greek National Tourism Organization. 1999. *Tourism Statistics.* Athens: GNTO (in Greek).

Green, R. H. 1979. 'Towards Planning Tourism in Developing Countries', pp. 79–100 in *Tourism: Passport to Development?*, Emanuel de Kadt (ed.). New York: Oxford University.

Grenon, M. and Batisse, M. (eds). 1989. *Futures for the Mediterranean Basin: The Blue Plan.* Oxford: Oxford University Press.

Grissa, A. 1991. 'The Tunisian State Enterprises and Privatization Policy'. Pp. 109–27 in *Tunisia: The Political Economy of Reform*, I. William Zartman (ed.). Boulder: Lynne Riener.

Gruber, H. 1992. *The Influence of Temporary Hotel Workers on the Social Texture and Quality of Life of Eilat's Residents.* An M. A. thesis. Department of Geography, University of Haifa.

Guidicini, P. and Savelli, A. con altri (M. Boyer, Nadji Safir). 1988. *The Mediterranean Area as a Complex Tourism System.* Sociologia urbane e rurale.

Hardy, S., Hart, M., Albrechts, L. and Katos, A. (eds). 1995. *An Enlarged Europe – Regions in Competition?* London: Jessica Kingsley Publishers.

Harrell-Bond, B. 1978. 'A Window on the Outside World: Tourism and Development in the Gambia.' *American Universities Field Staff Reports*, 19. Hanover, New Hampshire.

Harrison, D. and Price, M. F. 1996. 'Fragile Environments, Fragile Communities? An Introduction', pp. 1–18 in *People and Tourism in Fragile Environments*, M. F. Price (ed.). Chichester: John Wiley & Sons.

Heikel, R. 1990. *Mediterranean Cruising Handbook.* Imray.

Hermans, D. 1981. 'The Encounter of Agriculture and Tourism: A Catalan Case.' *Annals of Tourism Research*, 8: 462–79.

Hill, T. and Shaw, R. N. 1995. 'Co-marketing Tourism Internationally: Bases for Strategic Alliances.' *Journal of Travel Research*, 34: 25–32.

Houliaras, N. 1998. *One Day Before, Two Days After.* Athens: Nefeli (in Greek).

Inskeep, E. 1991. *Tourism Planning: an Integrated and Sustainable Development Approach.* New York: Van Nostrand Reinhold.

Institut National de la Statistique. 1990. *La Tunisie en Chiffres.* Tunis: Author.

Instituto de Estudios Turísticos. 1995. Las Vacaciones de los Españoles, 1995. Unpublished research, Secretaría General de Turismo, Ministerio de Comercio y Turismo.

Instituto de Estudios Turísticos. 1997. Dimension Regional de Jos Impactos Macroeconomicos del Turismo. Unpublished research, Secretaría de Estado de Comercio, Turismo y Pequeña y Mediana Empresa, Ministerio de Economía y Hacienda.

Ioannides, C. 1991. *Environmental Policy Support Study.* Nicosia. Cyprus: Cyprus Development Bank.

Ioannides, D. 1992. 'Agents of Tourism Development and the Cypriot resort Cycle.' *Annals of Tourism Research*, 19(4): 711–31.

Ioannides, D. 1994. The State, Transnationals, and the Dynamics of Tourism Evolution in Small Island Nations. Unpublished Ph.D. dissertation, Rutgers University. New Brunswick, NJ.

Ioannides, D. 1995. 'A Flawed Implementation of Sustainable Tourism.' *Tourism Management*, 16(8): 583–92.

Ioannides, D. and Debbage, K. G. 1998. *The Economic Geography of the Tourist Industry: A Supply-side Analysis.* London: Routledge.

Ioannides, D. and Apostolopoulos, Y. 1999. 'Political Instability, War, and Tourism in Cyprus: Effects, Management, and Prospects for Recovery.' *Journal of Travel Research*, 38: 51–6.

Ivandić, N. and Radnić, A. 1996. 'The War's Indirect Damage to Tourism in Croatia.' *Turizam 44* (1–2): 3–13.

Jenkins, C. L. and Henry, B. M. 1982. 'Government Involvement in Tourism in Developing Countries.' *Annals of Tourism Research*, 9: 499–521.

Jenner, P. and Smith, C. 1993. *Tourism in the Mediterranean.* London: The Economist Intelligence Unit.

Jensen, L. 1989. 'Cyprus: Boom in Tourism, Battle on the Environment.' *World Development*, 2: 10–12.

Jensen-Butler, C., Shakhar, A. and van den Weesep, J. (eds). 1996. *European Cities in Competition.* Aldershot: Avebury.

Jurdao, F. 1979. *Espana en Venta.* Madrid: Ediciones Ayuso.

Kalogeropoulou, H. 1993. 'Greece – Prospects for the Tourism Industry Within the Context of the European Single Market.' *Tourism Review*, 48: 2–4.

Kariotis, T. (ed.). 1992. *The Greek Socialist Experiment: Papandreou's Greece (1981–1989).* New York: Pella.

Karmon, Y. 1963. 'The Geographical Characteristics of Eilat.' *Proceedings of the Eighteenth Archeological Convention.* Jerusalem: The Israel Exploration Society.

Kassimati, K., Thanopoulou, M. and Tsartas, P. 1994. *Women's Employment in the Tourism Sector: Study of the Greek Labor Market and Identification of Future Prospects.* Athens: Pandio University and European Commission (in Greek).

Katohianou, D. 1995. 'Economic and Spatial Development of Tourism in Greece: A Preliminary Picture.' *Contemporary Issues*, 55: 62–71 (in Greek).

Khader, B. 1994. *La Ville Arabe d'hier a Aujourd'hui Jalons pour une rèflexion.* Citta e Societe Urbana nel Mondo Arabo Trasformazioni, Ffide, Prospettive.

Khelladi, A. 1992. *Les Islamistes Algériens Face au Pouvoir.* Algiers: editions Alfa.

King, B. 1994. 'Research on Resorts: A Review', pp. 165–80 in *Progress in Tourism Recreation and Hospitality Management – Volume Five*, C. P. Copper and A. Lockwood (eds). Chichester: John Wiley & Sons.

King, B. 1994. 'What is Ethnic Tourism? An Australian Perspective.' *Tourism Management*, 15.3: pp. 173–6.

King, R. 1991. 'Italy: "Multi-faceted Tourism"', pp. 61–83 in *Tourism and Economic*

Development: Western European Experiences, A. Williams and G. Shaw (eds). London: Wiley.

King, R. L. 1984. 'Population Mobility: Emigration, Return Migration and Internal Migration.' pp. 145–78 in *Southern Europe Transformed,* A. Williams (ed.). London: Harper & Row.

King, R. L. (ed.). 1993a. *Mass Migrations in Europe: The Legacy and the Future.* London: Belhaven Press.

King, R. L. (ed.). 1993b. *The New Geography of European Migrations.* London: Belhaven Press.

King, R. L., Proudfoot, L. and Smith, B. (eds). 1997. *The Mediterranean: Environment and Society.* London: Edward Arnold.

King, R. L. and Rybaczuk, K. 1993. 'Southern Europe and the International Division of Labor: from Emigration to Immigration', pp. 175–206 in *The New Geography of European Migrations',* R. L. King (ed.). London: Belhaven Press.

King, R., Warnes, T. and Williams, A. 2000. *Sunset Lives: British Retirement Migration to the Mediterranean.* Oxford: Berg.

Kinnaird, V. and Hall, D. (eds). 1994. *Tourism: a Gender Analysis.* New York: John Wiley & Sons.

Klemm, M. 1992. 'Sustainable Tourism Development: Languedoc-Roussillion Thirty Years On.' *Tourism Management,* 2 (June): 169–80.

Klemm, M. 1996. 'Languedoc-Roussillion: Adopting the Strategy.' *Tourism Management,* 2 (March): 133–9.

Kliger, Y. and Shmueli, V. 1997. *Israeli Tourism and Hospitality: a Sector Analysis* (in Hebrew). Tel Aviv: Globes Publishing.

Kofas, J. 1989. *Intervention and Underdevelopment.* University Park: The Pennsylvania State University Press.

Komilis, P. 1987. *The Spatial Structure and Growth of Tourism in Relation to the Physical Planning Process: the Case of Greece.* Ph.D. Dissertation, University of Strathclyde.

Komilis, P. 1992. The Spatial Structure of Tourism and Physical Planning Practices in Greece. Paper presented at the international conference on 'Tourism and the Environment', University of the Aegean, Lesvos, Greece.

Komilis, P. 1994. 'Tourism and Sustainable Regional Development', pp. 65–73 in *Tourism: The State of the Art,* A. V. Seaton (ed.). New York: Wiley.

Komilis, P. 1995a. 'Tourism Policy and Areas of Integrated Tourism Development.' *Contemporary Issues,* 55: 77–80 (in Greek).

Komilis, P. 1995b. Tourism's Growth and Sustainable Development in Coastal Areas: a Policy Framework for Local Authorities. Paper presented at the 4th International Conference on 'Coastal Cities, Tourism and Environment', Eliat, Israel, May 14–17.

Kopp, J. 1996. 'The Mediterranean as a Product.' Paper presented at the inter-ministerial tourism conference, 'Marketing the Mediterranean as a Region.' Sliema, Malta, November 7–9.

Kousis, M. 1989. 'Tourism and the Family in a Rural Cretan Community.' *Annals of Tourism Research,* 16: 318–32.

Koutsopoulos, K. and Nijkamp, P. (eds). (n. d.) *Regional Development in the Mediterranean.* Athens: Phebus.

Krippendorf, J. 1992. *The Holiday Makers: Understanding the Impact of Leisure and Travel.* Oxford: Butterworth-Heinemann.

Kritz, M. M., Lim, L. L. and Zlotnik, H. (eds). 1992. *International Migration Systems.* Oxford: Clarendon.

Kypros-Net. 1997a. General Information: http://www.kypros.org/Cyprus/root.html (July 1 1997).

Kypros-Net. 1997b. 'Holbrooke Says EU Accession Talks Facilitate Cyprus Solution.' http://www.kypros.org/Cyprus/root.html (July 7 1997).

Kyriazi, D. 1996. 'Greek Tourism Is Seeking the Correspondence of Cost to Quality', in *Hellenews.* Interview with S. Kokotos, President of the Association of Hellenic Tourist Enterprises. Athens: Hellenews (January).

LaFranchi, H. 1993. 'Algeria Buys Some Time as Hostage Crisis Ends.' *Christian Science Monitor,* November 13: 3.

Lakshmanan, T. R. 1983. 'The Regional Development Prospects in the Mediterranean: Some Reflections', in *Regional Development in the Mediterranean,* K. Koutsopoulos and P. Nijkamp. (eds) Athens: Phebus.

Lanquar, R. 1995. *Tourisme et Environnement en Mediterrannee.* Paris: Les Fascicules du Plan Bleu 8, Economica.

Lardies Bosque, R. 1997. Migration and Tourism Entrepreneurship: North-European Immigrants in Mediterranean Coasts. Unpublished Fellow's Report in Leontidou (coord.), 24 pp.

Larson, B. K. 1991. 'Rural Development in Central Tunisia: Constraints and Coping Strategies', pp. 143–52 in *Tunisia: The Political Economy of Reform,* I. William Zartman (ed.). Boulder: Lynne Riener.

Lash, S. and Urry, J. 1994. *Economies of Sign and Space.* London: Sage.

Lea, J. 1988. *Tourism and Development in the Third World.* New York: Routledge.

Leheta, W. 1998. *Infrastructure Development for Regional Integration.* Cairo: American Chamber of Commerce, 1–3 (www.amcham.org.eg).

Leno, F. (coord). 1995. *Turismo y Medio Ambiente: La Sostenibilidad como Referencia.* Madrid: Secretaría General de Turismo, Ministerio de Turismo y Comercio.

Leontidou, L. 1990. *The Mediterranean City in Transition: Social Change and Urban Development.* Cambridge: Cambridge University Press.

Leontidou, L. 1991. 'Greece: Prospects and Contradictions of Tourism in the 1980s', pp. 84–106 in *Tourism and Economic Development: Western European Experiences,* A. Williams and G. Shaw (eds). London: Wiley.

Leontidou, L. 1994. 'Gender Dimensions of Tourism in Greece: Employment, Sub-cultures and Restructuring', pp. 74–105 in *Tourism: a Gender Analysis,* V. Kinnaird and D. Hall (eds). New York: John Wiley & Sons.

Leontidou, L. 1995. 'Repolarization in the Mediterranean: Spanish and Greek Cities in Neoliberal Europe.' *European Planning Studies,* 3(2): 155–72.

Leontidou, L. 1997. 'Five Narratives for the Mediterranean City', in King, R. *et al.* (eds), pp. 181–93.

Leontidou, L. (coord) 1997. Migration and Tourism Development in Marginal Mediterranean Areas. Four unpublished Fellows' Reports for the EU DG XII Human Capital and Mobility Project, King's College London.

Leontidou, L. 1998. 'Greece: Hesitant Policy and Uneven Tourism Development in the 1990s', pp. 101–23 in *Tourism and Economic Development: European Experiences,* A. M. Williams, and G. Shaw (eds). Chichester: Wiley.

Leontidou, L., Gentileschi, M. L. Aru, A. and Pungetti, G. 1998. 'Urban Expansion and Littoralization', pp. 92–7 in *Atlas of Mediterranean Environments in Europe: the*

Desertification Context, P. Mairota, J. B. Thornes and N. Geeson (eds). London: J. Wiley & Sons.

Litersdorff, T. and Goldenberg, Y. 1976. *Tourism Master Plan for Israel – Final Report* (in Hebrew). Tel Aviv: Tourism Master Plan Integrated Team.

Lockhart, D. G. 1993. 'Tourism and Politics: The Example of Cyprus', pp. 228–46 in *The Development Process in Small Island States*, Douglas G. Lockhart, David Drakakis-Smith and John Schembri (eds).

Lockhart, D. G. 1994. 'Tourism in Northern Cyprus: Patterns, Policies, and Prospects.' *Tourism Management*, 15(5): 370–9.

Lombardo. 1995.

Lorch, J. and Bausch, T. 1995. *Sustainable Tourism in Europe. Tourism and the Environment in Europe*. DG XXIII.

Loukissas, P. J. 1977. 'Tourism's Regional Development Impacts: A Comparative Analysis of the Greek Islands.' Ph.D Thesis, Cornell University. Ithaca NY: 335 pp.

Loukissas, P. 1982. 'Tourism's Regional Development Impacts.' *Annals of Tourism Research*, 9: 4. 523–41.

Loukissas, P. 1983. 'Public Participation in Community Tourism Planning: A Gaming Simulation Approach.' *Journal of Travel Research*, XXII, 1: 18–23.

Loukissas, P. (ed.). 1993. *Marine Tourism*. Conference Proceedings, The University of Thessaly and the Technical Chamber of Greece, Magnesia Chapter, Volos (in Greek).

Loukissas, P. (coord.) 1997. *Strategic Management Actions Related to Tourism* (SMART), unpublished report for the EC DG XXIII – Tourism Unit. University of Thessaly: Volos.

Loukissas, P. (coord.) 1999. *Regional Planning for Tourist Development of Coastal Zone in Magnesia*, unpublished report for the EC LIFE 96ENV/GR580. University of Thessaly: Volos: Author (in Greek).

Loukissas, P. (coord.) 2000. *Regional Planning for Tourist Development of Coastal Zone of Magnesia*, EC, Life 96 ENV GR 580, University of Thessaly, Volos (in Greek).

Loukissas, P. and Skayannis, P. 1999. 'Tourism, Sustainable Development, and the Environment', in *Mediterranean Tourism: Facets of Socioeconomic Development and Cultural Change*, Y. Apostolopoulos, P. Loukissas and L. Leontidou (eds). London and New York: Routledge.

Loukissas, P. and Triantafayllopoulos, N. 1997. 'Competitive Factors in Traditional Tourist Destinations: The Cases of the Islands of Rhodes and Myconos.' *Papers de Turisme*, 22: 214–18.

Loukissas, P. and Triantafayllopoulos, N. 2000. 'Community Involvement in Tourism Planning', in Apostolopoulos, Y. and Gayle, D. J. (eds), *Tourism, Sustainable Development and Natural Resource Management Experience in Caribbean*, Greenwood (forthcoming).

MacCannell, D. 1976. *The Tourist. A New Theory of the Leisure Class*. New York: Shocken Books.

Magnusson, D. K. 1991. 'Islamic Reform in Contemporary Tunisia: Unity and Diversity', pp. 143–52 in *Tunisia: The Political Economy of Reform*, I. William Zartman (ed.). Boulder: Lynne Riener.

Mairota, P., Thornes, J. B. and Geeson, N. (eds). 1998. *Atlas of Mediterranean Environments in Europe: The Desertification Context*. London: J. Wiley & Sons.

Manning, E. and Dougherty, T. 1995. 'Sustainable Tourism: Preserving the

Golden Goose.' *Cornell Hotel and Restaurant Administration Quarterly*, 36(2): 29–42.

Mansfeld, Y. 1992. 'Group-differentiated Perceptions of Social Impacts Related to Tourism Development.' *Professional Geographer*, 44: 377–92.

Mansfeld, Y. 1994. 'The Middle East Conflicts and Tourism to Israel 1967–90.' *Middle Eastern Studies*, 30(3): 646–67.

Mansfeld, Y. 1996. The Israeli Government's Role in Financing Tourism Promotion in Europe. An Internal Memo Prepared for the Eilat Hotel Association (in Hebrew).

Mansfeld, Y. and Czmanski, D. 1995. *Guidelines for Sustainable Development of Eilat* (in Hebrew). Eilat: Municipality of Eilat.

Mansfeld, Y. and Ginosar, O. 1994. 'Determinants of Locals' Perceptions and Attitudes toward Tourism Development in their Locality.' *Geoforum*, 25: 227–48.

Mansfeld, Y. and Kliot, N. 1996. 'The Tourism Industry in the Partitioned Island of Cyprus', pp. 187–202 in *Tourism, Crime, and International Security Issues*, A. Pizam and Y. Mansfeld (eds). New York: Wiley.

Marmaras, E. 1996. Migration and Tourism Development in Marginal Mediterranean Areas: Foreign Second-home Owners in Spain and Greece. Unpublished Fellow's Reportin Leontidou (coord.) 1997, King's College London, 112 pp.

Marmaras, E. 1997. 'The Human Geography of the Foreigners in the Cyclades Islands.' *Synchrona Themata*, 63: 139–41 (in Greek).

Martin-Munoz, G. 1994. *Familles Arabes en Milieu Urbain*. Modification de l'ordre Social Traditionnel, Citta e Societa Urbana nel Mondo Arabo Trasformazioni, Sfide, Prospettive.

Mawforth, M. and Munt, I. 1998. *Tourism and Sustainability: New Tourism in the Third World*. London and New York: Routledge.

Mazor, A. 1987. *Tourism and Recreation Master Plan 1985–1995* (in Hebrew). Tel Aviv: Urban Institute and Tahal Consultants.

McHugh, K. E. 1990. 'Seasonal Migration as a Substitute for, or Precursor to, Permanent Migration.' *Research on Aging*, 12: 229–45.

McKinney, B. 1998. *Encouraging Investment in Egypt's Private Sector*. Cairo: American Chamber of Commerce, 1–2 (www.amcham.org.eg).

Mendonsa, E. L. 1983. 'Search for Security, Migration, Modernisation and Stratification in Nazare, Portugal.' *International Migration Review*, 6: 635–45.

Merchant, C. 1992. *Radical Ecology*. New York: Routledge.

Mies, M. 1986. *Patriarchy and Accumulation on a World Scale: Women in the International Division of Labor*. New Jersey: Zed Books.

Mikačić, V. 1996. 'Research into the Domestic Tourist Trade in Croatia.' *Turizam* 44, (3–4): 61–70.

Mingione, E. 1991. *Fragmented Societies: a Sociology of Economic Life Beyond the Market Paradigm*. Oxford: Basil Blackwell.

Mingione, E. 1995. 'Labor Market Segmentation and Informal Work in Southern Europe.' *European Urban and Regional Studies*, 2(2): 121–43.

Ministere de L'Environnement et de L'Amenagement du Territoire. 1993. *Rapport National L'Etat de L'Environnement*. Tunis: Author.

Ministerio de Industria, Comercio y Turismo (MICyT). 1992. *Futures*. Plan Marco de Competitividad del Turismo Español, Madrid, Ministerio de Industria, Comercio y Turismo, Secretaría General de Turismo.

Ministry of National Economy. 1985. 'Economic and Social Development Plan.' MoNE/KEPE. Athens: Author (in Greek).

Ministry of the Environment Regional Planning and Public Works. on-going 1997. *Regional Plan of Pelion*. Athens: Author (in Greek).

Ministry of Tourism and Culture. 1986. *Turkey: Tourism Opportunities for Investors*. Ankara: Nurol Matbaasi.

Ministry of Tourism and Culture. 1987. *Turkey: Tourism Opportunities for Investors*. Ankara: Nurol Matbaasi.

Ministry of Tourism of Egypt. 1998. *Tourism Indicators*. Cairo: Ministry of Tourism 1–5 (www.idsc.gov.eg/tourism).

Miossec, J. M. 1976. 'Elements pour une Theorie de l'espace Touristique.' *Les Cahiers du Tourisme*, C–3. Aix-en-Provence: CHET.

Miossec, J. M. 1994. *Tourisme et Loisirs de Proximite dans le Monde Arabe*. Monde Arabe Maghreb Machrek.

Miranda, M. T. 1985. *La Secunda Residencia en la Provincia de Valencia*. Universidad de Valencia.

Monfort, V. M. 1995. 'Estrategias de Competitividad del Sector Hotelro Español: Especial Referencia al Arco Mediterráneo y Canarias.' Economía de los Servicios. V Congreso Nacional de Economía, Congress Proceedings, Las Palmas de Gran Canaria, Vol. 6, -rea Economía del Turismo, pp. 51–9.

Montanari, A. 1995. 'The Mediterranean Region: Europe's Summer Leisure Space.' In Montanari, A. and Williams, A. M. (eds). 1995. *European Tourism: Regions, Spaces and Restructuring*. Chichester: John Wiley & Sons.

Montemagno, G. 1985. 'Turismo e Beni Culturali: Dalla Relazione Teorica all'integrazione Sistemica', in Quaderni ANIEST n. 5 *Lo Sviluppo del Turismo e la Protezione dell'ambiente*, Roma: ANIEST.

Montemagno, G. 1986. 'Funzione e Valutazione dei Beni Culturali Negli Itinerari Turistico-culturali del Mezzogiorno', in Quaderni ANIEST n. 6 *Turismo e Beni Culturali*. Roma: ANIEST.

Montemagno, G. 1987. 'Il Ruolo degli Enti Locali nello Sviluppo Turistico: Il Prodotto Turistico Locale (PTL)', in *Sicilia, Quale Turismo per lo Sviluppo*, G. Lo Re (ed.). Palermo: Lega delle Autonomie.

Montemagno, G. 1989. 'La Valutazione del Rapporto Turismo/Ambiente e la Proposta del Turismo Rurale', in *Turismo e Ambiente nella Società Post-industriale*, FAST–TCI Milano (ed.): FAST–TCI.

Montemagno, G. 1994. 'Tourisme Rural et Agrotourisme pour le Développement d'une Région du 'Mezzogiorno' Italien', in *Territoires d'Europe*, vol. II, Strasbourg: CEDRE (FEDER).

Montemagno, G. 1998. 'Nuove Strutture Organizzative per Nuove Prospettive del Turismo Siciliano', in AAPIT *Riordino del Settore Turismo: Quali Prospettive?* Catania: Maimone.

Montemagno, G. 1997. 'Settimo Rapporto sul Turismo Italiano.' Firenze: Turistica-Mercury.

Moore, H. C. 1991. 'Tunisian Banking: Politics of Adjustment and the Adjustment of Politics', pp. 67–97 in *Tunisia: The Political Economy of Reform*, I. William Zartman (ed.). Boulder: Westview.

Morello, C. 1998. Egypt *USA Today*, Egypt, 10 March, 2. (www.usatoday.com/life).

Mouzelis, N. 1978. *Modern Greece: Facets of Underdevelopment*. London: Macmillan.

Mouzelis, N. 1986. *Politics in the Semi-periphery: Early Parliamentarism and Late*

Industrialisation in the Balkans and Latin America. Basingstoke, Hampshire: Macmillan Education.

Nash, D. 1979. 'The Rise and Fall of an Aristocratic Tourist Culture, Nice 1763–1936.' *Annals of Tourisme Research,* 4(1).

Nathan, S. 1973. 'First Steps', pp. 38–48 in *The Second Million – Israel Tourist Industry Past-present-future,* C. H. Klien (ed.). Jerusalem: Amir Publishing.

National Technical University of Athens. 1982. *Plan for a National System of Marinas in Greece.* National Tourist Organization of Greece (in Greek).

Navez-Bouchanine, F. 1992. *Citadinite et Urbanite: Le Cas des Villes Marocaines.* Journees d'etudes sur la Citadinite, Tours, URBAMA.

Nebel, E. III (ed.). 1983. *Tourism and Culture, a Comparative Perspective.* New Orleans: University of New Orleans Press.

Noin, D. and Woods, R. (eds). 1993. *The Changing Population of Europe.* Oxford: Basil Blackwell.

North Cyprus Homepage. 1997. 'General Information.' http://www.emu.edu.tr/trnc/index.html (June 29, 1997).

Oberg, S., Scheele, S. and Sundstrom, G. 1993. 'Migration among the Elderly: The Stockholm Case.' *Espaces, Populations, Societes,* 3: 503–14.

Office National du Tourisme Tunisien. 1987.

Office National du Tourisme Tunisien. 1992. *Le Tourisme Tunisien en Chiffres.* Tunis: Author.

Office National du Tourisme Tunisien. 1995.

Office National du Tourisme Tunisien. 1996.

Ogden, P. E. 1989. 'International Migration in the 19th and 20th Centuries', pp. 34–59 in *Migrants in Modern France,* P. E. Ogden and P. E. White (eds). London: Unwin Hyman.

Ogden, P. E. and White, P. E. (eds). 1989. *Migrants in Modern France.* London: Unwin Hyman.

Olali, H. 1984. *Turizm Dersleri.* Izmir: Istiklal Matbaasi.

Panayiotou, T. 1989. Tourism and Environmental Impact Assessment: Towards a New Tourist Policy for Cyprus. Unpublished paper.

Panayotou, G. 1993. *Green Markets.* San Francisco: Institute for Contemporary Studies.

Papayannis, T. 1994. 'Tourism under the Perspective of Sustainability.' Proceedings of the Day-conference on *Tourism and the Environment: Selections for Sustainable Development.* Athens: Technical Chamber of Greece (in Greek).

Pearce, D. 1989. *Tourism Development,* 2nd edn, New York: Longman Scientific & Technical.

Pearce, D. 1995. 'Planning for Tourism in the 1990s: an Integrated, Dynamic, Multiscale Approach', pp. 229–45 in *Change in Tourism: People, Places and Processes,* R. Butler and D. Pearce (eds). London and New York: Routledge.

Pearce, D. 1995. *Tourism Today: a Geographical Analysis.* New York: Longman.

Pedrini, L. 1984. 'The Geography of Tourism and Leisure in Italy.' *Geojournal,* 9: 55–7.

Peisley, T. 1992. 'The World Cruise Ship Industry in the 1990s.' *The Economist Intelligence Unit* Special Report No. 2104, January 1992.

Peisley, T. 1995. 'The Cruise Ship Industry to the 21st Century.' *EIU Travel and Tourism Analyst,* 2: 4–25.

Perry, A. and Ashton, S. 1994. 'Recent Developments in the UK's Outbound Package Tourism Market.' *Geography*, 79: 313–21.

Perry, R., Dean, K. and Brown, B. 1986. *Counterurbanisation*. Norwich: Geo Books.

Petras, J. 1992. 'The Contradictions of Greek Socialism', in *The Greek Socialist Experiment: Papandreou's Greece (1981–1989)*, T. Kariotis (ed.). New York: Pella.

Phileleftheros. 1992. 'Tourism and the Economy' (Special Report) (January 31): p. 6.

Pierre P. Y. 1996. *Le Tourisme, un Phénomène économique*. La Documentation Française.

Piou, M. 1995. 'The World Tourism and Greece: Tourism Figures, Trends, Challenges and New Opportunities from Arthur Andersen.' *Money and Tourism*, 1: 34–8 (in Greek).

Poirier, R. A. 1995. 'Tourism and Development in Tunisia.' *Annals of Tourism Research*, 22(1): 157–71.

Poirier, R. A. and Wright, S. 1993. 'Tunisia: Political Economy of Tourism.' *The Journal of Modern African Studies*, 31(1): 149–62.

Poon, A. 1989. *Financing Tourism in the Caribbean: International Trends and Caribbean Perspectives*. Christ Church, Barbados: CTO.

Poon, A. 1993. *Tourism, Technology and Competitive Strategies*. Wallingford, Oxon: C.A.B. International.

Prefecture of Magnesia. 1995. *Economic and Social Development Plan of the Magnesia Prefecture*. Volos: Author (in Greek).

Preglau, M. 1983. 'Tourism Kills Tourism: Environmental, Cultural, and Economic Consequences of Tourism', pp. 35–63 in *Tourism and Culture, A Comparative Perspective*, Nebel, E. III. (ed.) New Orleans: University of New Orleans Press.

Radnić, A. 1994. 'Kamping Turizam u Novoj Razvojnoj Strategiji Turističkog Sektora Hrvatske.' *Turizam 42*, (1–2):15–23.

Reed, D. (ed.). 1992. *Structural Adjustment and the Environment*. Boulder: Westview.

Rees, P. *et al.* (eds). 1996. *Population Migration in the EU*. Chichester: John Wiley.

Richter, L. 1989. *The Politics of Tourism in Asia*. Honolulu: University of Hawaii Press.

Riviera in AUTREMENT, *Hors-série 1987 no. 2*, 230pp.

Rodriguez Martinez, F. 1991. 'Conocimiento Geograficos de la "Politica de Costas" en el Litoral Surmediterraneo Andaluz', pp. 53–9 in *Ordenacion y Desarrollo del Tourismo en Espana y en Francia*, F. Fourneau and M. Marchena (eds). Madrid: Ministerio de Obras Publicas y Transportes.

Rogers, A. 1993. *The Earth Summit*. Los Angeles: Global View.

Rowland, J. 1992. 'Algeria Aims to Revive Dormant Tourist Trade.' *MidEast Markets*, Vol. 2, No. 11.

Ryan, C. 1991. *Recreational Tourism: A Social Science Perspective*. London: Routledge.

Salt, J. 1996. 'Migration Pressures on Western Europe', pp. 92–126 in *Europe's Population in the 1990s*, D. Coleman (ed.). New York: Oxford University Press.

Scherbina, A. A. 1980. 'Mathematical Economics.' *Moscow: Science Academy*, Vol. XVI (6).

Scherbina A. A. 1982. 'Mathematical Economics.' *Moscow: Science Academy*, Vol. XVIII (2) pp. 346–8.

Schwaninger, M. 1986. 'Strategic Business Management in Tourism.' *Tourism Management*, 7(2): 77–85.

Scott, J. 1995. 'Sexual and National Boundaries in Tourism.' *Annals of Tourism Research*, 22(2): 385–403.

Sebstad, J. 1989. 'Towards a Wider Perspective on Women's Employment.' *World Development*, 17: 937–52.

Secretaría de Estado de Comercio, Turismo y de la Pequeña y Mediana Empresa (SECTyPYMEs). 1997. *Plan de Estrategias y Actuaciones de la Administracion General del Estado en Materia Turística.* Madrid: Ministerio de Economía y Hacienda.

Secretaría General de Turismo (SGT). 1994. *Nota de Coyuntura Turistica*, Enero. Madrid: Author.

Sekulić, B. and Pečar-Ilić, J. 1997. 'Climate, Morphological and Ecological Characteristics of the Eastern and Western Coast of the Adriatic.' *Turizam 45*, (1–2): 3–12.

Selwyn, T. (1996). 'Tourism, Environment, and Society in the Mediterranean.' In the Proceedings of the WTO Interministerial Tourism Conference on *Marketing the Mediterranean as a Region* (Malta, November). Madrid: WTO.

Shaari, Y. 1973. 'Regional Development', pp. 130–45 in *The Second Million – Israel Tourist Industry Past-present-future*, C. H. Klien (ed.). Jerusalem: Amir Publishing.

Shaw, G. and Williams, A. M. 1990. 'Tourism, Economic Development and the Role of Entrepreneurial Activity', pp. 67–81 in *Progress in Tourism, Recreation and Hospitality Management*, vol. 2, C. P. Cooper (ed.). Chichester: John Wiley.

Shaw, G. and Williams, A. M. 1997. 'The Private Sector: Tourism Entrepreneurship – A Constraint or Resource?', pp. 117–36 in *The Rise and Fall of British Coastal Resorts: Cultural and Economic Perspectives*, G. Shaw and A. M. Williams (eds). London: Pinter.

Shaw, G. and Williams, A. M. (eds). 1997. *The Rise and Fall of British Coastal Resorts: Cultural and Economic Perspectives.* London: Pinter.

Shawki, A. 1998. *Investment Incentives in Egypt.* Cairo: American Chamber of Commerce, 1–3 (www.amchan.org.eg).

Shucksmith, D. M. 1983. 'Second Homes: A Framework for Policy.' *Town Planning Review*, 54(2), 174–93.

Smaoui, A. 1979. 'Tourism and Employment in Tunisia', pp. 101–10 in *Tourism: Passport to Development?*, Emanuel de Kadt (ed.). New York: Oxford University Press.

Smith C. and Jenner, P. 1995. 'Marinas in Europe.' *EIU Travel and Tourism Analyst*, 6: 56–72.

Sonmez, S. and Apostolopoulos, Y. 1999. 'Conflict Resolution Through Tourism Cooperation? The Case of the Partitioned Island-State of Cyprus.' *Journal of Travel and Tourism Marketing* (forthcoming).

State Institute of Statistics. 1997. Web Page, Various Years.

State of Israel. 1996. *Statistical Abstracts.* Jerusalem: Israel Bureau of Statistics.

State Planning Organization. 1964. *First Five Year Development Plan 1963–1967.* Ankara: Central Bank of Turkey.

Steavenson, W. 1997. 'Cruises to the Future.' *Time*, 16 June 1997.

Steigenberger Consulting. 1997. *Competition of Croatia's Tourism in the European Markets: Findings, Conclusions and Recommendations.* Frankfurt-Zagreb: Author.

Sullivan 1985. 'The Ties That Bind.' *Research on Ageing*, 7: 235–60.

SVIMEZ. 1998. 'Rapporto 1996 sull'economia del Mezzogiorno.' Bologna: Il Mulino.

Syrimis, G. 1990. The Long-Term Development Strategy for the Cypriot Economy: Modernization and Technological Upgrading. Unpublished paper. Nicosia: Ministry of Finance.

Tahi, M. S. 1992. 'The Arduous Democratisation Process in Algeria.' *The Journal of Modern African Studies*, 30: 397–419.

The European. 1996. 'Cyprus: Special Report,' (April 11–17): 8–10.

Timothy, D. J. and Ioannides, D., in press.

Touloupas, P. 1996. 'Greece as a Tourist Destination for the British Tour Operators: an Analysis from a Marketing Perspective in the Particular Context of the United Kingdom.' Master's Thesis, University of Surrey, Guildford.

Touring Club Italiano. 1998. *Agenda 1998.* Milano: TCI.

Tourism and Economy. 1993. Special Anniversary issue (in Greek). *Tourism and Economy*, 1997.

Travel and Tourism Intelligence. 1997.

Travis, A. S. 1980. 'Tourism Development and Regional Planning in East Mediterranean Countries.' *International Journal of Tourism Management*, 2: 207–18.

Triantafyllopoulos, N. 1999. 'Les Mechanismes Fonciers de L'Urbanisation Touristique de Littoral en Grece: Etude de Cas – L'Ile de Rhodes.' These de Doctorat Universite Paris I Pantheon-Sorbonne (in French), 344 pp.

Troin, J. F. 1994. *Les Souks et le Bazar: Quel Avenir pour la Ville de l'histoire?* D' Ville et Societe Urbaine dans le Monde Arabe Transformations, Enjeux, Perspectives.

Tsartas, P. 1989. *Social and Economic Impacts of Tourism.* Athens: NCSR (in Greek).

Tunisia Digest. 1993. 2 (1 & 3).

Tunisian External Communication Agency. 1993. *Women of Tunisia.* Tunis: Author.

Tuppen, J. 1991. 'France: The Changing Character of a Key Industry', pp. 191–206 in *Tourism and Economic Development: Western European Experiences*, A. Williams and G. Shaw (eds). London: Wiley.

United Nations Development Programme. 1988. *Comprehensive Tourism Development Plan: Cyprus.* Madrid: World Tourism Organization.

United Nations. 1992. Conference on the Environment and Development, Rio de Janeiro.

University of Thessaly. 1997. *Strategic Management Actions Related to Tourism* (SMART) EC DG XXIII – Tourism Unit. Volos: Author.

URBAPLAN. 1996. *Etude de la strategie d'amenagement touristique.* Rapport final. Ministre du Tourisme, Rabat.

Valenzuela, M. 1991. 'Spain: the Phenomenon of Mass Tourism', pp. 40–60 in *Tourism and Economic Development: Western European Experiences*, A. Williams and G. Shaw (eds). London: Wiley.

Valero Escandell, J. R. 1992. *La Inmigracion Extranjera en Alicante.* Diputacion de Alicante: Instituto de Cultura 'Juan Gil-Albert.'.

Van den Berg, J. C. J. M. and Nijkamp, P. 1994. 'Modeling Ecologically Sustainable Economic Development in a Region: a Case Study in the Netherlands.' *Annals of Regional Science*, 28: 7–29.

Van der Heiden, Johan. 1996. *Population Counter, Top 25 Index*, 9 June. (johan.van.der.heijden@tip.nl).

Vassiliou, G. 1995. 'Tourism and Sustainable Development Lessons from the Cyprus Experience', in The United Nations University (World Institute for Development Economics Research) *Small Islands, Big Issues: Crucial Issues in the Sustainable Development of Small Developing Islands.* Helsinki: UNU/WIDER.

Vera, F. and Marchena, M. 1996. 'El Modelo Turístico Español: Perspectiva Economica y Territorial', in *Introduccion a la Economía del Turismo en España*, A. Pedreño and M. Monfort (eds). Madrid: Civitas.

308 *Bibliography*

Vera, J. F. 1993. 'Actividad y Espacios Turisticoso', in *Geografla de España*, R. McEndez and F. Molinero (eds). Barcelona: Ariel.

Vera, J. F. 1994. 'El Modelo Turístico del Mediterráneo Español: Agotamiento y Estrategias de Reestructuracion.' Papers de Turisme, no. 14–15, Valencia, Institut Turístic Valencia, pp. 133–47.

Veronese, B. 1963. 'Out of Brindisi to the Islands of Greece.' *Yachting World*, (January): 2–5.

Vries, H. J. M. de. 1989. *Sustainable Resource Use: an Inquiry Into Modelling and Planning*. Groningen: Rijks Universiteit.

Vukonić, B. 1997. *Turističke gencije*. Zagreb: Mikrorad.

Wahab, S. 1997. 'Sustainable Tourism in the Developing World', in *Tourism Development and Growth*, S. Wahab and J. J. Pigram (eds). New York: Routledge.

Walz, S. 1986. 'Islamist Appeal in Tunisia.' *The Middle East Journal*, 40: 651–70.

Warnes, A. M. (ed.). 1982. *Geographical Perspectives on the Elderly*. London: Wiley.

West Africa. 1991. January 14–20: 6–9.

Wilkinson, P. F. 1989. 'Strategies for Tourism in Island Microstates.' *Annals of Tourism Research*, 16: 153–77.

Williams, A. 1997. 'Tourism and Uneven Development in the Mediterranean', pp. 208–26 in *The Mediterranean: Environment and Society*, R. L. King, L. Proudfoot and B. Smith (eds). London: Edward Arnold.

Williams, A. M. (ed.). 1984. *Southern Europe Transformed*. London: Harper & Row.

Williams, A. M. 1997. 'Tourism and Uneven Development in the Mediterranean', pp. 208–26 in *The Mediterranean: Environment and Society*, R. L. King, L. Proudfoot and B. Smith (eds). London: Edward Arnold.

Williams, A. M. and Shaw, G. (eds). 1991. *Tourism and Economic Development: Western European Experiences*, 2nd edn, London: Wiley.

Williams, A. and Shaw, G. (eds). 1998. *Tourism and Economic Development: European Experiences*, 3rd rev. edn, London: Wiley.

Williams, A. M., Shaw, G. and Greenwood, J. 1989. 'From Tourist to Tourism Entrepreneur, from Consumption to Production: Evidence from Cornwall, England.' *Environment and Planning A*, 21: 1639–53.

Wilson, R. 1992. *Cyprus and the International Economy*. New York: St. Martin's Press.

Witt, S. F. 1991. 'Tourism in Cyprus: Balancing the Benefits and Costs.' *Tourism Management*, 12: 37–46.

World Bank. 1989. *Sub-Saharan Africa. From Crisis to Sustainable Growth*. Washington, DC: World Bank.

World Bank. 1992. *Republic of Cyprus: Environmental Review and Recommendations*. Washington, DC: World Bank.

World Commission on Environment and Development. 1987. *Our Common Future: Report of the World Commission on Environment and Development*. Oxford: Oxford University Press.

World Tourism Organization. 1980. *The Manilla Declaration*. Madrid: Author.

World Tourism Organization. 1994. 'International Incentives in the Cyprus Tourism Plan', pp. 181–6 in *National and Regional Tourism Planning*. London: Routledge.

World Tourism Organization. 1994. 'Tourism Planning Approach of Malta', pp. 81–5 in *National and Regional Tourism Planning*. London: Routledge.

World Tourism Organization. 1997. *Compendium of Tourism Statistics, 1991–1995*, Seventeenth Edition, Madrid.

World Tourism Organization. 1998. *Yearbook of Tourism Statistics.* Madrid: WTO.

World Tourism Organization. 1999. *Yearbook of Tourism Statistics.* Madrid: WTO.

World Tourism Organization. 1996. *Yearbook of Tourism Statistics.* Madrid: WTO.

World Tourism Organization. 1996. *Marketing the Mediterranean as a Region.* Madrid: WTO.

World Travel and Tourism Council and Travel Business Roundtable. 1999. *Travel and Tourism: ' White Paper.* Brussels: WTTC & TBR.

World Tourism Organization. 1999. *Yearbook of Tourism Statistics.* Madrid: WTO.

World Travel and Tourism Council. 1997. *European Union Travel & Tourism Creating Jobs.* London: World Travel & Tourism Council.

World Travel and Tourism Council. 1997. *Travel and Tourism in the World Economy.* Brussels: WTTC.

Wright, S., and Poirier, R. A. 1991. 'Tourism and Economic Development in Africa.' *TransAfrica Forum* 8: 13–27.

Zacharatos, G. 1986. *Touristic Consumption.* Athens: GCPER (in Greek).

Zlotnik, H. 1992. 'Empirical Identification of International Migration Systems', pp. 19–40 in *International Migration Systems*, M. M. Kritz, L. L. Lim, and H. Zlotnik (eds). Oxford: Clarendon.

Zoreda, J. L. 1996. 'How to Make an Umbrella Promotion of the Mediterranean – The Spanish Perspective.' Paper presented at the inter-ministerial tourism conference, 'Marketing the Mediterranean as a Region.' Sliema, Malta, November 7–9.

Zoubir, Y. H. 1995. 'Stalled Democratization of an Authoritarian Regime: the Case of Algeria.' *Democratization*, 2(2).

Zoubir, Y. H. 1999. 'State and Civil Society in Algeria'. in *North Africa in Transition: Socio-Economic and Political Change in the 1990s*, Yahia H. Zoubir (ed.). Jacksonville, FL: University Press of Florida.

Index